Language and the Rise of the Algorithm

∴

Language
and the
Rise of the
Algorithm

∵

Jeffrey M. Binder

THE UNIVERSITY OF CHICAGO PRESS

CHICAGO AND LONDON

The University of Chicago Press, Chicago 60637
The University of Chicago Press, Ltd., London
© 2022 by The University of Chicago
Published 2022
Printed in the United States of America

31 30 29 28 27 26 25 24 23 22   1 2 3 4 5

ISBN-13: 978-0-226-82253-2 (cloth)
ISBN-13: 978-0-226-82254-9 (e-book)
DOI: https://doi.org/10.7208/chicago/9780226822549.001.0001

Library of Congress Control Number: 202201812

♾ This paper meets the requirements of ANSI/NISO Z39.48-1992
(Permanence of Paper).

Contents

Introduction

```
/*
 * If the new process paused because it was
 * swapped out, set the stack level to the last call
 * to savu(u_ssav). This means that the return
 * which is executed immediately after the call to aretu
 * actually returns from the last routine which did
 * the savu.
 *
 * You are not expected to understand this.
 */
if(rp->p_flag&SSWAP) {
        rp->p_flag =& ~SSWAP;
        aretu(u.u_ssav);
}
```

—LIONS' COMMENTARY ON UNIX 6TH EDITION, WITH SOURCE CODE

THE COMPROMISE

In May 2020, as much of the world was focused on the COVID-19 pandemic and as racial justice protests took place across the United States, a technical development sparked excitement and fear in narrower circles. A computer program called GPT-3, developed by the OpenAI company, produced some of the best computer-generated imitations of human writing yet seen: fake news articles that were, according to the authors, able to fool human readers nearly half the time, and poems in the style of Wallace Stevens.[1] The program is based on a statistical model that does one thing: given a sequence of words, it tries to predict what word will come next. The model was trained on more than 570 gigabytes of compressed text scraped from the internet in addition to the contents of Wikipedia and a large number of books.[2] The system's creators describe it as a "task-agnostic" learner—that is, a machine learning model that can perform a wide range of cognitive tasks without having to be fine-tuned for any particular one.[3] This new approach to artificial intelligence (AI) aspires to transform the practice of computer programming: instead of designing an algorithm to solve a given problem, one tells the machine its goal in English, and it works out (one hopes) the correct answer.

From a humanistic standpoint, a striking aspect of this claim is how it locates knowledge in language. GPT-3's input and output consist of text, and it is trained on nothing but text; it has no experience, even in the loosest notional sense, of anything whatsoever.[4] Yet its apparent capabilities are not limited to such language-oriented tasks as rewriting paragraphs in different styles; to the extent that it really is a multitask learner, it unites the functions of writing aid, programmable calculator, and search engine. Skeptically viewed, the machine is acting like a parrot, saying things it cannot understand. But the idea that a language model can form the basis for a universal method could also suggest something like a deconstructive insight: that learning language cannot be distinguished from learning to think, that there is no limit to the sorts of cognitive operations that go into choosing words. If we are to believe the researchers—which we certainly should not do uncritically—then natural language is the essential ingredient needed to create the elusive artificial general intelligence (AGI).

The rise of large language models such as GPT-3 has unsettled the categories in which people have long understood the relation of computation to language. Computers are often described as symbol-manipulating machines; they work by rearranging electrically represented ones and zeros through mechanical rules that do not depend on the symbols' meanings. GPT-3 has rekindled a long-standing philosophical debate over whether such a machine can really be said to understand a language.[5] But even before this development, computers have seldom been used as purely uninterpreted symbol manipulators. In modern interfaces, screens are festooned with words—*save, submit, like*—that serve to mediate between computational logic and the social conventions by which people communicate. Engineers have long treated the communicational elements of computer systems as superficial ornaments when compared to the data structures and algorithms that form the real core of a computer program. Language models such as GPT-3 have blurred the lines. Since these systems depend, through and through, on data about people's linguistic practices, they make it harder than ever to judge where algorithm ends and language begins.

The term *algorithm*, as it is used in computer science, is notoriously easier to illustrate than to define. While the word has recently become associated with machine learning, textbooks typically explain algorithms, quite simply, as precisely defined procedures for solving problems. These procedures often take the form of sequences of steps, as in the following algorithm for finding the length of the longest sentence in a book:

> Write the number 0 on scrap paper
> For each sentence in the book, repeat the following:
> > Count the number of words in the sentence
> > If the result is greater than the number on the scrap paper:
> > > Replace the number on the scrap paper with the result

Although similar instructions occur in a wide range of contexts—a typical example is cooking recipes—calling a procedure an *algorithm* evokes a more specific set of disciplinary practices. Programming languages provide a way of describing procedures with the extreme precision demanded by machines. (To make the foregoing procedure a true algorithm, we would have to clarify what *words* and *sentences* are—not a straightforward matter.) Computational complexity theory provides methods for gauging and improving the efficiency of these procedures. More broadly, algorithmic thinking (in the expansive sense of thinking about algorithms) invites abstraction.[6] The technical theory of algorithms encourages the development of general solutions that can be reused for different purposes and in different contexts; the procedures are thought of as mathematical entities that exist apart from the complexities of the languages in which they are described and the concrete situations in which they are used.

This book is about how this form of abstraction came into being. It focuses on one thread in the prehistory of algorithms: the use of symbols in numerical calculation, algebra, calculus, logic, and, eventually, computer science. Standard programming languages such as Python and R draw (among other sources) on the symbolic notations of algebra and logic as ways of precisely defining operations. Yet these notations, like programming languages, have long combined computation with another function that is harder to reduce to mechanical rules: communication. A symbolic formula such as $F_s = kx$ provides both instructions for how to compute something—in this case, the force required to extend or compress a spring by a given length—and a way of conveying a proposition about the world.[7]

It is my contention that the modern idea of *algorithm,* as the term is used in computer science, depends on a particular way of disentangling computation from the complexities of communication that first took shape in the pure mathematics of the nineteenth century.[8] Although machine learning systems are often called (confusingly) by the same name as the precisely defined procedures dealt with in the theory of algorithms, I hope to show that machine learning represents a break from this technical concept that places centuries-old epistemological boundaries in jeopardy.

The history of algorithms has been told in both long and short versions. In a broad sense, algorithmic thinking goes back at least as long as the written record.[9] On clay tablets, the ancient Babylonians wrote down rule-based procedures for numerical computation in which the computer scientist Donald E. Knuth perceived the rudiments of his discipline.[10] The word *algorithm* (early on spelled a range of ways, such as *algorism, algorithmus, algram,* or *augrym*) is less ancient but still very old—it was formed in the twelfth century from the name of the Arabic mathematician Muḥammad ibn Mūsā al-Khwārizmī, who described techniques for computing with Hindi–Arabic numerals in the ninth century.[11] These techniques—including the familiar addition, subtraction, multiplication, and division procedures one still learns in school—made up the original "algorithm." As early as the sixteenth century, the word *algorithm* came to encompass a range of other techniques beyond these original ones, often involving symbolic algebra. In searching for precursors to the totalizing ambitions that now attend computation, popular histories commonly single out the German polymath Gottfried Wilhelm Leibniz. Starting in the 1660s, Leibniz attempted to create what he called a *calculus ratiocinator*—a system of symbolic "calculation" that could resolve disputes about virtually any topic. The science writer Martin Davis describes the modern computer as a fulfillment of "Leibniz's Dream" of extending mathematical symbol manipulation into a universal method that can be applied to anything whatsoever.[12]

More focused scholarship by historians including Michael Mahoney and Lorraine Daston has shown that such sweeping narratives overlook the ways computational practices have changed over the centuries.[13] Mark Priestley has argued persuasively that computer programming has no intrinsic relation to other fields such as symbolic logic but rather came to relate to them through intentional choices made by computer scientists.[14] Matthew L. Jones and Maria Rosa Antognazza have placed Leibniz into historical context and showed that his work was not exactly algorithmic in the modern sense.[15] This more historicist perspective has led to a contrasting narrative in which the concept of algorithm is very new. Venerable as the word *algorithm* may be, its meaning arguably did not reach its modern form until the 1960s, when computer science emerged as an academic discipline. The six authors of the book *How Reason Almost Lost Its Mind* have argued that algorithms were not a model of rationality until the Cold War period, when think tank researchers sought to replace human judgment with strictly rule-based decision-making.[16] The algorithm's rise to the status of a social concern is even more recent, stemming from a confluence of technical developments in machine learning with entrenched structures of inequality and discrimination.[17]

This historicization of the idea of algorithm should serve as a warning against uncritically identifying the symbolic methods of the past with modern algorithms. The algorithm as we know it is a complex amalgam whose prehistory encompasses a range of practices, including astronomical and statistical computation, bureaucratic procedures, market economics, and governmental data-gathering efforts such as the US census. As a background to modern algorithms, symbolic methods are important less on account of their intrinsic relevance than because of the role they came to play in technical discourse. In the 1960s and '70s, the discipline of computer science came to view algorithms as abstract processes that maintain a stable identity even as they are implemented, explained, applied, and interpreted in a range of ways. As I show in this book, this way of thinking is implicated in a long series of debates about the relation of symbols to language. Should the same symbols be used both to compute results and to present them to others? To what extent can their meanings be chosen at will, and to what extent does the establishment of meaning require social agreement? If a symbol is defined using words, does that entail that it inherits the imprecision of natural language?

Such issues would now be seen as extrinsic to computation, involving the significance people assign to algorithms, not the algorithms themselves. But this boundary has not always been in place, as one can see by examining how what counted as an *algorithm* has changed over time. The Indian computational techniques have always involved instructions, taught either through direct imperative statements or by example, for what to do with symbols: if the sum is greater than 9, write a 1 above the digit to the left. As people recognized long before the computer age, this type of procedure can potentially be performed by machines.[18] Yet the original "algorithm" also involved another, less obviously mechanical sort of rule: 9 means *nine*. The practice, that is, included rules not just for how to manipulate the symbols but also for how to interpret them. While mathematicians long recognized that these semantic rules differed from calculating procedures, they were a part of the "algorithm" just the same.

Symbolic algebra complicated these matters by introducing letters to indicate unspecified values, as in $ax + b$. This use of letters, introduced by François Viète in the 1590s, laid the groundwork for the modern algorithm by enabling procedures to be described in an abstract form that leaves the inputs unspecified. But these letters were linked together with operators such as + and − that were, at least early on, supposed to have fixed meanings. Establishing these meanings may not have posed a major problem in simple cases, but things became trickier as symbolic methods extended into theoretically fraught fields such as the infinitesimal calculus, and they became yet worse in utopian schemes like Leibniz's attempt to develop symbolic methods for politics. Suppose, for instance, we introduce a symbol to denote *equity*. How can we be sure that everyone using this symbol agrees about what equity is? The importance given to conceptual clarity made it difficult to ignore the question of what it takes to make a symbol mean something, and disparate answers to this question had strong implications for what symbolism could do.

The expulsion of meaning from algorithms did not so much resolve these issues as divest them of epistemological significance. An early phase of this process may be discerned in the nineteenth century, when algebraists like George Boole granted formal rules a newly foundational role in their science. The boundary solidified in the twentieth century with the development of programming languages. Early programming languages such as ALGOL, first introduced in 1958, provided at once a way to control computers and a standard medium for publishing algorithms. As means of communication, programming languages do not, in general, work autonomously from the languages people speak; code typically uses words, both in built-in keywords like `if` and `for` and the user-defined names of functions and variables, to make its workings easier to understand. The received explanation of these linguistic inclusions is that they are mere conveniences that aid comprehension without affecting the algorithm itself, which is defined in terms of a formal semantics. This division between "hard" algorithmic logic and "soft" communicational matters—a division that came to pervade the discourse of computer science—gives programmers license to push ahead in the design of computational systems without worrying about what it would take to establish an accord about meaning, if, indeed, this accord is ever established at all.

Historicizing the relation of symbolic methods to language shows that this way of thinking is not inherent to symbolic methods; things have been otherwise in the past, and they could be otherwise in the future. Language-based AI systems like GPT-3, with their admixture of computational logic and collectively produced linguistic data, push the distinction between computation and communication to its utmost limits, and they thus provide an occasion to reconsider fundamental assumptions about how computational processes relate to language. The central claim of this book is that the modern idea of algorithm depends on a particular sort of subject–object divide: the separation of disciplinary standards of rigor from the complex array of cultural, linguistic, and pedagogical factors that go into making systems comprehensible to people. In the discipline of computer programming, these standards provide a way of thinking about computational procedures—of creating them and judging them—that grants these procedures an objective existence as mathematical abstractions, apart from concrete computer systems. This subject–object divide is deeply embedded not just in textbook definitions of *algorithm* but also in the design of modern programming languages, which generally make algorithmic logic as independent as possible from matters of communication; this abstraction facilitates the transfer of algorithms across computer systems and across application domains. This way of thinking was not firmly in place until the nineteenth century, and revisiting the conditions that produced it can help us better understand the implications of language-based machine learning systems like GPT-3. The idea of algorithm is a levee holding back the social complexity of language, and it is about to break. This book is about the flood that inspired its construction.

FROM FORMULAE TO SOURCE CODE

In broaching linguistic issues in relation to mathematics, this book joins a long tradition in the historiography of science. In the 1990s, scholars such as Peter Dear and Robert Markley drew attention to the role of language in the emergence of experimental science in the seventeenth century.[19] More recent scholarship has explored the influence of linguistic disciplines from the past on mathematics and computation. There has, in particular, been a great deal of research on the intersection of linguistics with early computer history, including the importance of theories of syntax for programming languages and the emergence of machine translation as a research program.[20] Looking at earlier time periods, scholars such as Kevin Lambert and Travis D. Williams have discussed mathematicians' engagements with philology, which long concerned itself with the histories of mathematical symbols, and rhetoric, whose techniques can be discerned in mathematical proofs.[21]

With some exceptions, histories of mathematical symbolism have focused primarily on epistemological matters such as changing standards of mathematical proof, the new modes of thought opened by notations like a^b, and how mathematical constructs relate (or do not relate) to reality. This book considers these matters, but it places more emphasis on the relatively neglected communicational side of symbolism. Communication, as the form of the word suggests, requires a common ground between people, and it is not self-evident that this common ground works the same way with words and symbols. For centuries, it has been recognized that the use of words is to some extent constrained by convention. As the seventeenth-century philosopher Bernard Lamy put it, "We might, if we please, call a Horse a Dog, and a Dog a Horse; but the Idea of the first being fixt already to the word Horse, and the latter to the word Dog, we cannot transpose them, nor take the one for the other, without an entire confusion to the Conversation of Mankind."[22] To communicate effectively in English, one must, at least broadly, follow the usages of others. The meanings of algebraic symbols, on the other hand, appear to bend to the individual will: one can write, "let $a = 5$," and that is what a will mean.[23] To many observers, such individualistically defined symbols have seemed, paradoxically, to convey ideas with a level of transparency that words could not match. Historical thinkers have addressed this apparent paradox in a range of ways, reflecting changing precepts about language, knowledge, and the formation of thought.

An attention to these issues complicates received thinking about the role of algebraic symbolism in the origin of modern science. It has long been a common narrative that the Scientific Revolution of the seventeenth century involved the "mathematization" of the physical sciences. The trend more recently has been toward recognizing that the category of mathematics itself changed in the period. The 2016 edited collection *The Language of Nature: Reassessing the Mathematization of Natural Philosophy in the Seventeenth Century* works toward a more nuanced view of what it means for a science to be mathematized.[24] The present study contributes to this nuancing by examining the changing ways people made sense of mathematical symbols from the early modern period to present. While algebraic notation inspired a great deal of excitement in the seventeenth century, this excitement was not, as I hope to show, always tied to a conception of "mathematics" at all. Early on, the excitement had more to do with the visual nature of the symbols, which promised a mode of communication fundamentally different from spoken languages such as English and Latin. Understanding the place of symbolic methods in the history of science thus requires historicizing not only the category of mathematics but also the category of language; in particular, we must consider changing opinions on the relation of writing to speech and on how the common ground of communication should be established.

Attempting to find an absolute beginning for this history would be hopeless. Practices that look to us like algorithms have existed at least as long as writing itself, developing independently in a range of cultures. Aside from some background about ancient and medieval mathematics, my account starts in the sixteenth century, when the modern form of algebraic notation began to be codified. As I discuss in chapter 1, between the mid-sixteenth century and the mid-seventeenth, this notation revolutionized the practice of algebra: whereas equation-solving procedures had previously been expounded largely through words, one could now express them in compact formulae. Amid a general climate of suspicion toward language, such symbols came to be seen as a superior alternative, a way of presenting ideas directly to the eye without the mediation of words. This confidence in the transparency of symbols rested, I argue, on a belief that certain universal ideas were divinely etched onto all human minds, thus enabling perfect communication independently of the contingencies of language.

Leibniz's work was both a culmination of this early modern obsession with symbols and an inflection point. In chapter 2, I discuss the role of symbols in both his mathematical work and his attempt to create a *calculus ratiocinator*. Leibniz was one of the earlier writers (although not the first) to extend the word *algorithm* to something other than variants of the Indian calculating techniques: he used its Latin and French cognates to refer to the differentiation procedure of his version of calculus. Yet his meaning was not quite the modern one, and an attention to this semantic nuance reveals an aspect of the history of algorithmic thinking that has often been overlooked. Leibniz modeled his "algorithm" not on common arithmetic but on symbolic algebra; it consists not of a precisely defined procedure that determines the correct manner of proceeding at each step but rather of a collection of equations for use in transforming expressions. This algebraic sense of *algorithm*, which had widespread and enduring influence, placed the idea in an intimate relation to the development of new symbolic notations.

Leibniz experimented with such notations in a wide range of contexts, from the ∫dx notation for integrals to binary numerals to attempts to develop symbolic methods for politics and law. Shifting ideas about language were, however, undermining the grounds of this project. In Hannah Dawson's account, a pivotal figure in linguistic thought was Leibniz's intellectual rival John Locke.[25] Leibniz's dispute with Locke is commonly interpreted as epistemological, dealing with the legitimacy of nonempirical forms of knowledge. The debate also, as I show in the chapter, had implications for symbolic methods. Leibniz assumed that concepts already existed in the mind at birth, so that stabilizing the meanings of symbols would not be a major problem. Locke troubled this assumption and, by doing so, called into question whether the symbols really were so different from words. In Locke's long shadow, mathematicians paid a heightened attention to conceptual definitions; clarity, it was now believed, stemmed not from the notation itself but from the way mathematical concepts were formed in the mind.

Although the rise of Lockean views of language spelled doom for Leibniz's more extreme claims about symbols, it disrupted neither the development of symbolic methods nor the desire to turn algebra into a universal language. In chapter 3, I focus on a relatively little-discussed successor to Leibniz's universal characteristic developed in the 1790s by Nicolas de Condorcet. At the height of the Reign of Terror following the French Revolution, Condorcet sketched out a system that would provide algebra-like notations for all manner of subjects. Like Leibniz, Condorcet was out to resolve people's political and cultural differences by means of symbols. Yet his method was very different. Unlike Leibniz, Condorcet did not presume that the ideas expressed by symbols were already universal; rather, he wanted to make them universal through a program of education. This approach rendered his system overtly politicized, dependent on a particular vision of what society should look like.

Although Condorcet's scheme can be assigned little direct influence, it typifies a contention over the politics of symbolism that deserves a larger place in the historiography of computation. Standard accounts of eighteenth-century mathematics emphasize a division between national traditions: Continental mathematicians embraced Leibniz's notation and method, whereas the English followed Isaac Newton in rejecting them. I argue in chapter 3 that eighteenth-century mathematics was cut across by ideological as well as national divides. As Sophia Rosenfeld has shown, language became a divisive topic during the French Revolution, as people blamed the Revolution's splintering on a failure to agree on the meanings of such terms as *liberty*, *equality*, and *fraternity*.[26] A central thinker in the linguistic thought of the period, the Abbé de Condillac, held up algebra as a model of the clarity needed to resolve such disagreements. Viewing algebra as, in Condillac's terms, a "well-formed language" led to a number of debates over whether the symbols really did have clear definitions, as in a notorious controversy over the existence of negative numbers. How symbolic methods worked hinged, in this moment, on an issue regarding the politics of language: whether the meanings of signs ought to be governed collectively by the people or decided on by the learned.

This conflict was less resolved than it was abandoned. In the early nineteenth century, algebraists turned their attention from conceptual definitions to formal rules, which provided a new standard of mathematical rigor. In chapter 4, I focus on the work of George Boole, the English Irish mathematician who described the system that would eventually become Boolean logic. Boole's work has seldom been considered part of the universal language tradition exemplified by Leibniz and Condorcet, being typically positioned at the intersection of algebra and logic. But in the 1847 book in which he first introduced his system, Boole describes symbolic logic as "a step toward a philosophical language."[27] Taking this claim seriously, I contend that Boole's project was enabled by another major shift in linguistic thought. While Boole was just an enamored with symbols as his precursors, he lacked their hostility toward words; instead, he espoused a respect and even a reverence toward the languages people inherit from their ancestors. This attitude enabled the two factions that clashed in the eighteenth century to arrive at a truce: instead of replacing language, the symbols were supposed to work together with it, at once drawing rigor from mechanical rules and meaning from words.

The old antagonism toward language would soon enough return in the work of Gottlob Frege, Ernst Schröder, and Rudolf Carnap, who once again envisioned replacing words with symbols. But even their work did not undo the epistemological divisions that formed in Boole's time. In chapter 5, I consider the early programming language ALGOL, whose name means "algorithmic language"—a choice that heralds the widespread adoption, starting around 1960, of the word *algorithm* as a general term for precisely defined computational procedures. ALGOL's creators described it as a "universal language" that could specify algorithms in a form both readable by humans and executable by machines.[28] But as with Boolean logic, ALGOL's claims to universality are narrow. Rather than replacing the vernacular all the way down to the formation of actual human thought, ALGOL employs words (often in English) to help people understand programs. What was supposed to be universal in ALGOL was only the algorithmic "essence" of a program, which was distinguished sharply from issues in which ordinary language still had to play a role, such as communication and education—in short, from the aspects of computation that were coming to be known as "human factors."

The example of ALGOL shows that the algorithm as we now know it depends on a particular way of drawing disciplinary lines. When computer scientists started giving theoretical heft to the term *algorithm*, they were trying to identify essential elements of computational systems that could be analyzed mathematically, in isolation from the messiness of how those machines worked in their social contexts. This division between "hard" algorithmic matters and "soft" social ones remains deeply ingrained in the technical design of programming languages and the discourse surrounding them. But it is not inevitable. Before the late nineteenth century, "algorithms" were not usually understood to exclude issues of communication; through Boole's time, computational procedures typically included rules not just for what to do with symbols but also for what the symbols meant. How to establish this meaning was a matter for philosophical contention, and disparate views about language entailed divergent visions for what universal computation would be.

It is primarily these earlier ways of thinking—the ones that are noticeably different from modern computation—that I emphasize in this book. In the history of science, it is a methodological precept to avoid falling into the style called "Whig history"—to avoid, that is, describing historical developments through linear narratives of progress that implicitly side with the positions that won. Histories of mathematical symbols tend to be extremely Whiggish, complimenting authors who use notations that later became standard and chastising those who do not. I certainly do not mean to deny the advantages of symbolism, but my purpose is less to celebrate it than to understand it, and I accordingly hope to describe what was lost with the adoption of symbols as well as what was gained. I also hope to show that the symbolic method is not a fixed category. The ways people have understood symbols changed multiple times over the centuries, and the modern idea of algorithm is a product of particular circumstances and epistemological commitments.

Signs of another such change began to appear in the early twenty-first century. Over the course of the 2010s, the word *algorithm* came increasingly to refer not to the precisely defined procedures ALGOL was designed to represent but to machine learning systems like GPT-3. While the idea of machine learning has existed since the early computer era, this shift in the meaning of *algorithm*, as I argue in the coda, represents more of a break from twentieth-century conceptions than has generally been recognized.[29] Text generators like GPT-3 promise a new programming paradigm in which, instead of designing a computational procedure, programmers give the computer orders in English. Even for those (perhaps a minority) who are fully comfortable with this idea, it is hard to deny that its widespread adoption would give a renewed importance to the flaws of language—to the possibility that words are not actually clear or stable enough to form an adequate medium for technical knowledge. With the widespread adoption of machine learning, the division between "hard" logic and "soft" communicational matters has become troubled, and algorithms have become a site of contestation.

New as these developments are, they in some ways mark a return to the situation in the eighteenth century, before Boole and his contemporaries threw up a barrier between symbols and language. Mathematicians in the eighteenth century did not view the meanings of words as irrelevant to symbolic methods; instead, they heartily debated whether symbols had to correspond to received definitions of words or whether they could be defined anew. Nor did they set computational systems apart from politics. Some viewed symbols as a way of challenging received ways of thinking, an idea that came to be associated with the rationalizing reforms of the French Revolution. Others took the opposite view, cherishing words as a precious inheritance whose influence was needed to keep mathematical knowledge in line with the culture of a country. Attending to these earlier discourses, as this book aims to do, can provide us with a better sense of the possibilities and problems that exist at the intersection of computation and language.

It may be helpful to think of this history as a succession of guiding terms—ideas that, in particular historical contexts, set the standards by which symbolic methods were judged. In the seventeenth century, Europeans typically described computation as an *artifice* or *art*, meaning a systematically developed set of skills. What made computation an art was its transmissibility: one could physically demonstrate, articulate, or write down the correct way of doing it, thus enabling people to develop and practice the skill in a controlled fashion. In the eighteenth century, the valuing of artifice largely gave way to the cult of *natural reason*—a guiding principle that valued the mind's inborn faculties. This way of thinking encouraged a deemphasis of explicit rules in favor of conceptual explanations that were supposed to make the correct way of performing a computation intuitively obvious. In Boole's time, the reaction against Enlightenment thought led to a turn away from natural reason to the quite contrary valuing of *culture*. Under this star, the mechanical had to be balanced with the organic, and thus abstract mathematical systems and human thought, as fostered by the languages that develop in communities, formed two halves of a whole.

While the idea of culture continues to influence computation, the idea guiding the modern algorithm is, if anything, technology. *Technology* is a very old word, but it once meant something very different from its present sense, referring either to a treatise about a skilled practice or to the set of technical terms used in discussing it.[30] The modern meaning, which became dominant in the late nineteenth century, has more to do with the practical application of scientific knowledge. Viewing computation as technology encourages defining problems precisely so as to isolate aspects of systems that can be subjected to rigorous engineering methods—a perspective that motivated early computer scientists to theorize algorithms as abstract procedures that may be analyzed apart from the specific contexts in which they are used. The full ramifications of this divide-and-conquer strategy did not become apparent until the early twenty-first century, when techniques that were developed within an intellectual framework that abstracted out almost all human experience became a force that runs much of the world.

The history of symbolic methods is in some ways remote from the political contentions that now surround algorithms. This book largely deals with a time when the idea of universal computation was more a matter of starry-eyed speculation than a social reality. But many of the issues that arose from this speculation have remained with us in the computer age. Questions like whether symbolic methods can or should be politically neutral have come up again and again over the centuries at moments when these methods were venturing into new territory. The terms of debate, however, have varied widely, and attending to earlier moments can be revealing about the assumptions of the present discourse. I begin in the early modern period, when excitement about symbolic methods was widespread—but for reasons quite opposed to those that have inspired the hype surrounding twenty-first-century AI.

CHAPTER ONE

Symbols and Language in the Early Modern Period

The alphabet is really now superfluous
for in this sign all men can find salvation.

—GOETHE, *Faust, Part II* (trans. Atkins)

IDOLS AND HIEROGLYPHS

In the scientific circles of the seventeenth century, words had a bad reputation. In the 1623 version of his book *The Advancement of Learning*, Francis Bacon warned against what he called the "idols of the market"—the "vulgar" notions that, in everyday speech, tend to "insinuate themselves into the understanding" by means of words.[1] As a protection against "the seducing incantation of names," he tentatively suggests definitions and "terms of art," but even these are not enough; truly preventing words from "doing violence to the understanding," he states, will require "a new and deeper remedy."[2] At almost exactly the same time, there was an explosion of new mathematical symbols.[3] In the mid-1500s, algebra often took the form of words, with even equations, which we now think of as made out of symbols, appearing in knotty prose. By the mid-1600s, this logorrhea had given way to compact symbolic expressions like $ax + b = c$. Although Bacon himself had little interest in mathematics, scholars have long noted an alliance between these new symbols and his followers' hostility toward language.[4] Algebraic notation, brought into something like its modern form by Thomas Harriot and René Descartes in the early decades of the 1600s, came to be associated with a philosophical ideal of clarity, and numerous thinkers, G. W. Leibniz among them, envisioned developing analogous symbols for all manner of subjects.

This chapter gives an overview of the symbolic methods that existed before Leibniz's arrival on the scene in the 1660s. It focuses on two practices that would eventually form major sources for the modern idea of algorithm. The first is the set of techniques to which the word *algorithm* originally referred. This word (then more commonly spelled *algorism*) generally referred to the procedures of numerical computation that probably originated on the Indian subcontinent in the medieval period.[5] The second is the algebraic symbolism that solidified in the early 1600s. Whereas it is now a cliché to call mathematical notation a universal language, early modern textbooks presented numerals and algebraic symbols less as language than as forms of writing comparable to the alphabet.[6] Alphabetical writing, as the linguist Amalia E. Gnanadesikan explains, "is a transformation of language, a technology applied to language, not language itself."[7] To their early modern advocates, symbols promised a way of improving the technology of writing so as to free it from the uncertainty of words. This view raised theoretical problems that would ultimately explode in the debate between Leibniz and John Locke, and that would render symbolic methods philosophically contentious for centuries.

The early reception of symbolic algebra reflected a clash between conceptions of mathematical knowledge. As numerous scholars have shown, the question of what constituted "mathematics" was far from settled at the time; the category traditionally encompassed not just geometry and arithmetic but also astronomy and music, and some writers extended it to other practices such as the construction of machines.[8] For many thinkers in the period, the heart of mathematics was Euclidean geometry. For instance, when Galileo Galilei made his famous statement—in his 1623 book *The Assayer*—that God wrote the book of the world in the language of mathematics, he was explicitly referring to geometric diagrams, not to any sort of symbolic notation.[9] Throughout the sixteenth and seventeenth centuries, Europeans held algebra in lower esteem than geometry, since it was not one of the traditional liberal arts and was perceived to lack rigorous standards of proof.[10] Symbolic algebra transformed a range of practices in the seventeenth century, but its methods were widely regarded as practical rather than truly scientific, and they would long be hounded by conceptual difficulties.

Going back to G. H. F. Nesselmann's work in the nineteenth century, historians of mathematics have explained the development of algebraic symbolism with a three-stage model.[11] First is the rhetorical phase, in which equations are presented entirely in words: "Three unknowns plus five equals twenty." Next is the syncopated phase, in which some symbols are used as ligatures or abbreviations of words: "3 co. p. 5 eq. 20." Finally, in the symbolic phase, the symbols replace words altogether and take on an epistemological role: "$3x + 5 = 20$." This model captures the gradualness of the process by which words gave way to symbols. Some of the basic algebraic symbols originated as abbreviations: in his 1557 book *The Whetstone of Witte*, Robert Recorde explains the = sign as a way "to auoid the tediouse repetition of these woordes: is equalle to."[12] Such symbols, according to Nesselmann, eventually took on uses beyond merely shortening texts, making it possible to solve complex problems by transforming arrangements of symbols on a page.

This three-stage account places much of early modern algebra in a gray area. As Albrecht Heeffer has argued, Nesselmann's chronology is unclear; the syncopated phase includes both ancient mathematicians such as Diophantus and early modern ones such as François Viète, ignoring the variety of mathematical practices that existed between them.[13] Nesselmann's account also muddles the issues of what symbols people used—special signs such as + and = versus words such as *plus* and *equals*—and how they used the symbols. Viète employed a notation that mixed symbols with Latin words, which he even inflected in accordance with the rules of grammar—instead of =, he used *æquatur*. Yet he subjected these semiverbal equations to rule-based transformations much like the ones now employed in symbolic algebra. If we are looking for the origins of the style now known as symbol manipulation, then the transitions Nesselmann describes are not necessarily pivotal. As far as problem-solving methods go, it makes little difference whether one manipulates words, abbreviations, or symbols.

This revision of Nesselmann's account, however, leaves an explanatory gap. Even if trading words for symbols had little effect on the procedures of algebra—on what one would now call the algorithms—symbolic notation was not viewed as a minor development in the seventeenth century. Leibniz was far from the only one to see symbols as a basis for a universal method; numerous thinkers, including Descartes and Isaac Newton, considered the possibility of doing for other fields what numerals and algebraic symbols had done for mathematics. Symbolic notation has an obvious advantage in its compactness, but this fact alone cannot explain the degree of the fervor. Advocates represented symbols as a way of putting thoughts directly on the page without mediation; algebraic notation was widely viewed as a way of circumventing Bacon's idols of the market and granting knowledge a degree of certainty that words could not match. To understand these attitudes, we must contextualize the development of the notation not only in terms of the mathematical thought of the time, but also in terms of early modern ideas about language and writing.

To do so, one must step into the mindset of a population for whom reading and writing were not nearly as pervasive as they are today. Prior to the late seventeenth century, the word *language* primarily referred to spoken communication, and writing still seemed to many people, as Jonathan Hope puts it, "a strange technology," a sometimes unreliable means of recording spoken words so that they could be recited later.[14] The sixteenth century witnessed a number of attempts to make the technology of writing more efficient and dependable. A 1588 book by Timothie Bright describes an art of "characterie" that provides a means of "shorte, swifte, and secrete writing by Character."[15] Bright's system consists of a large number of "Characters," each of which has a "value, or signification" defined by a word (figure 1.1).[16] By this means, one could write a whole word using no more space than a single letter. The shorthand movement was, as James Dougal Fleming has noted, entirely confined to England in the early modern period, but shorthand-like practices existed in a range of languages.[17] Alchemy and astrology, for instance, employed complex systems of symbols that were viewed as secret forms of writing akin to cryptographs and hieroglyphs.[18]

The fascination with these "characters" stemmed in part from the fact that they placed writing in a different relation to spoken language compared to alphabetical writing. In the early modern period, Europeans tended to discuss reading as if it involved a voice, be it literal or imagined.[19] This way of thinking had a classical warrant, albeit one that was increasingly viewed as unsatisfactory. In *On Interpretation*, Aristotle describes the signification of written language as a multistage process: letters signify (spoken) words, words signify concepts, and (at least in the interpretation of some medieval readers) concepts signify things.[20] Whereas the last step was endlessly controversial among the Scholastics, the first step was often glossed over. In early modern linguistic thought, it was common to use the word *letter* (in Latin, *littera* or *litera*) to refer indifferently to both alphabetical characters and the speech sounds they represent.[21] Nonphonetic symbols would seem to change this situation: Recorde's "=," for instance, does not in any obvious way represent the sounds of the words *is equalle to*. The Aristotelian model provides no clear guidance as to such symbols—the equals sign could be taken to signify the words in the manner of Bright's "Characterie," or else it could be seen as bypassing words and cutting straight to the concept of equality.

In the case of numerals and algebraic symbols, there was a major mark in favor of the latter. Unlike alphabetical writing, these symbols could be read aloud in multiple languages: English speakers read "9 – 1" as *nine minus one*, whereas French speakers read it as *neuf moins un*, and the meaning

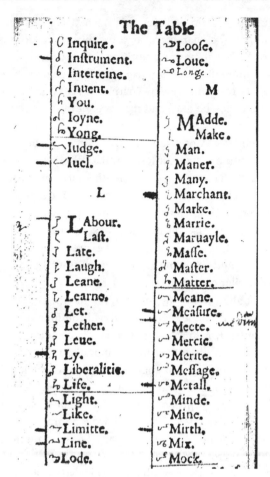

Figure 1.1. An example of an early modern shorthand notation, from Timothie Bright's 1588 book *Characterie*. The Bodleian Libraries, University of Oxford, Douce W 3 (Weston Stack), sig. ¶3v. Images produced by ProQuest as part of *Early English Books Online*. www.proquest.com. Images published with permission of ProQuest. Further reproduction is prohibited without permission.

appears to be the same in both cases. Borrowing from the Scholastic termi-nology, early modern thinkers explained such translinguistic symbols by distinguishing between *nominal characters*, which represented the sounds of words, and *real characters*, which referred directly to ideas or things. The idea of a real character appears most famously in Bacon's 1623 *Advance-ment of Learning*, where it had particular reference to kanji—a subset of the Chinese han characters that can be read in either the Japanese or the

Chinese language.[22] Bacon explains this quality as a departure from the Aristotelian model: the characters "express, not their letters [i.e., speech sounds] or words, but things and notions; insomuch, that numerous nations, though of quite different languages, yet, agreeing in the use of these characters, hold correspondence by writing."[23] As a result, "a book written in such characters may be read and interpreted by each nation in its own respective language."[24]

Although Bacon does not discuss mathematical symbols in the passage, it later became routine to cite numerals and algebraic symbols as examples of real characters.[25] For instance, the mathematician John Wallis wrote that, like Chinese characters, algebraic signs "so little need the intervention of Words to make known their meaning, that, when different persons come to express, in Words, the sense of those Characters, they will as little agree upon the same Words, though all express the same sense, as two Translators of one and the same Book into another Language."[26] Robert Hooke made a similar comparison with regard to "Arithmetical Figures."[27] Since it does not align with how modern linguistics understands writing, a number of modern scholars have dismissed the idea of a real character as mistaken or absurd; Jaap Maat goes so far as to call the idea a "myth."[28] But the idea is not wholly senseless when applied to mathematical symbols. Calling these symbols real characters amounts to claiming that they express universal ideas that are accessible to all people, regardless of what words one chooses when pronouncing them—that the English word *minus* and the French word *moins* really share a common core of meaning by means of which the minus sign can be used to communicate the idea of subtraction clearly across languages.[29]

This is not to say that the real-character idea still holds water. The idea depended on a faith that the human mind was divinely constructed to mirror the world, which precludes any serious recognition of cultural diversity. Not everyone thought this way in the seventeenth century. But numerals and algebraic symbols really are, in a sense, more like an alternative to the alphabet than a language. As the next section shows, seventeenth-century textbooks taught numerals in a way that emphasized physical pen skills and speaking numbers aloud. This pedagogy had more in common with learning to read and write (or with learning a shorthand like Bright's "characterie") than with learning a second language. There was also a difference in regard to gender: vernacular tongues were largely learned from women—from mothers and nurses—whereas writing and mathematics alike were both typically learned from male teachers.[30] Examining how numerals were taught sheds light on why early modern philosophers such as Leibniz put so much stock in the power of symbols—and on how the algorism of their time was different from algorithmic thinking as we now know it.

THE MEANING OF ALGORISM

That computation has anything to do with writing is not a given. Pre-modern cultures developed a wide array of calculating implements, from abaci to the intricate, multilevel counting tables developed by the Inca. The rise of the Hindi–Arabic numeral system, however, gave computation a strong link to writing that would long implicate it in philosophical debates about language. This system probably originated on the Indian subcontinent; the astronomer Brahmagupta described it in Sanskrit verse around 628 CE, although there is evidence that it was already in use prior to his work.[31] The numerals later spread to the Arabic world, where they were described by the mathematician and astronomer Muḥammad ibn Mūsā al-Khwārizmī.[32] While al-Khwārizmī did not invent these methods, it was his name that inspired the word *algorithm*, and so it is worth considering the contents of his work.

Probably born in what is now Uzbekistan, al-Khwārizmī secured a position at the House of Wisdom, a library in Baghdad, where he wrote on a number of topics.[33] His c. 820 *Compendious Book on Calculation by Completion and Balancing* is largely about solving equations; the word *algebra* is derived from the word *al-jabr* (usually translated as "completion") in the Arabic title of this book.[34] His work on arithmetic, unfortunately, survives only in unreliable Latin translations.[35] The most famous of these translations is sometimes called "Dixit Algorizmi" ("Algorizmi said") because the translator inserted that phrase at the beginnings of the first two paragraphs; this bit of scribal happenstance is thought to be the origin of the Latin word *algorithmus* and thus, ultimately, of the English word *algorithm*.

From what we can gather from the surviving translations, al-Khwārizmī's arithmetic book presented procedures for addition, subtraction, multiplication, division, halving, and doubling.[36] In the "Dixit Algorizmi" version, the explanation of addition begins like this:

> You will add each place to the place that is above it with regard to its own kind, i.e., units to units and tens to tens. When ten has been collected in one of the places, i.e., in the place of the units or tens or in some other place, put a one instead of it and elevate it to a higher place, i.e., if you have ten in the first place which is the place of the units, make a one of it and raise it to the place of the tens and there it will signify ten. But if there remains something from the number that is less than X or the num-

ber itself is less than X, leave it in the same place. And if nothing remains, put a circle (i.e., o), so that the place may not be empty.[37]

This procedure contains many of the hallmarks of the intellectual style that eventually came to bear al-Khwārizmī's name, including the use of conditional, if–then logic and even what we might think of as a loop: "Do likewise," we are instructed, "also in all the places."[38] This is an archetypal algorithm: early computer scientists took it as a model for the sort of procedure that a machine can be programmed to perform.

Through the seventeenth century, the word *algorism* or *algorithm* still referred primarily to variants of this particular set of procedures. (While the spelling with a *th* appeared in Latin as early as the 1480s and in English in the 1650s, I will refer to these historical practices as *algorism* so as to avoid confusion.[39]) Some authors distinguished algorism from arithmetic, which was one of the seven liberal arts set out by the medieval philosopher Boethius. Arithmetic, in the Boethian sense, was about types of numbers: squares, primes, perfect numbers, and a range of others that are less well remembered.[40] As the Elizabethan polymath John Dee put it, the purpose of this study was "arise, clime, ascend, and mount vp (with Speculatiue winges) in spirit, to behold in the Glas of Creation, the *Forme* of *Formes*, the *Exemplar Number* of all thinges *Numerable*: both visible and inuisible, mortall and immortall, Corporeall and Spirituall."[41] In contrast to such lofty doctrines, algorism was seen as one of the lower branches of mathematics. While it was sometimes taught in Latin schools, the art of computation was primarily the business of the (mostly) men E. G. R. Taylor dubbed "mathematical practitioners"—people who taught mathematics independently of the university system through textbooks, lecturing, and tutoring.[42] The methods they taught had applications in navigation, trade, and finance, in artisanal trades such as bricklaying and construction, and in military practices such as ballistics.

In these practical fields, the Hindi–Arabic algorism competed with and sometimes worked together with a range of other forms of computation. The practitioners sold instruments such as the sector, which consisted of a hinged pair of rulers inscribed with scales that could be used to perform approximate calculations. Abaci and counting stones had a long history, and some people continued to prefer them; counting stones were especially popular in Germany, and abaci would remain in use for centuries in eastern Europe. Calculation could also involve numerical tables, in which one could look up certain values without having to compute them oneself. The early seventeenth century saw a major advance in such techniques with the development of logarithm tables, which were introduced in 1614 by John Napier.[43] Since adding the logarithms of two numbers produces the logarithm of their product, one could use logarithm tables to reduce multiplication to the much easier operation of addition. This technique was the basis of some of the period's most advanced mathematical instruments, such as William Oughtred's "circles of proportion," a precursor of the slide rule that he described in 1632.[44]

Apart from teaching and selling instruments, the practitioners also published books from which one could, at least in principle, learn the art of calculation. Most of these books started with "common algorism," meaning the use of ordinary counting numbers. This doctrine began with "numeration" or "notation," which meant learning the meanings of the digits; afterward came discussions of operations such as addition, subtraction, multiplication, and division.[45] The exact list varied. Some texts followed al-Khwārizmī in including special procedures for halving and doubling, and some added further operations such as extracting roots. Some textbooks also included special "algorisms" for calculations involving currency as well as more advanced ones for rational numbers and decimal fractions. Algorism also included additional procedures intended to verify results, since (no doubt) human computers would often make mistakes.

The procedures described in these books resemble modern algorithms in their use of rigidly defined steps that begin and end with arrangements of symbols. It is this rigidity that gives the procedures the mechanical quality that inspired Blaise Pascal and Leibniz to build calculating machines. But the first part of algorism—numeration—is different. According to Johann Lantz's 1616 arithmetic text, which Leibniz encountered in school, numeration is "the enunciation and expression of whatever number is set forth."[46] (The Latin *enunciatio* can mean either *pronunciation* or *proposition*, suggesting a conflation of words and ideas similar to that of the Greek *logos*.) The first step was to familiarize oneself with the nine digits 1, 2, 3, 4, 5, 6, 7, 8, and 9 as well as the "cipher" 0, which was viewed as a mere placeholder that had no inherent meaning. Students had to be able to recognize and inscribe these symbols dependably; algorism thus, as Jessica Otis has argued, required the basic pen skills that were a part of literacy.[47] They also had to learn the symbols' values, which were often taught through tables similar to the one by which Bright defined the "significations" of his shorthand characters (figure 1.2). They also had to learn the rules by which numerals are composed into numbers so that they could translate them into the number words of a language, as 84 becomes "eighty-four."

An extended discussion of this translation appears in another text Leibniz studied in detail: Johann Heinrich Alsted's 1630 *Encyclopaedia*.[48] Alsted was educated at the Herborn Academy, which was a center of pansophism,

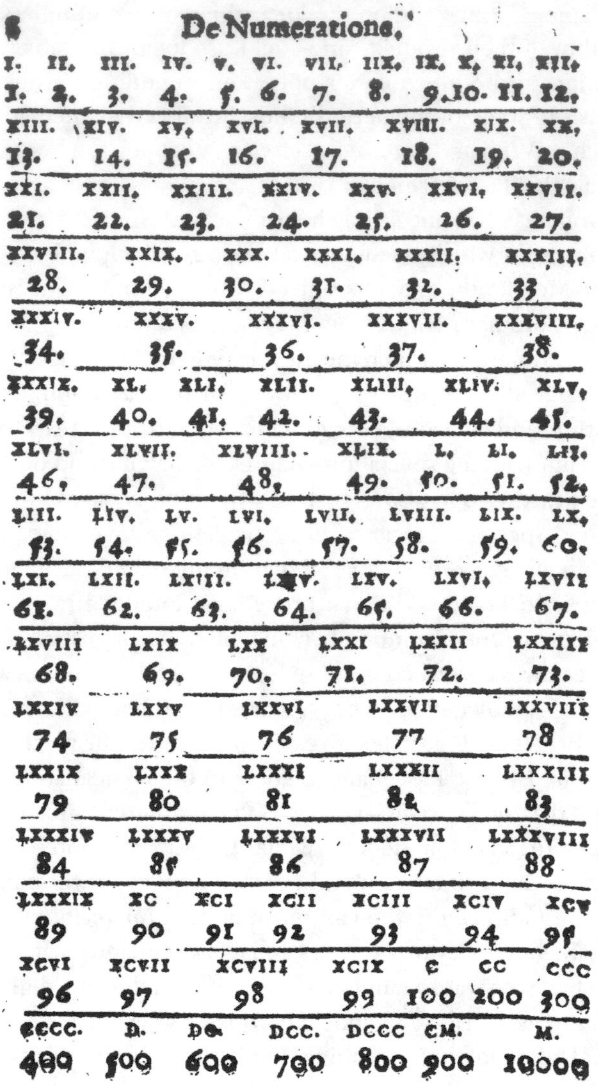

I. II. III. IV. V. VI. VII. IIX. IX. X. XI. XII.
1. 2. 3. 4. 5. 6. 7. 8. 9. 10. 11. 12.

XIII. XIV. XV. XVI. XVII. XVIII. XIX. XX.
13. 14. 15. 16. 17. 18. 19. 20.

XXI. XXII. XXIII. XXIV. XXV. XXVI. XXVII.
21. 22. 23. 24. 25. 26. 27.

XXVIII. XXIX. XXX. XXXI. XXXII. XXXIII.
28. 29. 30. 31. 32. 33

XXXIV. XXXV. XXXVI. XXXVII. XXXVIII.
34. 35. 36. 37. 38.

XXXIX. XL. XLI. XLII. XLIII. XLIV. XLV.
39. 40. 41. 42. 43. 44. 45.

XLVI. XLVII. XLVIII. XLIX. L. LI. LII.
46. 47. 48. 49. 50. 51. 52.

LIII. LIV. LV. LVI. LVII. LVIII. LIX. LX.
53. 54. 55. 56. 57. 58. 59. 60.

LXI. LXII. LXIII. LXIV. LXV. LXVI. LXVII
61. 62. 63. 64. 65. 66. 67.

LXVIII LXIX LXX LXXI LXXII LXXIII
68. 69. 70. 71. 72. 73.

LXXIV LXXV LXXVI LXXVII LXXVIII
74 75 76 77 78

LXXIX LXXX LXXXI LXXXII LXXXIII
79 80 81 82 83

LXXXIV LXXXV LXXXVI LXXXVII LXXXVIII
84 85 86 87 88

LXXXIX XC XCI XCII XCIII XCIV XCV
89 90 91 92 93 94 95

XCVI XCVII XCVIII XCIX C CC CCC
96 97 98 99 100 200 300

CCCC. D. DC. DCC. DCCC CM. M.
400 500 600 700 800 900 1000

Figure 1.2. Hindi–Arabic numerals explained by means of roman ones. From Nicolaus Kauffunger's 1647 German-language textbook *Plenaria Arithmetica*, p. 8. Kauffunger explains that numeration (*Numeriren*) teaches students how they "actually should correctly and tidily write and pronounce each number, just as, in grammar, orthography teaches correct writing" (2; translation mine).

an educational movement that emphasized making knowledge accessible to all; his encyclopedia provided a model for Leibniz's own encyclopedic endeavors. Leibniz judged Alsted's treatment of mathematics to be merely "average for his time," but it is useful as an example of what would have been considered typical in the mid-seventeenth century.[49] In his chapter on arithmetic, Alsted states that the digits are like an "Arithmetical alphabet."[50] Like Lantz, he associates understanding this alphabet with translating the symbols into words: numeration, he writes, is "the right enunciation of rightly written numbers."[51] He explains several techniques for this translation, including ways of marking the symbols up so as to make the translation easier: to help make sense of 89765878910, for instance, one can draw dots above or below every third digit after the first, going right to left, as in 89765878910.

While Alsted is concerned with the "right enunciation" of numbers, his point is not that there is only one correct way to do it. The Greeks and Romans, he observes, expressed numbers in various ways that are often much more verbose than the numeration of modern Latin. According to Alsted, Pliny the Elder might have expressed the number of soldiers in Xerxes's army, 5,283,220, as (to translate loosely) "fifty times and twice a hundred and eighty-three thousands, two hundred twenty," whereas one would now write "five double thousands, two hundred eighty-three thousands, two hundred twenty."[52] Likewise, markup procedures can produce different readings depending on how they are done: 10000000000 leads to "ten thousand thousand thousand," whereas 10000000000 leads to "a hundred hundred thousand thousand."[53] Grouping the symbols by threes is, Alsted tells us, the easiest way of doing it, but he makes it clear that, whichever words one chooses, the number itself remains the same.

This mediation between symbols and words has been overlooked in accounts of how language and mathematics related in the early modern period. Walter Ong and Robert Markley each have argued that early modern mathematics was antidialogic—that it suppressed the back-and-forth interchange that was valued in classical thought.[54] In *Phaedrus*, Plato has Socrates argue that speech is superior to writing because, in conversation, one can dynamically respond to questioning.[55] When one reads an explanation off a sheet of paper, the words are always the same, but when one explains something one genuinely understands, the words come out differently every time. Measuring intelligence by the ability to hold a conversation (an idea that persisted all the way to Alan Turing) implies that merely having a written text at hand is no guarantee that one knows anything. One might suppose that numerals, in their alienation from the phonetics of spoken languages, are even more inimical to this Socratic conception of knowledge than alphabetical writing. But in the pedagogical scene evoked by algorism texts, the opposite is the case. Contrary to Ong's association of mathematics with a "silent object world," numerals were not supposed to be contemplated mutely.[56] Instead, they enable the teacher to prod the student to speak: sure, it is that many thousands, but how many hundreds? How many tens?

Yet this proliferation of verbalizations had its limits. It may be equally acceptable to read 4206 as "four thousand two hundred and six" or as "forty-two oh-six," but reading it as "twelve" would simply be wrong. Teaching people to understand digits required establishing an accord about their values, which provided the common ground on which the dialogue took place. Alsted does not directly reference Bacon's idea of a real character in this passage, but his account of the "signification" of numerals reveals a similar way of thinking. Just like kanji characters, numerals enable people from many nations to "read and interpret," as Bacon put it, a text in their own languages; yet the numbers to which the symbols refer remain the same (we are supposed to assume) regardless of what language or what specific wording one chooses. The numerals thus offload, as it were, the signifying function that is ordinarily handled by languages like Latin onto the basic operating system of written communication: the alphabet itself.

This background makes it easier to understand Leibniz's confidence that the power of "calculation" could be extended to other areas. While the *calculus ratiocinator* is often discussed together with Leibniz's work on calculating machines, the two are distinct, and it is not clear that the "calculation" by which he hoped to settle disputes was supposed to encompass only those aspects of calculation that a machine could perform. Knowing how to numerate correctly—being able to choose the right digits at the beginning of the process and read off the results correctly at the end—was also part of what *calculation* meant in his time, and using a machine did not render this knowledge irrelevant. Crucially, the fact that numeration was less mechanical than the rules of operation did not imply that it was any less certain. Alsted's account of numeration suggests that the Hindi–Arabic system can express numbers in such a way that there can be no doubt about their true values. It thus provides reason to think that symbols could extend a similar level of certainty to other areas.

This confidence in the power of symbols depends, however, on the stability of the ideas or things those symbols are supposed to signify. One might pronounce the symbol 8 with either the English *eight* or the Latin *octo*, but are the meanings of these two words really the same? If we place the languages into their historical contexts, they are arguably not: the Romans did not understand numbers the same way we do today. This may seem a pedantic distinction, but the problem is more glaring in domains that are more obviously contentious than common arithmetic. If we devise a character, for instance, to signify *grace*, it is far from a given that everyone will understand its meaning in the same way. As the next section shows, such issues were not specific to utopian schemes like Leibniz's. They also arose in another major part of al-Khwārizmī's mathematical work: algebra. While it had clear practical use, algebra also presented answers that could not be "said" in the way the results of algorism could. Its adoption thus unsettled the peaceful relationship numerals instated between symbols and language, raising theoretical difficulties that would recur in discussions of symbolic methods for centuries.

UNSPEAKABLE NUMBERS

The fact that the words *algebra* and *algorithm* derive from the work of the same person should not be taken to mean they have any necessary connection. If algebra means a way of solving equations—which is what the word almost always meant before the nineteenth century—then its history goes back millennia. Geometric equation-solving methods existed in Mesopotamia and ancient Greece; Chinese mathematicians began developing numerical techniques for equation solving around 200 BCE.[57] Diophantus of Alexandria anticipated some elements of symbolic algebra in the third century CE.[58] That such practices are algorithmic is far from a given, and modern algebra, indeed, contains much that is not. But algebra did play an important role in the development of algorithmic thinking. Al-Khwārizmī opens his account of algebra by placing it under the scope of calculation: "When I considered what people generally want in calculating, I found that it always is a number."[59] Algebra, in this form, is an art of calculation that produces numbers—not one that is identical to algorism but one that can work together with it.

Al-Khwārizmī's version of algebra survives in the work titled *The Compendious Book on Calculation by Completion and Balancing*. His book begins with definitions and general techniques for equation solving along with geometric proofs, which he performed in the style of Mesopotamian geometry.[60] The two basic techniques are *al-jabr* (الجبر), which means *completion* or *restoration*, and *al-muqābala* (الْمُقَابَلَة), which is translated in this context as *balancing*. While the word *algebra* derives from *al-jabr*, it now encompasses both of these techniques—moving terms from one side of an equation to the other (*al-jabr*) and canceling terms (*al-muqābala*). Employing these two techniques, al-Khwārizmī describes procedures for solving six types of equation; he then discusses mercantile calculations, techniques for computing the areas of geometric figures, and calculations involving inheritance. This final section is indicative of the uses to which algebra was put in al-Khwārizmī's context. The Abbasid Caliphate had complex rules for inheritance that required fulfilling certain equation-like conditions; al-Khwārizmī's discussion of these issues takes up almost half the book.[61]

Like the surviving translations of al-Khwārizmī's arithmetic, the algebra is oriented toward teaching the reader a set of practical methods. In contrast to the direct instructions of the algorism, he explains his algebraic methods through specific problems, such as "half of a square and five roots are equal to twenty-eight dirhems [a unit of currency]."[62] That is, in modern notation, $x^2/2 + 5x = 28$. His solution for this problem begins as follows: "Your first business must be to complete your square, so that it amounts to one whole square. This you effect by doubling it. Therefore double it, and double also that which is added to it, as well as what is equal to it. Then you have a square and ten roots, equal to fifty-six dirhems."[63] What he is instructing us to do here—doubling the coefficients—is specific to this instance of the problem: if the first number is something other than a half, one has to multiply by some other number. Yet this example is meant to stand in for a broader class of problems. "Proceed in this manner," he concludes the section, "whenever you meet with squares and roots that are equal to simple numbers: for it will always answer."[64] Al-Khwārizmī's compiling of these instructions marks the beginning of what Victor J. Katz and Karen Hunger Parshall call the "algorithmic stage" in the history of algebra—a stage in which algebra consisted primarily of procedures for solving particular classes of problem.[65]

Like algorism, algebra was long regarded as a practical matter in Europe. Al-Khwārizmī himself would have had no strong reason to classify his work as either practical or theoretical—Islam, as it was interpreted by the Abbasids, encouraged a continuum between secular and holy knowledge.[66] European universities had comparatively rigid hierarchies of prestige, and algebra did not self-evidently deserve the status of a learned discipline. Unlike arithmetic and geometry, algebra had no place among the seven liberal arts; its diffusion in Europe was largely due to merchants such as Leonardo of Pisa (Fibonacci), who encountered Arabic mathematics while accompanying his father on a trading expedition to North Africa and described algebraic methods in his 1202 *Book of Calculation*.[67] During the Renaissance, the Italian mercantile academies known as "abacus schools" taught algebra alongside a range of calculating practices.[68] While algebra was sometimes taught at universities, its academic status long remained less secure than those of geometry and arithmetic.[69] Even in Leibniz's lifetime, powerful figures such as Isaac Barrow, who became the first Lucasian professor of mathematics at Cambridge at 1663, were dismissing algebra as a mere problem-solving technique lacking the force of demonstration.[70]

Algebra's initial lack of prestige resulted not just from its association with commerce but also from conflicting ideas of number. In the early modern period, European mathematics was heavily under the sway of Euclid, along with other classical Greek thinkers such as Archimedes and Plato.[71] Whereas we now tend to think of cardinal number as a singular concept, Euclid employs two number-like concepts that he treats as entirely distinct: quantity and magnitude. A quantity, for Euclid, is a number of things, such as *four apples*, whereas a magnitude is a length, area, or volume. Euclidean magnitudes can also be compared by means of ratios, which he treats as distinct from the magnitudes themselves; this distinction persists in the use of the different notations $a/b = c/d$ and $a : b :: c : d$. Euclid also differentiated magnitudes by dimension, which algebra violated in its use of square numbers: the expression $x^2 + x$ would seem to add an area to a length, which, in Euclidean terms, is nonsense.[72]

Arabic mathematicians had extensive access to Greek sources, so their work should not be seen as a wholly separate tradition. But the *al-jabr* and *al-muqābala* did in some ways clash with Euclidean ideas of number. When solving a quadratic equation, one has to take square roots. The situation was clear enough when the root was rational: $\sqrt{16} = 4$. Yet one could also end up with a root whose exact value cannot be pinned down. In book 10 of the *Elements*, Euclid proves the existence of lines whose lengths cannot be expressed as multiples of any common unit.[73] The proportion of a square's diagonal to one of its sides, for instance, is $\sqrt{2}$, whose value is not quite equal to any fraction. Such numbers arose frequently in solving quadratic equations and posed a problem for attempts to "calculate" an exact numerical solution.

Euclid provided a suggestive term for these numbers: *alogos* (ἄλογος), or, as it might be translated, unspeakable.[74] If the ratio between two magnitudes is irrational, this term suggests, then its true value cannot be said. This suggestion of a lack of speech was widely known in the medieval and early modern periods. While a number of thinkers, including the Persian astronomer Jamshīd al-Kāshī, had developed methods for calculating roots, this could never be done exactly in such cases.[75] A marginal note in a copy of al-Khwārizmī's algebra book states that one must be content with "an approximation, and not the exact truth: for God alone knows what the exact root is."[76] Such attitudes were common among Islamic mathematicians as well as some Christians such as Nicolas of Cusa and Jacques Peletier, who both connected mathematics to a notion of divine mystery.[77] Others, however, were more disturbed by the apparent unspeakability of irrational quantities, whose use in algebra did not seem to measure up to Euclidean or Archimedean standards of knowledge.

One potential resolution was to forget about speaking the numerical value and be satisfied with saying "the square root of two." A suggestive, albeit equivocal, example appears in Michael Stifel's 1544 book *Arithmetica integra*.[78] "It is justly disputed of irrational numbers," Stifel writes, "whether they are true numbers, or fictions."[79] In favor of the existence of irrationals is their utility in calculation, on account of which "we are moved and compelled to confess, that they truly exist, namely by their effects, which we perceive to be real, certain, and constant."[80] On the other hand, when we "try to subject them to numeration, and proportion them by rational numbers, we find that they flee perpetually."[81] The idea that the numbers "flee" (a subtly violent metaphor) exemplifies a characteristically Protestant emphasis on clear apprehension as a standard of mathematical truth. A personal friend of Martin Luther who was repeatedly imprisoned for his bold expressions of support, Stifel was not content to take matters on authority; he wanted knowledge whose force one could perceive.[82]

In spite of these reservations, Stifel developed a method, which he calls an "algorithm" (*Algorithmus*), for manipulating irrationals using a notation somewhat like the modern $\sqrt{2}$.[83] For instance, in modern notation, to compute $\sqrt{18} + \sqrt{8}$, one first determines the ratio between the two roots, which in this case turns out to be rational: $\sqrt{18/8} = \sqrt{9/4} = 3/2$. The sum of the roots, then, must stand in proportion to $\sqrt{8}$ as $3 + 2$ is to 2; hence, the sum, as we would now write it, is equal to $\sqrt{8(3+2)^2/2^2}$. Based on this, Stifel determines that $\sqrt{18} + \sqrt{8} = \sqrt{50}$.[84] He also includes a more advanced "algorithm" for working with composites of different types of number, such as $6 + \sqrt{12}$.[85] This work (which has roots in Euclid's discussion of "unspeakable" numbers) indicates that the term *algorithm* was already, in the sixteenth century, beginning to expand beyond its original reference to the Hindi–Arabic methods of computation into a broader category of symbolic method. In this case, it is notable that the methods in question deal with exact relations of irrationals, not numerical approximations—a quality that aligns them more with algebra than with computation.

As the example of Stifel indicates, European mathematicians were, by the sixteenth century, feeling the limitations of the classical Greek number theories that had long reigned in universities. One of the earliest explicit repudiations of these theories appeared a few decades later in the work of the Flemish mathematician Simon Stevin. In 1585, Stevin published a pamphlet whose title is variously translated as *The Tenth* or *The Tithe*, in which he introduced decimal notation (versions of which were already known in the Middle East and in China) to Europe.[86] Rather than a decimal point, he used circled numbers to indicate the significance of each digit: 3①7②5③9④ means three tenths, seven hundredths, five thousandths, and nine ten-thousandths, corresponding to what we would write today as 0.3759.[87] He goes on to show that a method much like common algorism may be used to perform operations with these numbers; he provides applications in surveying, measurement, and mercantilism.

Although Stevin's circled numbers may have been cumbersome compared to the modern decimal point, his work led to a major shift in conceptions of number. For Stevin, a number is constructed not by measuring lines or counting indivisible units, but through the digits themselves—the tools of the commoners who practiced algorism. As a result, Stevin maintains that "there are no absurd, irrational, irregular, inexplicable, or surd numbers."[88] Perhaps one cannot write $\sqrt{2}$ as a fraction, but one can write it (in modern notation) as 1.41421356 . . . and continue the expansion as far as one likes. In 1594, Stevin described a procedure that can do just that: pinning down the root of an equation digit by digit by progressively dividing the number line into tenths.[89] (Centuries later, Turing would prove the existence of "uncomputable" numbers, as Gregory Chaitin put it, whose digits cannot be generated through any clearly defined procedure; Stevin was not on as steady ground as he thought.[90]) While there is some debate about the exact nature of Stevin's numbers, his work points in the direction of what is now called the real number continuum, a number concept that breaks entirely with classical theories.

Stevin's redefinition of number does not, however, encompass every numerical entity algebra can produce. A procedure like Stevin's can approximate the roots of positive numbers, including compounds such as $\sqrt{6 - \sqrt{6}}$, but it cannot account for the roots of negatives, which can readily emerge from algebraic methods. If, for instance, one were to apply al-Khwārizmī's equation-solving method to the equation we would now write as $x^2 + (10 - x)^2 = 48$, one gets $5 \pm \sqrt{25 - 26} = 5 \pm \sqrt{-1}$.[91] Unlike irrationals, the square root of negative one cannot even be approximated by decimal fractions, since no positive or negative number has a negative square. This result could simply be rejected as a sign that the equation has no solution, and for centuries this would remain the typical reaction. Yet the fact that the procedures of algebra could produce such results seemed to imply that the procedures themselves were ill founded, and the problems worsened as mathematicians attempted to extend their range.

The best-known instance of this issue appears in the work of the Italian polymath Gerolamo Cardano. In his 1545 book *The Great Art, or, the Rules of Algebra*, Cardano described a complete solution for cubic equations—that is, in modern notation, equations of the form $ax^3 + bx^2 + cx + d = 0$. The desire for such a solution had been long-standing; the Persian mathematician Omar Khayyam (best known as the presumed author of the poetic cycle the *Rubaiyat*) had analyzed cubic equations in the eleventh century, but neither he nor his immediate followers could find an algorithmic solution.[92] Cardano's solution was attended by a well-known scandal that gives a taste of what life as a mathematician was like in sixteenth-century Italy. Cardano learned part of the solution, as he acknowledges, from an unpublished poem shared by an acquaintance known as Tartaglia (which means "the stammerer"). Tartaglia swore him to secrecy about this result, and yet Cardano published it anyway. Cardano had an excuse. Tartaglia, he had discovered, was not the first to discover the result; another mathematician named Scipione del Ferro had discovered it over a decade before. Cardano took this to mean it was fair game to publish. Yet he failed to acknowledge the oath, which led to a bitter dispute culminating in a Renaissance alternative to a duel: a public mathematics contest between Tartaglia and one of Cardano's students, Ludovico Ferrari, who bested the stammerer and put the matter to an end.

The solution Cardano assembled breaks the problem down into more specific cases such as $x^3 + ax = b$ and $x^3 = ax^2 + b$. This division into cases is necessary because, like al-Khwārizmī, Cardano does not allow equations to have negative coefficients; it is thus impossible to combine all cubic equations into one general form. His "rules" take the form of knotty prose with the occasional use of an abbreviation, such as ℞ for "root." To solve $x^3 + ax = b$, for instance, one follows these instructions:

> Cube one-third the coefficient of the number of things, add it to the square of one-half the constant of the equation; & take the square root of the whole. You will duplicate this, and to one of the two you add the one-half of a number you have already squared and from the other you subtract the same. You will then have a *binomium* and its *apotome*. Then, subtracting the cube ℞ of the *apotome* from the cube ℞ of the *binomium*, the remainder [or] that which is left is the value of the thing.[93]

The historian of mathematics Helena M. Pycior characterizes these procedures as "prose algorithms," and they can readily be interpreted as algorithms in the modern sense.[94] Unlike al-Khwārizmī and Stifel, Cardano presents the rules in general forms rather than by means of specific examples, prefiguring the abstraction that would later come to characterize the programming language. But Cardano's theory is not wholly algorithmic. Along with each "rule," he presents a geometric demonstration of its validity, which he includes "so that, beyond mere experimental knowledge, reasoning may reinforce belief" in the results.[95] The goal of his book, then, was not just to compile procedures for solving practical problems; the reader was also supposed to come away with an understanding of why those procedures were right.

Such, at least, was the ideal. The geometric basis of Cardano's methods ran into trouble in the so-called irreducible case. As Pycior points out, certain cubic equations, such as the innocent-looking $y^3 = 8y + 3$, have rational solutions that one cannot find via Cardano's methods without encountering the square roots of negatives.[96] In this case, applying Cardano's rule gives (in modern notation) this rather cumbersome expression:

$$y = \sqrt[3]{\frac{3}{2} + \sqrt{-\frac{1805}{108}}} + \sqrt[3]{\frac{3}{2} - \sqrt{-\frac{1805}{108}}}$$

Using a procedure like Stifel's "algorithm of composite irrational numbers," one can reduce this value to $y = 3$, which clearly does satisfy the original equation.[97] This result is correct by the standards of twenty-first-century algebra, but the majority of mathematicians in the sixteenth century viewed such expressions as nonsense. Some of Cardano's early followers, such as Raphael Bombelli, managed to overcome their reservations; square roots of negative numbers, Bombelli wrote, initially "seemed to me to be based more on sophism than on truth, but I searched until I found the proof."[98] Yet the theoretical basis for such proofs remained highly uncertain.

Cardano's tentative foray into new realms of number, it should be emphasized, did not depend on the use of symbols. Cardano was undeniably practicing algebraic reasoning, but apart from numerals and a few abbreviations, he explained his procedures in words. This aspect of algebra was soon to change. Already in the mid-1500s, the convoluted prose of medieval equation-solving procedures was starting to give way to compact formulae. By the mid-1600s, symbols had taken over. As I discuss in the next section, this new notation gained much of its power from the use of letters: x, y, z for unknown values and a, b, c for known ones. Unlike numerals, with their fixed meanings, these symbols bear different values with every problem solved. In his 1650 textbook *Arithmetick*, Jonas Moore explains the difference this way: numerals are "*Notation certain*, and *determinate*," whereas algebraic symbols are "*Notation uncertain*, *undeterminate*, and *ar-*

bitrary."⁹⁹ These "uncertain" symbols seemed to many thinkers, paradoxically, to achieve a level of clarity words could not match—a development that almost immediately inspired dreams of a universal method.

FROM NUMBERS TO LETTERS

In today's primary schools, the transition from arithmetic to algebra is marked by the sudden appearance of letters. After years of learning about fixed values such as 4 + 5, the student is suddenly confronted with expressions such as 4 + x and must learn how to reason about a value that is not yet known. The need for some way of referring to unknown quantities is essential to algebra, but the use of letters to fill this office is a relatively recent development.[100] Modern algebraic notation is based on an idea that would later become central to the design of programming languages—the use of arbitrary symbols to represent values that are left unspecified for the purpose of generalizing a procedure. This use of symbols has become deeply entrenched in modern algorithmic thinking, but it was not there in the work of al-Khwārizmī himself, and it came at the cost of placing computational procedures in a fraught relation to meaning.

Early iterations of Arabic algebra most commonly represented unknown quantities with words. Medieval Arabic writers referred to the unknown as *shay* (شَيْءٌ), meaning *thing*; in Latin, this became *res*, which is the word Cardano uses. But long before Cardano's time, other ways of expressing unknowns had appeared. Robert of Chester's c. 1145 Latin translation of the *Al-jabr* includes a condensed summary of the "rules" (*regul[a]e*) of the art that uses symbols not in the Arabic original. In his critical edition, Barabas B. Hughes approximates these symbols as ø, 9, and ʒ. For instance: "When ø is equal to 9 and ʒ, ø and 9 must be divided by ʒ, 9 halved, the half drawn into itself, the product added to the number. The radix of the aggregated whole minus half 9 reveals what is sought."[101] The letters correspond to coefficients attached to terms of specific degrees: ø is a constant, 9 the unknown, ʒ the square of the unknown. In modern terms, then, the equation is $3x^2 + 9x = \text{ø}$.

Yet placing Chester's symbols in the company of modern notation like that is misleading. His ø, 9, and ʒ are not interchangeable tokens like the *a*s and *b*s of modern algebra; they function more like common nouns in that they have, to borrow a pair of terms from Gottlob Frege, both sense and reference.[102] In Frege's terms, reference is what is singled out: the reference of the phrase *the liar*, for instance, is the person being called a liar. Sense is the semantic freight the words carry: in this case, all it is to be a liar. Chester's symbols employ both types of meaning, at once referring to numerical values of the coefficients and conveying information about how those values fit into the equation—about, that is, which value is applied to the root and which to the square of the root. There is, to be sure, something of sense in the symbols of modern algebra; René Descartes instated a convention of using *x*, *y*, and *z* for unknowns and *a*, *b*, and *c* for knowns, thus using the letters as hints as to the purposes different values play. But the rules of symbol manipulation do not depend on these conventions; *a* and *x* obey identical rules. Chester's "rule," on the other hand, is incomprehensible without a recognition of the different senses of the symbols.

The transition from Chester's ø, 9, and ʒ to our *a*s and *x*s was not a linear process. In his 1494 *Summa de arithmetica*, Luca Pacioli presented equations in a compact form that abbreviated unknowns as "co.," after the Italian *cosa* (thing).[103] This word inspired a type of symbol that came to be known as the "cossic character," which was placed next to a number to indicate a certain power of the unknown. These symbols first appeared in Christoff Rudolff's c. 1525 algebra textbook, the shortened title of which—*Die Coss*—led a generation of Germans to refer to algebra with a term that means, etymologically, "the thing."[104] Rudolff's symbols were later adopted by Stifel and Recorde, while others employed similar notations with different symbols (figure 1.3). Whereas Chester's symbols represent unspecified coefficients, cossic symbols accompany coefficients to indicate

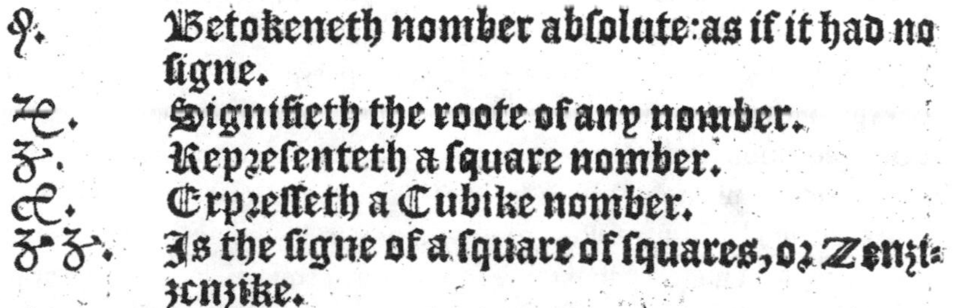
Betokeneth nomber abſolute: as if it had no ſigne.
Signifieth the roote of any nomber.
Repreſenteth a ſquare nomber.
Expreſſeth a Cubike nomber.
Is the ſigne of a ſquare of ſquares, or Zenzizenzike.
Standeth for a Surſolide.
Doeth ſignifie a Zenzicubike, or a ſquare of Cubes.
Doeth betoken a ſeconde Surſolide.
Doeth repreſent a ſquare of ſquares ſquared.

Figure 1.3. Robert Recorde's explanation of cossic characters from *The Whetstone of Witte* (1557), sig. S.i.v. Call no. 56546, Rare Books, The Huntington Library, San Marino, California. The first character is a unit indicating a constant value; the second is a unit equal to the unknown, the third to the square of the unknown, and so forth. He continues the sequence further on the next page.

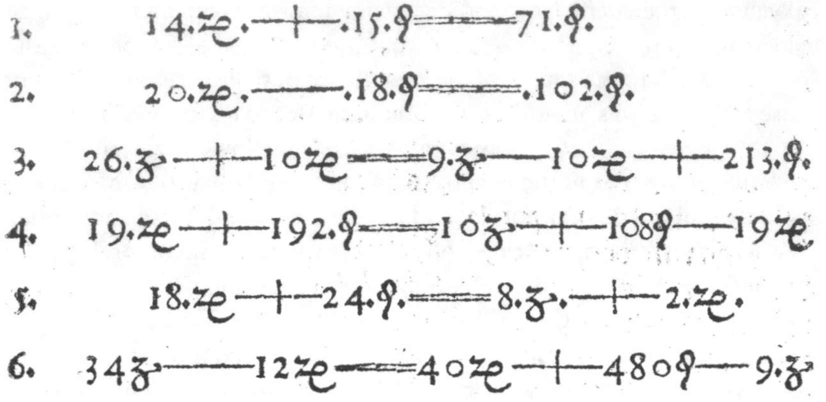

1. $\quad 14.\text{æ}.—|—.15.\text{9}==—71.\text{9}.$

2. $\quad 20.\text{æ}.——.18.\text{9}==.102.\text{9}.$

3. $\quad 26.\text{3}—|—10\text{æ}==9.\text{3}.—10\text{æ}—|—213.\text{9}.$

4. $\quad 19.\text{æ}—|—192.\text{9}==10\text{3}.—|—108\text{9}—19\text{æ}$

5. $\quad 18.\text{æ}—|—24.\text{9}.==8.\text{3}.—|—2.\text{æ}.$

6. $\quad 34\text{3}.——12\text{æ}==40\text{æ}—|—480\text{9}—9.\text{3}.$

Figure 1.4. The first set of equations presented in Robert Recorde's *The Whetstone of Witte*, sig. Ff.i.v. Call no. 56546, Rare Books, The Huntington Library, San Marino, California.

the degree of the term. The symbols are thus, in a sense, less abstract than Chester's ø, 9, and ʒ, providing a notation for the unknown but offering no way to leave the parameters of a problem unspecified so that the solution can be stated as a general rule.

Combined with Recorde's = sign, cossic characters enabled the construction of what are sometimes characterized as the first-ever equations (figure 1.4). On account of the absence of words in Recorde's version of cossic notation, Heeffer argues that his algebra was fully symbolic.[105] But Recorde's symbols are, from a semantic perspective, quite different from the ones we are used to. In modern notation, the left-hand side of the first equation in the figure would be $14x + 15$—an expression whose value is indeterminate until x is fixed. This indeterminacy is alien to Recorde's algebra.[106] Instead, Recorde explains cossic characters as units of measurement—one talks about five roots or five squares in the same way one talks about "20. shippes."[107] As in common algorism, the cossic characters have a "numeration" that involves translating them into words—in this case, we might say something like *fourteen roots more fifteen nombers*.[108] These words express not an indeterminate quantity but rather a "compounde nomber" that has, as far as Recorde is concerned, a fixed value as much as *twelve* does.[109] While he employed symbols, Recorde was still concerned primarily with representing numbers, which he understood as numbers of things.

As far as algorithmic thinking goes, a crucial turning point occurred in the work of the French lawyer, councillor to Kings Henri III and IV, and amateur astronomer François Viète. In his 1591 *Introduction to the Art of*

Analysis, Viète introduced a notation that kept the units but did away with numbers:

$$\frac{G \text{ in } A \text{ planum} + B \text{ in } Z \text{ quadratum}}{B \text{ in } G}$$

This would be, in modern notation, $\frac{ga+bz^2}{bg}$.[110] Yet the notations are not quite equivalent. The "planum" specifies that A is two-dimensional, which is necessary because, for Viète, lengths and areas are different types of value that may not be added together.[111] If these dimensions indicate an attachment to classical ideas of number, Viète's use of letters pointed in a new direction. In contrast to the cossic notations, Viète used letters to represent both the unknowns and the knowns. He used vowels to represent unknowns and consonants for knowns in order, as he wrote, "that this work may be assisted by some art."[112] These letters, which he called "species," enabled a practice known in English as "specious arithmetic" (specious was probably pronounced with a hard c, as in Latin). Instead of working on single equations such as $14x + 15 = 71$, one could now draw conclusions about general "species" of equation such as $Ax + B = C$.

There have been a number of differing accounts of where Viète's As and Bs came from, and it is not clear that they have any connection at all to the cossic characters used by Rudolff and Recorde.[113] In his 1685 *Treatise of Algebra*, John Wallis linked the letters to Viète's background in law, arguing that they originated from a way lawyers abbreviated people's names.[114] In his classic study of the history of number concepts, Jacob Klein treats them as wholly novel, arguing that Viète initiated the turn away from classical conceptions of mathematics toward *"symbolic formalism."*[115] The best recent account of Viète is by Jeffrey A. Oaks, who argues that Klein overlooked the geometric basis of Viète's method; in Oaks's account, Viète was drawing on the practice of using letters to refer to elements of geometric diagrams.[116] There is, however, a difference worth noting. Of necessity, geometric diagrams show a figure with particular proportions, even when the drawing is meant to stand in for a general class of figure. One can draw many scalene triangles with various proportions, but one cannot draw a scalene triangle with indeterminate proportions. Viète's notation, on the other hand, enables procedures to be described abstractly, without the need for any particular example.

Viète's idea emerged as a part of the humanist push to recover supposedly lost forms of ancient knowledge. There had, in particular, been a revival of interest in the third-century work of Diophantus of Alexandria, along with the later Alexandrians Theon and Pappus. Diophantus's work was preserved by Arabic scholars, but it was not widely read in Europe until the 1570s, when portions of it were translated into Latin by Raphael Bombelli.[117] Probably drawing on a combination of ancient Greek, Egyptian, and Babylonian mathematical traditions, Diophantus had developed methods for solving simple equations as well as indeterminate systems of equations—that is, systems that do not impose enough conditions to exclude the existence of infinitely many solutions.[118] His solutions, which he presents in a compact notation somewhat like cossic characters, involve making arbitrary assumptions about the relations of values and then working out their consequences.[119] For instance, to solve the problem of finding two squares that sum to 16, he assumed that their roots stood in the relation, as we would write it now, $y = 2x + 4$.[120] On this assumption, the original equation may be solved to produce a solution to the problem: $\frac{256}{25}$ and $\frac{144}{25}$.

Fairly or not (almost certainly not), Viète positioned himself as restoring the theory of equations to a pristine Alexandrian state. This art, he wrote, has been "spoiled and defiled by the barbarians," on account of which he must get "rid of all its pseudo-technical terms (pseudo-categorematis) lest it should retain its filth and continue to stink in the old way."[121] The unsubtle subtext is that he wishes to expunge Arabic sources in favor of Hellenistic ones, a chauvinism manifest in his dislike of the term *algebra*.[122] Viète hoped to replace al-Khwārizmī's techniques with what he took to be Diophantus's secret method, which he called *analysis*.[123] In Viète's definition, which he attributes to Theon of Alexandria, *analysis* starts by assuming what is sought and working out its consequences, whereas *synthesis* starts from what is true and deduces other truths.[124] Viète divided analysis into three phases. The first is the zetetic, which involves the use of specious arithmetic to derive a formula for the unknown in terms of what is known. The second phase, the poristic, involves the formulation and proof of the resulting solution. Last is the exegetic or rhetic phase, in which "from the equation set up or the proportion there is produced the magnitude itself which is sought."[125] Rhetic analysis, in other words, is the "let $A = 5$" moment: the moment in which one sets the values of the letters and resolves the formula into a number.

In his later work *Zetetics* (1593), Viète presents a large number of worked-out problems that give a better sense of how this method actually worked than the general explanations of the *Introduction*. For instance, he considers this problem: "Given the difference of lines, & difference of cubes: to find the lines."[126] That is, find two numbers that differ by B, and whose cubes differ by D. Viète solves it like so. Call the sum of the numbers E; then $E + B$ is twice the larger number and $E - B$ is twice the smaller number. Through some algebraic manipulation, he ends up with this:

$$\frac{D \, sol.4, - B \, cubo.}{B3} \; \text{æquatur } E \text{ quadrato}$$

That is, as we would now write, $\frac{4D - B^3}{3B} = E^2$. Next, he restates the formula in words: "The quadruple difference of cubes, minus the cube of the difference of lines, if applied to triple the difference of lines: appears the square of the aggregate line."[127] Finally, he plugs in the values $B = 6$ and $D \, solidum$ = 504 to get a numerical answer. In a fact often glossed over, he presents the result in an entirely different notation from the one he used to manipulate equations: "*summa laterum* 1*N*, 1*Q* æquatur 100."[128] Much like cossic characters, this notation treats the unknown value as a unit of measurement. In this case, N is a unit set to equal the "sum of the lines" and Q is another unit equal to its square; thus, the square of the sum is 100. From this, we can deduce that the sum itself is 10, and two numbers that satisfy the problem are 2 and 8.[129] The approach parallels that of Diophantus, but with a difference: whereas Diophantus uses numbers all along, Viète does most of the work with letters, only plugging in the numbers at the end.

Viète was grandiose in his ambitions for this art. He ends the 1591 *Introduction* with an imperialistic statement of purpose set in caps: "TO LEAVE NO PROBLEM UNSOLVED."[130] This statement signals the arrival—right alongside the first glimmerings of modern algebraic symbolism—of universalizing ambitions for what those symbols could do. There is a clear resonance between Viète's notation and programming languages and, indeed, a historical line of influence linking the two. But the road, again, is not linear. As the example in the foregoing paragraph shows, what Viète was doing was not quite symbolic algebra as we know it. The verbal statement of the formula seems, from a modern point of view, redundant—why bother with words if you already have the symbols? The rhetic phase, in which he plugs in the numbers to get an answer, raises an issue that is subtler but especially indicative of the problems that would come to face universal computation schemes. The verb *let*—in Latin, Viète uses the jussive subjunctive *sit*, as was common in geometric proofs—indicates a mediation between the undetermined Bs and Ds of specious arithmetic and the realm of numbers. Viète's self-consciousness about this mediation is apparent from the fact that the instant he "lets" the letters have values, he switches to a different notation more aligned with classical ideas of number. At the rhetic moment, then, the symbols change in nature, ceasing to be abstract instructions and coming to represent numbers of things.

All this complexity would soon vanish. While Viète's idea of analysis would have a lasting influence, the most consequential element of his work was the letters, which were soon extracted from their verbal settings. Among their earliest and most enthusiastic adopters was Thomas

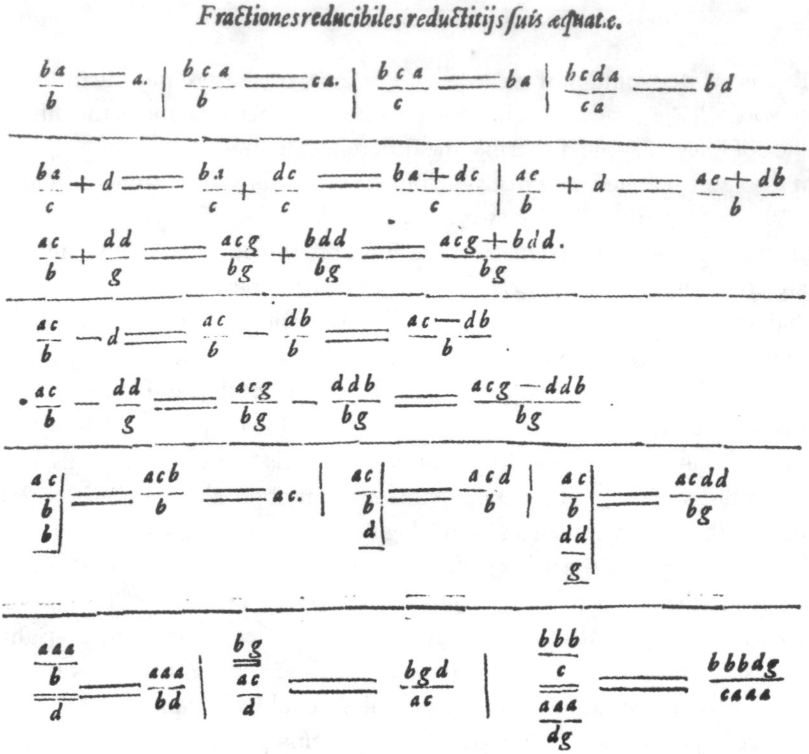

Fractiones reducibiles reductitijs suis æquat.e.

$$\frac{ba}{b} == a. \mid \frac{bca}{b} == ca. \mid \frac{bca}{c} == ba \mid \frac{bcda}{ca} == bd$$

$$\frac{ba}{c} + d == \frac{ba}{c} + \frac{dc}{c} == \frac{ba+dc}{c} \mid \frac{ac}{b} + d == \frac{ac+db}{b}$$

$$\frac{ac}{b} + \frac{dd}{g} == \frac{acg}{bg} + \frac{bdd}{bg} == \frac{acg+bdd}{bg}.$$

$$\frac{ac}{b} - d == \frac{ac}{b} - \frac{db}{b} == \frac{ac-db}{b}.$$

$$\frac{ac}{b} - \frac{dd}{g} == \frac{acg}{bg} - \frac{ddb}{bg} == \frac{acg-ddb}{bg}$$

$$\frac{\frac{ac}{b}}{b} == \frac{acb}{b} == ac. \mid \frac{\frac{ac}{b}}{d} == \frac{acd}{b} \mid \frac{\frac{ac}{b}}{\frac{dd}{g}} == \frac{acdd}{bg}$$

$$\frac{\frac{aaa}{b}}{d} == \frac{aaa}{bd} \mid \frac{\frac{bg}{ac}}{d} == \frac{bgd}{ac} \mid \frac{\frac{bbb}{c}}{\frac{aaa}{dg}} == \frac{bbbdg}{caaa}$$

Figure 1.5. Thomas Harriot's explanation of the rules for division in his algebra system, from the 1631 *Artis analiticae praxis*. © British Library Board, General Reference Collection C.74.e.4., p. 10. Images produced by ProQuest as part of *Early English Books Online*. www.proquest.com. Images published with permission of ProQuest. Further reproduction is prohibited without permission.

Harriot.[131] Harriot was best known for accompanying Sir Walter Raleigh to Roanoke Island and for writing a book about his experiences in Virginia. His algebraic work was not published until 1631, about a decade after his death, although it probably circulated in manuscript form earlier.[132] After a preface and a few pages of definitions in Latin, the 1631 volume consists almost entirely of symbols. Harriot developed a simplified version of Viète's notation that expunged the Latin words: in place of *A cubum*, he wrote *aaa* (figure 1.5). The preface to the book, probably written by Walter Warner, states that Viète had performed his analyses with "interpreted signs," whereas Harriot found it more convenient to use "a literal notation: that is, the letters of the alphabet."[133] Harriot also broke down the linguistic barriers by which Viète had so carefully demarcated the three phases of analysis. No more would cossic notations reappear once numbers enter the scene; now, the same symbols could be used for both algebraic manipulation and numerical computation.

It did not take long for admirers of these new symbols to conceive of extending them to other areas. In the 1630s, John Pell attempted to develop the ideas of Viète and Harriot into a more complete and rigorous theory of equations.[134] Pell connected this interest in algebra to a fascination with novel forms of communication such as shorthand, in which his father-in-law, Henry Reynolds, was deeply interested.[135] As did a number of others at the time, Pell thought specious arithmetic could provide a model for a new form of writing that could express anything in symbols.[136] The idea was to divide concepts up into simple components—he thought *fire*, for instance, combined "hot thing" and "shining"—and develop symbols for those simple components that could be placed together in various configurations.[137] The idea of such an "art of combinations" was very old, but Viète's use of the term *analysis* provided a new way of explaining it. Whether the analysis of concepts into parts really had anything to do with algebraic analysis would long be a topic of debate. Regardless, the example of Pell shows that Viète's early followers were interested in more than solving numerical problems—they also saw something in his method that could be applied to any number of topics of inquiry.

Harriot and Pell were both located in England, where the mania for symbols was strongest in the early seventeenth century. Across the English Channel, Pierre de Fermat, Marino Ghetaldi, and Jean-Louis Vaulezard applied Viète's methods to geometric problems.[138] In regard to symbols, however, the crucial Continental figure was René Descartes, whose 1637 book *Geometry* introduced a notation recognizably like the one we use today. It was Descartes who popularized the convention of using x, y, and z for unknowns as well as the exponent notation x^3. Apart from being more compact than Harriot's *aaa*, this notation eventually opened the possibility of exponents other than integers and, in particular, of a way of unifying exponents and roots: $\sqrt{x} = x^{1/2}$. Descartes claimed not to have been influenced by Viète at all, a statement that remains controversial among scholars.[139] Whatever the case may be, he contributed further to the establishment of a fully "literal" notation for expressing formulae.

One practical issue of the new notation was an important precursor to the modern algorithm: the formula. Viète's "species" made it possible to explain at least some types of computational procedure entirely through symbols. As an example, take Michael Dary's *Interest Epitomized, Both Compound and Simple* (1677). This text presents rules for computations regarding compound interest in a compact form (figure 1.6). The procedure shown in the figure, which gives the principal of a loan based on the

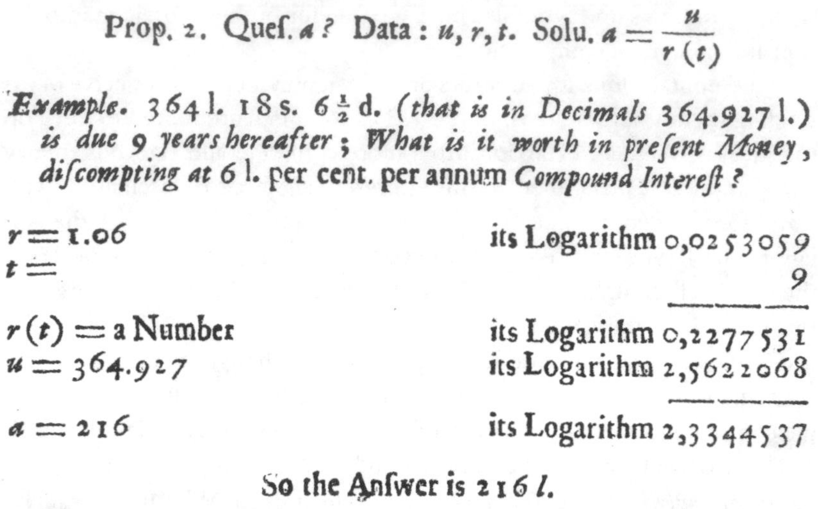

Prop. 2. Quef. *a ?* Data : *u, r, t.* Solu. $a = \dfrac{u}{r\,(t)}$

Example. 364 l. 18 s. $6\frac{1}{2}$ d. *(that is in Decimals* 364.927 l.) *is due* 9 *years hereafter* ; *What is it worth in prefent Money, difcompting at* 6 l. per cent. per annum *Compound Intereft ?*

$r = 1.06$	its Logarithm 0,02 5 3 0 5 9
$t =$	9
$r\,(t) =$ a Number	its Logarithm 0,2 2 7 7 5 3 1
$u = 364.927$	its Logarithm 2,5 6 2 2 0 6 8
$a = 216$	its Logarithm 2,3 3 4 4 5 3 7

So the Anſwer is 2 1 6 *l.*

Figure 1.6. A computational procedure for working with compound interest, first presented as a formula and then worked out as an example. Note that the notation $r(t)$ indicates what we would now write as r^t. The calculation is done using base ten logarithms; the logarithm used for u is slightly inaccurate, but the error does not affect the rounded answer. From Michael Dary's 1677 book *Interest Epitomized*. Cambridge University Library M.6.29, p. 2. Reproduced by kind permission of the Syndics of Cambridge University Library.

amount owed after the accrual of compound interest, is expressed in an abstract formula—$(a = \frac{u}{r(t)})$—that one can apply in short order. The compilation of such computational rules aligned with the Baconian attitudes of experimentalists like Robert Hooke, with whom Dary was friendly.[140] Bacon thought that science would be incomprehensible to the masses but would nonetheless produce new techniques that would be useful to them.[141] Symbolic formulae fit this model nicely: mathematical geniuses could develop theories that would trickle down to commoners in the form of practical operations that they could use without having to trouble their heads about how it all worked.[142]

Accounts of the emergence of algebraic notation have shown a strong tendency toward Whiggish narratives that presume the superiority of symbolic methods. If only Arabic algebraists had developed symbols, Katz and Parshall speculate, they could have surpassed their European counterparts.[143] But the superiority of symbolic methods is not self-evident. The Islamic mathematicians Viète denigrated as "barbarians" did gain access to Diophantus's work—indeed, they had more of it than Europeans would see until the 1970s—and yet they paid little attention to the symbols.[144]

Perhaps they were right in that. Viète's program of abstraction—solving problems by the hundred rather than one at a time—opened the way for the totalizing ambitions that now give the word *algorithmic* an increasingly ominous ring. That these ambitions should develop in the way they ultimately did, however, was not predestined. In pursuing the universal method Viète had promised, seventeenth-century mathematicians plotted out two divergent paths, neither of which quite aligns with algorithmic thinking as we know it.

OUT OF THE COVERT OF WORDS

Did the new symbols make algebra easier to understand? It depends on what one means by *understand*. Advocates emphasized the compactness of symbolic notation, which, as Robert Hooke wrote, "is of huge Use in the Prosecution of Ratiocination and Inquiry, and is of vast Help to the Understanding and Memory."[145] These cognitive advantages were not specific to mathematical fields: Hooke wishes that natural history could be expressed in a similar shorthand employing "as few Letters or Characters as it has considerable Circumstances."[146] William Oughtred, an influential mathematical practitioner who popularized specious arithmetic in England, went further. In the 1647 book *The Key of the Mathematicks New Forged and Filed* (a revised English translation of his earlier Latin textbook), he suggests that symbols can reveal the naked truth beneath the veil of language: "Wherefore that I might more cleerly behold the things themselves, I uncasing the Propositions and Demonstrations out of their covert of words, designed them in notes and species appearing to the very eye."[147] By claiming to be "uncasing" ideas from words, Oughtred positions the symbols as a neutral medium devoid of rhetoric—the naked truth, unadorned, all substance, no style. If so, they would seem a perfect remedy to the Baconian complaint about language.

Yet the symbols did not always appear clear or comprehensible. One of Oughtred's pupils, the Oxford astronomer Seth Ward, notes the potential opacity of symbols in a 1654 book on trigonometry: "It does not escape me that this little book, which abstains for the most part from words and strives to carry the things themselves to the understanding at once, will be seen by some as portentous and difficult: Truly it was produced chiefly for the sake of those to whom (provided they did not neglect themselves) designations of this kind have already become familiar."[148] This passage exhibits a tension that would continue to pervade discussions of symbolic methods for centuries, all the way to the emergence of the programming language. The symbols seemed to represent mathematical ideas with precision and transparency, yet to the untrained reader they could be altogether opaque.[149] To put it another way, notations that serve cognitive functions well do not necessarily work equally well for communication; inscriptions that seem clear to their writers might make no sense to others.

The major mathematicians in the period largely emphasized individualistic uses of symbols and gave little thought to matters of communication. In his 1637 *Geometry*, Descartes shows how geometric problems may be translated into algebraic ones and vice versa; it is here that he introduced his version of what is now known as the Cartesian coordinate system, although his system was rather different from the one we now know.[150] This work was supposed to be an example of the method of reasoning that he describes in *Discourse on Method*, in which he resolves "to comprise nothing more in my judgment than what was presented to my mind so clearly and distinctly as to exclude all ground of doubt."[151] As Matthew L. Jones has emphasized, Descartes was situated in an intellectual culture that devalued activities perceived as mechanical, including algebraic calculation.[152] Descartes accordingly embraced symbols less as an instrumental means of producing results than as a way of making problems easier to grasp mentally.

Descartes discusses symbols at length in his early text *Rules for the Direction of Our Native Intelligence* (c. 1628). This work was not published until 1701, although Leibniz acquired a manuscript copy of it in 1670.[153] Of all sciences, Descartes writes, "Arithmetic and Geometry alone are free from any taint of falsity or uncertainty" because they derive knowledge wholly through deduction, not from experience; he thus takes them as a model for a method he calls "universal mathesis."[154] Like Viète, Descartes identifies algebra with a secret art the ancients supposedly used in solving problems.[155] Descartes thinks the power of this art could be extended beyond the toy problems Diophantus was out to solve, producing "a more powerful instrument of knowledge than any other that has been bequeathed to us by human agency."[156] Here appears another instance of the universalizing ambitions that we already saw in Viète and Pell: the idea that a single method of "analysis" could be used to solve all problems.[157]

As a part of Descartes's universal mathesis, symbolic notation was primarily important because its compact form could help people visualize complex ideas. In *Rules*, Descartes emphasizes the importance of running ideas over "in a continuous and uninterrupted act of thought."[158] One way of doing so is to employ "compendious abbreviations" that leave out irrelevant aspects of objects.[159] As an example, he introduces a simple algebraic notation somewhat like that of *Geometry*, in which lowercase letters represent known quantities and uppercase letters unknowns.[160] As in the algorism texts of the time, he begins by establishing an equivalency between the symbols and words: the expression $2a^3$, he writes, "will be the equivalent of the words 'the double of the magnitude which is symbolized by the letter a, and which contains three relations.'"[161] The advantage of the symbols over these words, as he explains it, is that they present a problem in a way that contains nothing "superfluous" and nothing that would "exercise our mental powers to no purpose, by requiring the mind to grasp a number of things at the same time."[162]

This explanation of the cognitive function of symbols can only be fully understood in light of Descartes's commitment to mind–body dualism. Contrary to Klein's account, Descartes does not identify symbolic formulae with mathematical knowledge itself.[163] In the opening paragraph of *Rules*, Descartes states that it is an error to compare "the sciences, which entirely consist in the cognitive exercise of the mind, with the arts, which depend on an exercise and disposition of the body."[164] This dualism places philosophy in opposition to the computational practices that were taught at the time, in which matters of meaning were mixed up with physical pen skills. True science was about clear and distinct perceptions, which were not, for Descartes, tied to any particular form of writing. Before presenting his symbolic notation, he also suggests some other ways of representing quantities that have more in common with Boethius's theory of arithmetic, such as six points arranged in a triangle; he also shows that arithmetical operations can be thought of in terms of geometric diagrams.[165] Each of these "figures," he makes it clear, is one form of representation among many; the numbers themselves are objects of thought.[166]

Descartes's focus on conceptual clarity would continue to influence discussions of mathematical symbols for centuries, especially in the French context. But his account of the function of symbols was only one of a number of views that arose in the early modern period. An alternative appears in the work of his lesser-known contemporary Pierre Hérigone. Hérigone's only major work is the six-volume *Cvrsvs mathematicvs* (1634–42), which aims to give an overview of all of mathematics in both Latin and French. At the beginning of the book, he introduces abbreviations and symbols for both algebra and geometry, some of which are familiar today and some of which are not (figure 1.7). This extensive notation made Hérigone a particular target for critics of symbolic algebra, including Isaac Barrow; the twentieth-century historian Florian Cajori attributed to Hérigone "an almost reckless eagerness to introduce an exhaustive set of symbols."[167] His goal in developing these symbols, as he states, is to make mathematics "brief and intelligible, without the use of any language."[168]

Hérigone's work presents a more expansive vision for what symbols could do than that of Descartes.[169] Descartes situated his notations within

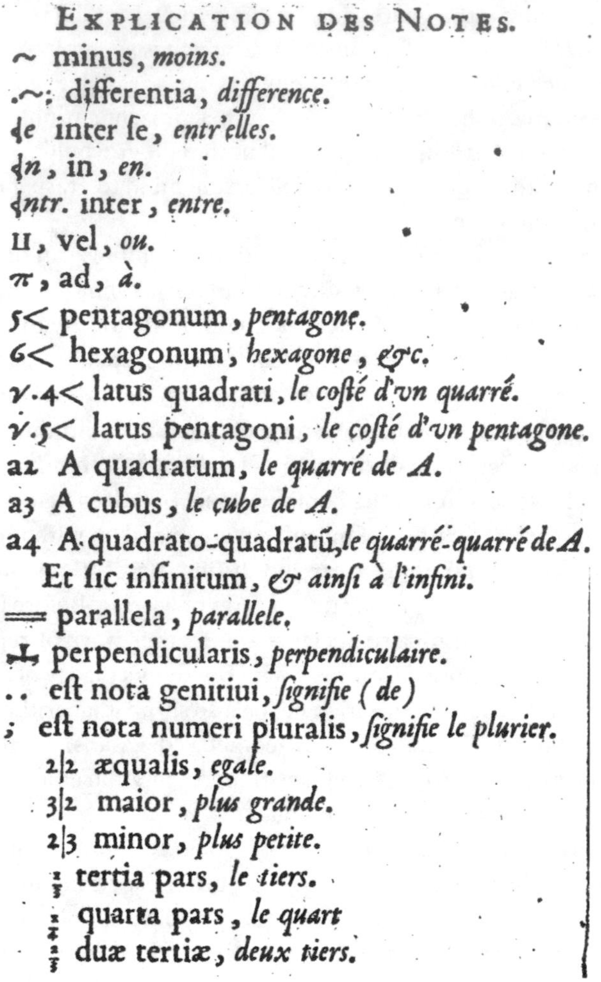

EXPLICATION DES NOTES.

~ minus, *moins*.

.~; differentia, *difference*.

⸙e inter fe, *entr'elles*.

⸙n, in, *en*.

⸙ntr. inter, *entre*.

ıı, vel, *ou*.

π, ad, *à*.

5< pentagonum, *pentagone*.

6< hexagonum, *hexagone*, *&c*.

ν·4< latus quadrati, *le cofté d'vn quarré*.

ν·5< latus pentagoni, *le cofté d'vn pentagone*.

a2 A quadratum, *le quarré de A*.

a3 A cubus, *le cube de A*.

a4 A quadrato-quadratū, *le quarré-quarré de A*.
    Et fic infinitum, *& ainfi à l'infini*.

⸗ parallela, *parallele*.

⊥ perpendicularis, *perpendiculaire*.

.. eft nota genitiui, *fignifie (de)*

; eft nota numeri pluralis, *fignifie le plurier*.

2|2 æqualis, *egale*.

3|2 maior, *plus grande*.

2|3 minor, *plus petite*.

⅓ tertia pars, *le tiers*.

¼ quarta pars, *le quart*

⅔ duæ tertiæ, *deux tiers*.

Figure 1.7. Some of the many symbols introduced by Pierre Hérigone; from his 1634 *Cvrsvs mathematicvs*, vol. 1, n.p.

paragraphs of text; even the symbol-besotted Harriot used Latin words like *sit* and *ergo* to stitch equations together into an argument. Hérigone invented symbols and abbreviations to replace these verbal elements of mathematical proofs, enabling (at least in principle) an entire proof to be developed and communicated without any words at all. On the title page, Hérigone states that his purpose of presenting mathematics in "notes real & universal."[170] What he means by "universal" is apparent from the polyglot nature of the text. When introducing the symbols, he explains them in both Latin and French: ~ is "minus, *moins*."[171] Even his abbreviations are designed so as to resemble (albeit imperfectly) both Latin and French words. "Pr." may be read as either *primus* or *premier*; "reg." is either *regula* or *reigle*.[172] The translingual nature of the symbols is crucial to the structure of the book. His prose explanations appear in both Latin and French, but in his symbolic proofs, he usually presents a single version that is supposed to be readable by both Latinate and Francophone audiences. He was trying to create a real character—to outfit mathematics with a set of symbols that, like kanji, could be read by speakers of multiple languages.

Hérigone's work was an early expression of a tendency that would re-cur time and again, from Leibniz in the later seventeenth century to Charles K. Bliss in the twentieth: a desire to improve communication by replacing words with symbols.[173] There was, however, a problem with the way Hérigone and others in the mid-seventeenth century approached this goal that would ultimately lead to a philosophical reckoning. Even though Hérigone's notations are supposed to enable "universal" communication, they are still defined by words. Their universality thus depends on the ex-istence of a common stock of ideas shared by Latin and French. If the two languages fail to align—if, say, the Latin *tangit* (it touches, it reaches, it af-fects) does not suggest quite the same thing as the French *elle touche* (she/it touches)—then clear translingual communication has not actually been established.[174] Seventeenth-century mathematicians were mostly con-fident that the concepts they dealt with were universally intelligible and that, as a result, explaining the meanings of symbols would not be a signifi-cant problem. Yet hints were bubbling up that communication was harder than the real-character view let on.

One of the earliest arguments along these lines came from Thomas Hobbes, whose eccentric views about geometry embroiled him in an ex-tended feud with the Oxford mathematician John Wallis.[175] In his 1656 book *Six Lessons to the Professors of the Mathematiques*, Hobbes criticizes Wallis's mathematical work along a number of lines, among them his over-reliance on symbols. Symbols, Hobbes wrote, might be useful in working out the details of a proof, but they "ought no more to appear in publique, then the most deformed necessary business which you do in your Cham-bers."[176] This scatological statement implies that the symbols are, in terms Hobbes laid out elsewhere, mere "marks" rather than true "signs"—they can aid the memory of the person who chose them, but they lack the so-cial warrant needed to convey meanings to others.[177] Hobbes later refers to algebraic notation as "a very narrow Language," suggesting that sym-bols were not as different from words as the advocates of real characters claimed.[178] For Hobbes, the idea that symbols could circumvent the need for language was illegitimate; all the algebraists were doing by adopting symbols was creating a secret cant comprehensible only to their exclusive clique.

Hobbes's antisymbolic polemics have generally been overlooked in accounts of the prehistory of computation. Much more has been made of his 1655 book *Elements of Philosophy*, in which he makes an unprecedented comparison that seems to anticipate much later developments in cognitive science: "By RATIOCINATION, I mean *Computation*."[179] According to Hobbes, all reasoning consisted of combining and splitting ideas in a manner analogous to addition and subtraction; he thus titles the first section of the book "Computation or Logique."[180] This striking comparison has led the philosophers Hubert Dreyfus and John Haugeland each to frame Hobbes as a founding figure of artificial intelligence.[181] Hobbes may indeed have led some of his early readers down this track; his statements were probably an inspiration for Leibniz's logical calculus. But as Hobbes's attack on symbolic algebra shows, viewing the mind as a symbol-manipulating machine does not entail that it would be a good idea to create an externalized system of symbolic reasoning, be it made of paper or of metal. If the phrase "artificial intelligence" would have meant anything to Hobbes, it would have meant the state—the complex aggregate created when many human beings work in consort. For its results to be authoritative from a Hobbesian perspective, a symbol-manipulating machine would have to be grounded in a common language, which algebraic symbols (at least in his opinion) were not.

Hobbes's polemics did little to stem the tide of symbols, at least at first. Textbooks confidently assumed that the "significations" of symbols would readily become transparent to all. As far as numerals went, this confidence may have been justified; establishing an agreement about what 7 meant was not, generally speaking, a significant problem. While algebra raised larger interpretive difficulties, the symbols had advantages over words that were (Hobbes's dissent notwithstanding) hard to deny, and the desire to extend these advantages to other areas was widespread. Yet there was a serious case to be made that symbols were less autonomous from words than the rhetoric of Oughtred and Ward let on. At the beginning of his trigonometry book, Ward maps his symbols onto Latin words, explaining *R*, for instance, as "Radius."[182] Thus, just after stating that his book "abstains" from the use of words in favor of symbols that express things directly, he defines those symbols with—words. The use of words in defining symbols went, for the most part, unnoted by advocates of real characters, but it posed a problem for attempts to develop symbolic notation from a mere specialized form of writing into a full-fledged alternative to language.

This tension reached its crisis in the work of Leibniz. Like the proposals of Pell and Hooke, Leibniz's universal characteristic was supposed to provide symbolic methods for reasoning about subjects ranging from theology to law. The characteristic was only one of a plethora of projects in which Leibniz explored the possibilities opened by symbols. Yet the theoretical concerns that had been mounting since the sixteenth century only worsened in Leibniz's wake, as his followers struggled to find a rationale for his version of calculus, and as Newton and others attacked him for employing "mechanical" methods. In pushing seventeenth-century ideas about symbols to their utmost limit, Leibniz's work marks the end of the somewhat naïve belief in the power of "characters" and the beginning of the long historical process by which, over the course of centuries, the boundaries of algorithmic thinking would be contested and ultimately set.

CHAPTER TWO

The Matter Out of Which Thought Is Formed

It is impossible to say just what I mean!
But as if a magic lantern threw the nerves in patterns on a screen:

—T. S. ELIOT, "The Love Song of J. Alfred Prufrock"

THE DREAM

To consider Gottfried Wilhelm Leibniz's position in the prehistory of computation is to peer through a thick fog of anachronism. Leibniz designed an early mechanical calculator, and later commentators have found in his work anticipations of numerous ideas that would later become important to computer science, from Boolean logic to binary numerals. In the mid-twentieth century, Norbert Wiener ensured Leibniz's place in the computer history canon by suggesting that he could be "a patron saint for cybernetics," the field that deals with systems of control and communication involving both humans and machines.[1] Leibniz has since become a fixture in histories of computation: the popular writer Martin Davis refers to the idea of a general-purpose computer as "Leibniz's dream," and in his account of the history of algorithms, Wolfgang Thomas considers whether the recent expansion of the category of algorithm means we are "Returning to Leibnizian Visions."[2] While Leibniz's thinking does overlap with the concerns of computer science, it also reflects a worldview quite alien to modern sensibilities, and this fact tends to get lost amid attempts to make Leibniz into a founding figure.

The belief that Leibniz imagined something like a modern computer stems in large part from his essays "Preface to the General Science" (1677) and "The Art of Discovery" (1685), along with similar remarks in a few other places.[3] In both texts, Leibniz expresses a dream that was, as I showed in chapter 1, common in the seventeenth century: extending the power of symbols to new areas of knowledge. This project, as Leibniz conceives it, would consist of a set of symbols called a *universal characteristic* along with a *calculus ratiocinator*—a rational calculus—by which any question can be resolved. In the 1685 essay, Leibniz explains:

> The only way to rectify our reasonings is to make them as tangible as those of the Mathematicians, so that we can find our error at a glance, and when there are disputes among persons, we can simply say: Let us calculate, without further ado, in order to see who is right.[4]

This passage is often quoted out of context with the implication that the "Let us calculate" (*calculemus*) refers to something like a programmable computer. But it is not clear that he understood this "calculation" to be algorithmic in the modern sense. At least some of Leibniz's efforts, as Matthew L. Jones has shown, involved inductive reasoning about patterns in tabular data.[5] Moreover, Leibniz's goals for this system were in some regards more ambitious than those of modern computing machines. The system Leibniz describes in "The Art of Discovery" is not just supposed to spit out answers; it is supposed to demonstrate them in a way that makes them impossible to doubt. This idea is worlds away from the black box algorithms of the twenty-first century. A Leibnizian computing machine, if such a thing were really possible, would only compute the truth.

This chapter gives a historically contextual account of Leibniz's role in the development of algorithmic thinking. It focuses, in particular, on two projects involving symbols: the universal characteristic and his work on mathematical notation. In chapter 1, I showed that seventeenth-century mathematicians described numerals and algebraic symbols as writing systems comparable to the alphabet or to the Chinese han characters. Viewing these symbols as writing rather than language highlighted the fact that they could be read aloud in multiple vernaculars while bearing (apparently) the same meanings for everyone; this made them, in period terms, *real characters*. While Leibniz did not intend the universal characteristic to be a real character in the strict sense, his work exhibits a similar way of thinking: whereas words were inextricable from the histories of particular communities and thus inevitably divisive, Leibniz's symbols were supposed to latch onto ideas about which everyone could agree. Leibniz used this universalism to explain the cognitive power of symbolic methods in mathematical fields such as calculus. But doubts about the conceptual foundations of these methods surfaced in the early reception of Leibniz's mathematical work, which contributed, ironically, to a collapse of the faith in symbols that had inspired his utopian project.

Leibniz's views about symbols can only be fully understood in the context of his philosophical and political commitments. In Leibnizian metaphysics, human souls, which he called *monads*, have no direct interactions with other beings but exist in a preestablished harmony with one another and with the material world.[6] This emphasis on harmony had a political meaning in Leibniz's context. In her biography of Leibniz, Maria Rosa Antognazza has emphasized the influence of Leibniz's upbringing on his philosophy.[7] Leibniz was born in Saxony, which was then part of the Holy Roman Empire. In contrast to the centralized monarchy of France, the empire consisted of autonomous and sometimes antagonistic principalities with varying cultural identities and religious affiliations. Politically and religiously, Leibniz tended strongly toward reconciliation—toward finding common ground between antagonistic factions and sects. Antognazza has demonstrated his lifelong commitment to *irenicism*, meaning the project of reuniting the Protestants and the Catholics. He viewed the universal characteristic as an integral part of this project, a basis for a unified body of knowledge that would bridge religious divides by grounding theology in indisputable principles.

Something similar may be said of his attitude toward language. English and French were moving toward standardization in the late seventeenth century, as dictionaries and grammars codified the dialects of the metropoles (London and Paris) and of the universities as defaults.[8] The German language, which was spoken across much of the Holy Roman Empire, was less standardized at the time, and the empire was multilingual, also including communities speaking Italian, Czech, Yiddish, and other languages.[9] Communication across regions would have been easier in writing than in speech, since written forms of German did not reflect all the differences between dialects; indeed, the standard German that would form in the eighteenth century was at first exclusively written.[10] The characteristic might be seen as an alternative to standardization that is more suited to this context—an attempt not to instate one dialect as the lingua franca but rather to base a new mode of communication on the harmony that already existed among all languages and dialects.

Famous as it eventually became, the universal characteristic amounted to little more than scattered notes and suggestive comments. It was, in modern terms, vaporware. In terms of influence, Leibniz's most important contribution to the history of computation was his work on the infinitesimal calculus, which greatly expanded the scope of what symbols could do.

The 1684 article in which Leibniz introduced his version of calculus, titled "New Method for Maximums and Minimums," called this method an *algorithm*.[11] Leibniz's mathematical innovations had two countervailing effects in the ensuing decades. The technique that Leibniz called an "algorithm" could do astonishing things, solving numerous problems that had puzzled the ancients; his work thus heightened, especially in Continental Europe, the excitement about symbols that had been building over the course of the seventeenth century. But Leibniz did not articulate (at least publicly) a satisfactory explanation of why his method worked, and so symbolic methods came to be subject to a greater degree of critical scrutiny and, in some circles, skepticism.

Leibniz hashed out these issues in a series of public debates.[12] His work on calculus embroiled him in an extended interchange with Isaac Newton, since both claimed credit for developing the idea first. Although he did employ notation, Newton was more skeptical of the value of symbolic methods than Leibniz; even a "bungler," he reportedly said, can use the mechanical methods of algebra.[13] Leibniz also had occasion to defend his use of symbols in a dispute with John Locke. In his *Essay Concerning Human Understanding* (1689), Locke had advanced a kind of protoanthropological relativism that undermined Leibniz's claims about the power of symbols to convey ideas transparently. At issue was nothing less than the possibility of the universal characteristic: Locke's position entailed that there was no getting around the need for semantic convention, that mere symbolic forms were not enough to ensure that people were genuinely thinking the same things. Broadly speaking, Locke's side won. In the eighteenth century, it would no longer be widely accepted that symbols could express prelinguistic ideas, as Leibniz maintained to the end of his life. Yet Leibniz's methods remained—the "algorithm" of his calculus apparently worked, even though people could not agree on its conceptual basis. For over a century, then, the status of "algorithms" hinged on a question about language: what does it take to make symbols meaningful?

A DESIGN AGAINST LANGUAGE

As I mentioned in chapter 1, a number of seventeenth-century thinkers, including John Pell and Robert Hooke, envisioned a system that could express anything whatsoever with the clarity of algebraic symbols. Another proposal along these lines (although referring to a different type of symbol) came from René Descartes. In a 1629 letter to his friend Marin Mersenne, Descartes observes that the Hindi–Arabic numerals can express infinitely many numbers using only ten symbols, compounded together according to simple rules. "In a single day," he writes, "one can learn to name every one of the infinite series of numbers, and thus to write infinitely many words in an unknown language. The same could be done for all other words necessary to express all the other things which fall within the purview of the human mind."[14] Leibniz's universal characteristic was, at least in some versions, similar to this proposal: the system was supposed to analyze ideas into components and so enable them to be expressed through combinations of a small set of basic symbols.

Although Descartes's proposal had specific reference to numerals, such ideas fit into a tradition that was more linguistic than mathematical—the idea of creating what is sometimes called a "philosophical language," an artificial form of communication more precise than the languages people presently speak.[15] This idea has roots in medieval thinkers such as William of Ockham, but it underwent a major resurgence in the seventeenth century.[16] Jan Amos Comenius, Cyprian Kinner, and Francis Lodwick tried to build such systems in the 1640s. In the early 1660s, Athanasius Kircher and Johann Joachim Becher made further attempts, both of which Leibniz cites as inspirations.[17] Around the same time, Isaac Newton tried his hand at it as well. The most full-fledged scheme, however, was attributable to John Wilkins, a scientific polymath and (eventually) Anglican bishop who was involved in the founding of the Royal Society of London. Scholars have generally dismissed the importance of Wilkins for Leibniz, and it is true that their aims were different: Wilkins was mainly trying to facilitate communication, whereas Leibniz wanted to create a method of discovery that could produce new knowledge. Leibniz did not encounter Wilkins's system until after he conceived the universal characteristic, so it cannot be considered a crucial influence. But Wilkins's work is a useful point of comparison because, as an uncommonly detailed example of the genre, it illustrates some misunderstood aspects of the intellectual movement in which Leibniz was participating. While Leibniz's project is habitually likened to later developments in formal logic, it existed within a seventeenth-century milieu that was worlds away from formalism as we now understand it. For Leibniz's contemporaries, the critical term was not *formal*; it was *artificial*.

Wilkins first discussed the idea of a universal character in his 1641 book *Mercury, or the Secret and Swift Messenger*. After explaining Francis Bacon's idea of a real character, Wilkins lists four examples of symbols that could be understood by speakers of multiple languages: numerals, astrological symbols, chemical symbols, and musical notes. He then considers the possibility of "a generall kinde of writing invented for the expression of every thing else" in this manner.[18] The symbols would be pronounced differently by speakers of different languages, as the astrological symbol ♉ is pronounced *toro* in Italian (or, one might add, Spanish) and *bull* in English; "yet the sense would be still the same" for everyone.[19] While scholars have often characterized this proposal as a "universal language," Wilkins makes it clear that he views it as a new form of writing, not a language. "The perfecting of such an invention," he writes, "were the only way to unite the seventy two Languages of the first confusion."[20] The point is not to replace the languages people already speak; it is to repair the rifts between those languages by recording ideas on paper in a way that is unaffected by the Confusion of Tongues imposed on humankind after the fall of the Tower of Babel.

Wilkins discussed this idea in detail with his friend Seth Ward, who published his own thoughts on the matter in his 1654 book *Vindiciae academiarum* (*Vindications of Academies*). For Ward, the best model for a universal character was symbolic algebra, which presented ideas in their full complexity while avoiding the "confusion or perterbati[on] of the fancy made by words."[21] Rejecting an opponent's suggestion that grammarians should study algebraic symbols, Ward stridently denies that the symbols constitute a language: specious arithmetic "was a designe perfectly intended against Language and its servant Grammar, and that carried on so farre, as to oppose the use of *numbers* themselves, which by the Learned, are stiled *Lingua Mathematicorum* [the language of mathematics]."[22] Algebraists, that is, replaced numbers with as and xs as a way to expunge the linguistic traces that persisted in Hindi–Arabic numerals. While this statement is sometimes cited as evidence of an opposition between mathematics and language in the period, the rest of the passage suggests otherwise.[23] "The use of Symbols," Ward continues, "is not confined to the Mathematicks only, but hath been applied to the *nature of things*"; he lists the Pythagoreans, cabbalists, and the *ars combinatoria* of Giovanni Pico della Mirandola as examples.[24] The tradition Ward assembles under the name "Symbolicall writers" includes many of the figures who would soon inspire Leibniz.[25] What he is celebrating in them is neither mathematics nor language; it is symbols, which Ward is careful to distinguish from both.

Ward goes on to propose extending this symbolical style into an ambitious project somewhat like the one Wilkins had previously described. After first learning about "the Symbolicall way, invented by *Vieta*, advanced by *Harriot*, perfected by Mr *Oughtred*, and *Des Cartes*," Ward writes, "I was put upon an e[ar]nest desire, that the same course might be taken in other things."[26] Doing so would provide a system in which "exact discources may be made demonstratively without any other paines then [*sic*] is used in the operations of specious Analytics."[27] Although Ward was almost certainly unaware of Descartes's letter to Mersenne, what he describes is quite similar: breaking ideas down into simple elements that could be represented by a small set of easily learned symbols. The advantage of symbols, according to Ward, is that they enable people "to discourse . . . freely without the trouble of words, upon which while the mind of man is intended, it neither sees the consequence so cleerely, nor can so swiftly make comparison as when it is acquitted of those obstacles."[28] The description of words as "obstacles" is characteristic of the attitudes of the scientific circles of the time: language appears as an annoyance, a shackle on humanity that holds back the advancement of knowledge.

Ward did not seriously attempt to create a universal character, but Wilkins, working together with collaborators such as George Dalgarno and John Ray, eventually did.[29] In 1668, he published *An Essay towards a Real Character and a Philosophical Language*, which, at over 450 folio pages, is one of the most detailed universal character schemes ever produced. Wilkins's project involved two parts: a "real character" that expressed ideas through written symbols, along with a "philosophical language" that was supposed to replace spoken languages like English.[30] The real character consists of special symbols assembled according to rules (figure 2.1). These symbols are supposed to be "legible by any Nation in their own Tongue" just as all of the inhabitants of China use the same writing system, each "reading it in his own Language."[31] The other part, the philosophical language, is meant to provide a substitute for existing languages and is printed in a phonetic alphabet loosely based on Latin and Greek characters. This system is supposed to differ from existing languages in that it was formed "according to the *rules of Art*" at a single founding moment rather than emerging through a long, haphazard process like the ones that produced English and Latin.[32] The symbols would thus (according to Wilkins) readily become comprehensible to everyone and, once established in use, resist the corruption to which words are liable.

The heart of Wilkins's system is a detailed, hierarchical classification scheme (figure 2.2). This hierarchy, Wilkins writes, provides "a just *Enumeration* and description of such things or notions as are to have *Marks* or *Names* assigned to them," arranged in a fashion inspired by the Aristotelian system of categories and Peter Ramus's diagramming method.[33] The symbols each indicate a particular set of coordinates within this hierarchy (figure 2.3). Loops and hooks can also be added on to characters to indicate part of speech or conjugation and to modify the meaning in various ways. Wilkins also provides smaller characters for pronouns and grammatical words; the way the characters are positioned relative to each other also plays a role in determining their meanings. The words of the philosophical language express more or less (although, as James Dougal Fleming has

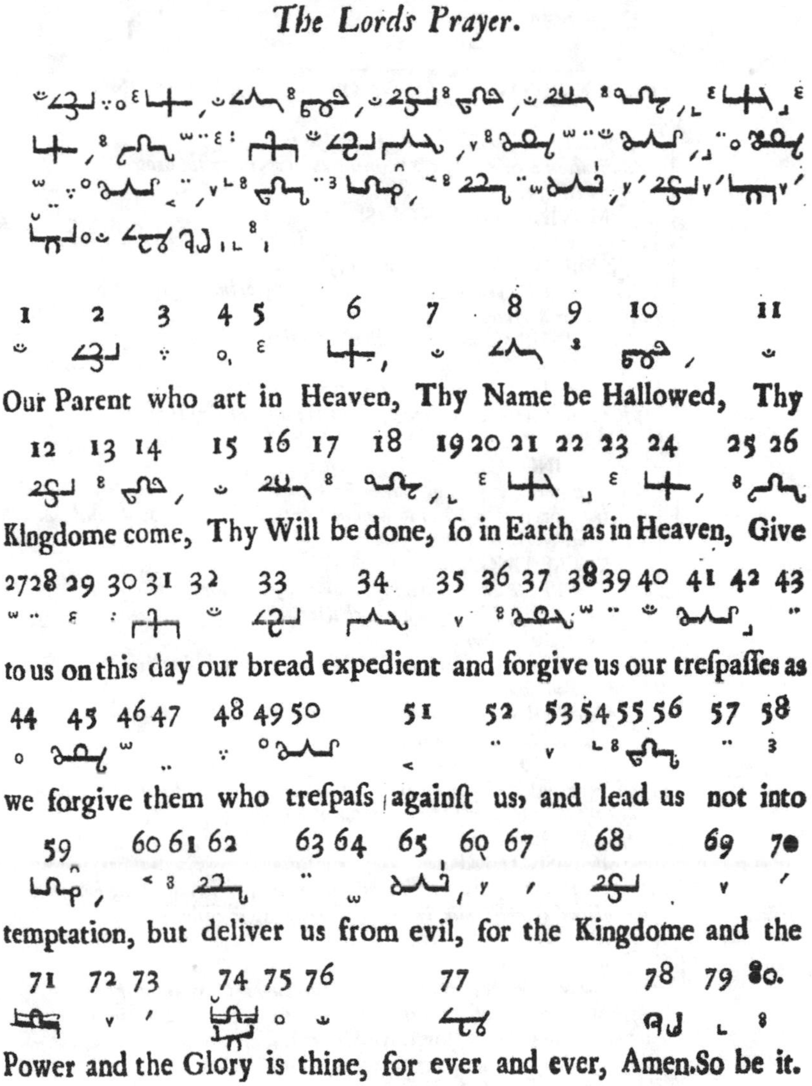

The Lords Prayer.

1 2 3 4 5 6 7 8 9 10 11

Our Parent who art in Heaven, Thy Name be Hallowed, Thy

12 13 14 15 16 17 18 19 20 21 22 23 24 25 26

Kingdome come, Thy Will be done, so in Earth as in Heaven, Give

27 28 29 30 31 32 33 34 35 36 37 38 39 40 41 42 43

to us on this day our bread expedient and forgive us our trespasses as

44 45 46 47 48 49 50 51 52 53 54 55 56 57 58

we forgive them who trespass against us, and lead us not into

59 60 61 62 63 64 65 66 67 68 69 70

temptation, but deliver us from evil, for the Kingdome and the

71 72 73 74 75 76 77 78 79 80.

Power and the Glory is thine, for ever and ever, Amen. So be it.

Figure 2.1. The Lord's Prayer in John Wilkins's real character. Houghton Library f *EC65.W6563.668e (A), p. 395. Images produced by ProQuest as part of *Early English Books Online*. www.proquest.com. Images published with permission of ProQuest. Further reproduction is prohibited without permission.

pointed out, not exactly) the same information, mapped onto syllables rather than shapes.[34] The purpose of this system, as Wilkins explains it, is in part to clear up scientific communication and in part to aid missionary work by presenting religious doctrines in a form that could (supposedly) be understood by speakers of any language.

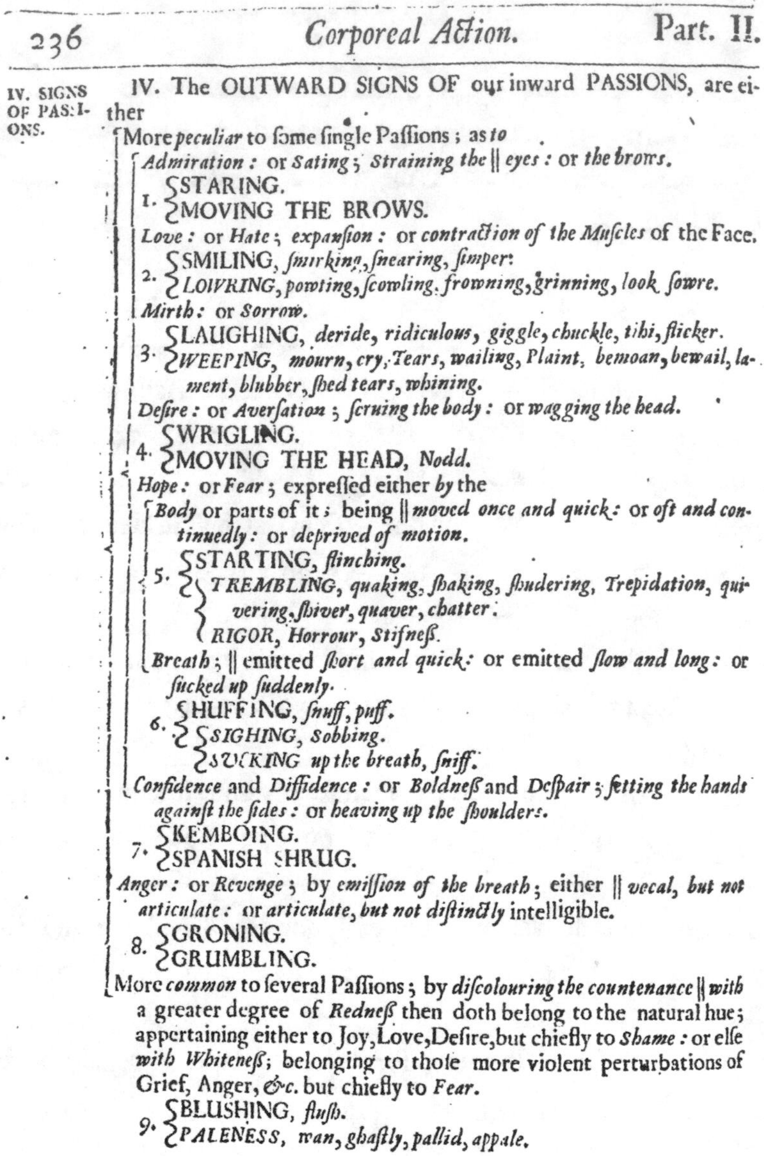

IV. SIGNS OF PAS-I-ONS.

IV. The OUTWARD SIGNS OF our inward PASSIONS, are either

More *peculiar* to some single Passions ; as *to*

Admiration : or *Sating* ; *Straining the* || *eyes :* or *the brows.*

1. { STARING.
{ MOVING THE BROWS.

Love : or *Hate* ; *expansion :* or *contraction of the Muscles* of the Face.

2. { SMILING, *smirking, sneering, simper:*
{ LOWRING, *powting, scowling, frowning, grinning, look sowre.*

Mirth : or *Sorrow.*

3. { LAUGHING, *deride, ridiculous, giggle, chuckle, tihi, flicker.*
{ WEEPING, *mourn, cry, Tears, wailing, Plaint, bemoan, bewail, lament, blubber, shed tears, whining.*

Desire : or *Aversation* ; *scruing the body :* or *wagging the head.*

4. { WRIGLING.
{ MOVING THE HEAD, Nodd.

Hope : or *Fear* ; expressed either *by* the

Body or parts of it ; being || *moved once and quick :* or *oft and continuedly :* or *deprived of motion.*

5. { STARTING, *flinching.*
{ TREMBLING, *quaking, shaking, shudering, Trepidation, quivering, shiver, quaver, chatter.*
{ RIGOR, *Horrour, stifness.*

Breath ; || emitted *short and quick :* or emitted *slow and long :* or *sucked up suddenly.*

6. { HUFFING, *snuff, puff.*
{ SIGHING, *Sobbing.*
{ SUCKING *up the breath, sniff.*

Confidence and *Diffidence :* or *Boldness* and *Despair* ; *setting the hands against the sides :* or *heaving up the shoulders.*

7. { KEMBOING.
{ SPANISH SHRUG.

Anger : or *Revenge* ; by *emission of the breath* ; either || *vocal, but not articulate :* or *articulate, but not distinctly* intelligible.

8. { GRONING.
{ GRUMBLING.

More *common* to several Passions ; by *discolouring the countenance* || *with* a greater degree of *Redness* then doth belong to the natural hue ; appertaining either to Joy, Love, Desire, but chiefly to *Shame :* or else with *Whiteness* ; belonging to those more violent perturbations of Grief, Anger, &c. but chiefly to *Fear.*

9. { BLUSHING, *flush.*
{ PALENESS, *wan, ghastly, pallid, appale.*

Figure 2.2. A page of John Wilkins's categorization scheme. Houghton Library f *EC65. W6563.668e (A), p. 236. Images produced by ProQuest as part of *Early English Books Online.* www.proquest.com. Images published with permission of ProQuest. Further reproduction is prohibited without permission.

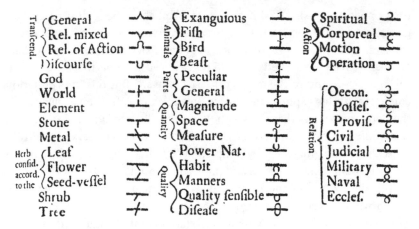

The Differences are to be affixed unto that end which is on the left fide of the Character, according to this order;

The Species fhould be affixed at the other end of the Character according to the like order.

Figure 2.3. John Wilkins's explanation of the basic structure of the real character. In addition to what is shown here, the system contains small symbols for grammatical words, notations used to accommodate larger categories, and extra strokes that modify the meanings of characters. Houghton Library, Harvard University, f*EC65.W6563.668e (Λ), p. 387. Images produced by ProQuest as part of *Early English Books Online*. www .proquest.com. Images published with permission of ProQuest. Further reproduction is prohibited without permission.

From our perspective, Wilkins's system may seem to have little in common with the algebra-like system that Leibniz would ultimately create. The overall structure of Wilkins's system is based on grammar, which (as Ward would surely remind us) was an entirely separate discipline from algebra. Moreover, rather than analyzing ideas into their simple components as Leibniz proposed, Wilkins employed a top-down taxonomy. But Wilkins's work does resemble algebraic notation in its use of real characters—visual, nonphonetic symbols that can (supposedly) be read in multiple languages. Wilkins's evident fixation on these symbols has long perplexed historians of linguistic thought. As Jaap Maat has observed, the real character appears to be unnecessary, since, with its phonetic alphabet, the philosophical language can serve for both written and spoken communication.[35] As a way of explaining the dual nature of Wilkins's system, Hans Aarsleff and Michael Isermann each have argued that Wilkins believed symbols to have a "mystical" power to capture the true natures of things.[36] But this interpretation conflicts with his explicit statements on the matter. In the last section of the book, he asserts that the symbols signify "by *Institution*" rather than "*Naturally*"—a clear indication that he saw the real characters as arbitrary signs chosen pragmatically by people.[37] Why, then, did he go to so much effort to create separate visual and phonetic symbols?

This choice makes more sense if we recognize that he was working within what was still, to a large extent, an oral culture. In explaining the real character, Wilkins drew on a long-standing intellectual tradition that thought of language primarily as speech and viewed alphabetical writing as an encoding thereof. Following Aristotle's theory of signs in *On Interpretation*, Wilkins states that "*Names*" are "arbitrary *sounds* or *words*" that people have agreed to use as signs of "their Mental notions"; "The *Written word*," he continues, "is the figure or picture of that Sound."[38] In the taxonomic tables, he classifies "READING" as one of the "*particular kinds of speaking*"; he defines it as a variety of articulate sound that refers to "*such words as we see before us*."[39] The idea that reading ordinarily involved the ability to speak was common at the time, as one can infer from the way people discussed the consequences of deafness: it was generally presumed that people could not learn to read without first learning a spoken language, although John Wallis disagreed.[40] Real characters promised to alter this situation by making writing fully independent of speech, much as numerals can convey the same meanings to speakers of any language.

This interpretation still leaves some puzzles regarding how the real character was supposed to work. Certainly, Wilkins's symbols are not, in spite of his claims, nearly as easy to learn as numerals. With regard to Leibniz, what is important is less the specifics of Wilkins's plan than the general way of thinking that made the idea of a universal character plausible. As Fleming has argued, Wilkins's project is premised on the early modern belief that the human mind is a mirror of the world, divinely outfitted with ideas that suit the true natures of things.[41] Wilkins states this principle explicitly in the first part of the *Essay*: "As men do generally agree in the same Principle of Reason, so do they likewise agree in the same *Internal Notion* or *Apprehension of things*."[42] That is, ideas such as *horse* and *heretic* are already there in every human mind, waiting to be uncovered. A similar assumption may be discerned in William Oughtred's statement about "uncasing" mathematical propositions from the "covert of words" by means of algebraic symbols. If one believes that the ideas one wishes to express are already waiting in the reader's mind, then all it takes to create perfect clarity is to cut away the verbal thickets that obscure the light of reason.

Opposition to this way of thinking was already mounting by the time Wilkins's *Essay* was published in 1668. A pointed critique came from Wilkins's erstwhile collaborator George Dalgarno. Dalgarno had worked together with Wilkins in the early phase of the project, but the two fell out as a result of a difference of opinion about how the system should be designed. Like Ward, Dalgarno thought it would be better to analyze concepts into parts rather than place them in a hierarchy. He also became skeptical of the distinction between real characters and languages. In his 1661 book *The Art of Signs*, Dalgarno argues at length that "mute characters" are not as different from words as his contemporaries tended to assume; "the art of *characters* and *sounds*," he maintains, "is one and the same."[43] Dalgarno's work, as Maat has argued, marks an important step toward the idea of language as an abstract system rather than a set of concrete practices, and thus to a de-emphasis of the difference between writing and speech.[44] An implication of this conceptual shift was that written symbols were not inherently immune to the Confusion of Tongues; since they depended on arbitrary significations, they were subject to the contingencies of communication just as much as English and Latin were.

Dalgarno's attack on the division between languages and characters was not the only threat that the real-character view faced in Leibniz's time. A more epistemological line of criticism emerged from the Cartesian tradition. Although Descartes entertained the idea of a universal symbolism in his letter to Mersenne, he was skeptical that the dream could be realized. Descartes's philosophy emphasized the importance of what he called clear and distinct perceptions, meaning perceptions that are accessible to the mind and sharply distinguished from other perceptions.[45] As a result, Descartes concludes in his letter to Mersenne that a universal symbolism could not be created without "la vraie philosophie" (the true philosophy)—that is, a complete and correct theory of everything in the world.[46] This argument, when accepted, put a damper on the idea of extending the clarity of mathematical symbols to other areas. From a Cartesian perspective, a science can only be expressed with the clarity of numerals when it rests on principles as certain as those of arithmetic; until then, the universal character would have to wait.

As we will see, Descartes's "vraie philosophie" argument would recur frequently in discussions of symbolic methods. Again and again, people who attempted to extend the power of these methods to new domains would face the criticism that their ideas were insufficiently clear. Yet Leibniz, at least early in his career, thought that the Cartesian objection could be overcome. To be sure, Leibniz made less of the language–character distinction than Wilkins did; he at times described the universal characteristic as a type of "language," albeit one that was supposed to work differently from words. But he shared Wilkins's belief in a divinely instituted correspondence between mind and reality. Like his English predecessor, he located the problems of language primarily in the arrangement of signifiers, presuming that a properly designed system of symbols would readily become transparent to everyone. Leibniz's confidence that he could have it both ways, embracing what looks to us like formalism while still maintaining that the symbols had conceptual meanings, is what differentiates his approach from algorithmic thinking as we know it—in short, what makes him not modern.

INTO THE INTERIOR OF THINGS

Leibniz's earliest attempts at a philosophical language predate the publication of Wilkins's *Essay*. The *Dissertation on the Art of Combinations* (1666), which he wrote while still a teenager, deals with what is now called combinatorics: the problem of determining how many ways simple elements can be compounded together, as ABC can (if we disregard order) be combined into pairs in three ways: AB, AC, or BC.[47] Borrowing a notation from Mersenne, he called these assemblages "com2nations"; when three elements are combined, they make up "con3nations," and so forth.[48] His ambitions for this method are broad. Leibniz wanted to turn combinatorics into a general "art" that could be applied to a wide range of fields, from theology to jurisprudence. This work marks the beginning of the lifelong project that would eventually lead to his famous statements about a universal method of calculation.

The idea of an "art of combinations" is often traced back to the medieval thinker Ramon Llull.[49] Born around 1232 in Majorca, Llull had a spiritual vision at the age of thirty that led him to abandon his family and devote the rest of his life to evangelism. To this end, he developed what he called the *Ars Magna* (Great Art), a general method of reasoning that is supposed to encompass both logic and metaphysics. Some versions of this method involved a rotatable device called a *volvelle* that could generate every combination of a certain set of symbols; others involved intricate tables and diagrams. Using Llull's art, one can iterate through all possible questions and statements based on certain fixed terms, such as, "What is great difference of eternity?" and "What is great and eternal concordance?"[50] For

Llull, the point of this endeavor was religious: he intended to use his art to produce arguments that convinced people of the truth of Christianity. Llull's later admirers, however, turned his ideas to a variety of ends. Johann Heinrich Alsted, for instance, employed Llullian ideas in the organization of his 1630 encyclopedia, which Leibniz later used as a starting point for his own encyclopedic endeavors.

In the *Dissertation*, Leibniz attempts to extend the Llullian art into a general "Logic of invention"—a method for discovering new knowledge in the spirit of Francis Bacon's *Advancement of Learning*.[51] The method begins with the identification of primitive terms that are to be left undefined, which he represents by numbers. He links this numerical representation to Hobbes's statement that reasoning is computation.[52] He also introduces a shorthand by which complex ideas may be expressed as pseudo-fractions: $\frac{3}{2}$ is the third in the sequence of com2nations, as enumerated in some particular order.[53]

Using this notation, Leibniz attempts to develop a method for enumerating all possible logical inferences. As an example of how this system would work, he lists twenty-seven primitive terms and then derives some definitions from them. For example, here is how he defines *infinite*.[54] The relevant primitive terms are as follows: 9 is part, 11 is same, 14 is number, 15 is plurality, 17 is possible, 18 is all, and 19 is given. To get from here to *infinite*, we must follow a series of combinations of increasing complexity. First, he defines $\frac{1}{2}$: "*Quantity* is 14 of 9s (15)." Next, $\frac{2}{3}$: "*Equal, A* of 11, $\frac{1}{2}$." (That is to say, something of *same quantity*.) Then, $\frac{1}{4}$: "*A* is *Greater* when it has a 9, $\frac{2}{3}$ more than *B*." (Note that $\frac{2}{3}$ means *equal*, not *two-thirds*!) Finally, we can understand the definition of $\frac{2}{7}$: "*Infinite*, $\frac{1}{4}$ than 18, 19, 17." While these definitions include some Latin words linking things together, Leibniz suggests that, by reducing grammatical categories to their simple components, it would be possible to define new ideas entirely through primitive terms represented by numbers.

Although he initially presents this system as a means of generating syllogisms, he then proceeds to another idea that "flows like a corrolary" from it: the possibility of developing "a Universal Notation [*Scriptura Universalis*], that is, intelligible to whatever reader, no matter what his language."[55] He describes this notation as a "universal polygraphy" comparable to the Egyptian hieroglyphs and the Chinese characters.[56] Leibniz's polygraphy is supposed to consist of pictorial symbols for primitive terms, such as groups of points for numbers and lines for relations; these symbols would provide "a kind of alphabet" out of which more complex terms could be assembled.[57] The result would be "a universal script that will be as easy as it is general, readable without the aid of a dictionary, and allowing fundamental knowledge of everything to be absorbed at once."[58] The idea of constructing symbols from simple components resembles the proposals of Descartes, Ward, and Dalgarno, who all preferred this analytical approach to Wilkins's taxonomic one. The result would be what, later in his life, he would come to call a "universal characteristic," a system of symbols whose construction reflected the true natures of things.

Leibniz later came to view the *Dissertation* as a piece of juvenilia, but he continued to revisit the idea of a universal writing throughout his career. In the 1677 essay "Towards a Universal Characteristic," he recounts the development of his thinking on the project up to that point. Early on, he writes, his hope was to create an "alphabet of human thoughts" to which the combinatorial art could be applied: "through the connection of its letters and the analysis of words which are composed out of them, everything else can be discovered and judged."[59] More recently, he had moved away from the alphabetical model toward the idea of representing ideas by "characteristic numbers," which would be determined based on a comprehensive system of knowledge akin to an encyclopedia.[60] These numbers would make it possible to subject metaphysics and morals to "an infallible method of calculation."[61] He sums up the consequences in another statement that is often quoted in relation to computation: "Once the characteristic numbers are established for most concepts, mankind will then possess a new instrument which will enhance the capabilities of the mind to a far greater extent than optical instruments strengthen the eyes, and will supersede the microscope and telescope to the same extent that reason is superior to eyesight."[62] This project, as he describes it here, is far more ambitious than that of Wilkins. Leibniz did not want to produce a fixed language for use in international communication; instead, he wanted to create a method of discovery that could be used to advance scientific knowledge.

Leibniz's efforts from 1677 and 1678 exhibit an obsession with characteristic numbers that verges on numerology. In 1677, he began writing a detailed commentary on Wilkins's real character that might be seen as an attempt to construct the sort of encyclopedic knowledge that would be necessary to realize the characteristic.[63] Much of the commentary consists of Latin definitions for concepts in Wilkins's hierarchy; Leibniz also briefly tried to convert a simple definition into a numerical form, criticized some of Wilkins's assumptions, and constructed his own skeletal version of the hierarchy. In 1678, he made two abortive attempts at a characteristic, one titled "Lingua Generalis" and the other "Lingua Universalis." Probably inspired by the previous work of Dalgarno, he constructs a set of rules for converting between words and numbers:

a	e	i	o	u
1	10	100	10000	100000

[...]

b	c	d	f	g	h	l	m	n
1	2	3	4	5	6	7	8	9^{64}

The word is constructed from consonant–vowel pairs, with the consonant indicating the digit, the vowel indicating its place, and the order being arbitrary; thus, to use his example, the word *"bodifalemu"* has the number 81374. The idea is to represent the compounding of ideas through multiplication; the mathematical statement *"48 est 6ⁿᵃʳⁱᵘˢ"*—that is, 48 is divisible by 6—could thus, for instance, translate into the logical statement *a human is rational*.[65] Or, by Leibniz's rules, one could say that *fema* is *ha*.

Leibniz's work in these two texts exhibits a close attention to the musical harmonies of language. As he explains, his system gives no significance to the order of syllables, so that one can indifferently choose to write either *bodifalemu* or *mubodilefa*. This flexibility, he writes, enables a "pleasing, apt language to be restored to Music and poetry, and to all other delights of discourse."[66] Leibniz goes further in the second fragment, claiming that a line of verse could be composed in this language "as if by a sure demonstration" with "everything determined."[67] While Louis Couturat takes this passage to show that Leibniz drew on mathematical methods even in the domain of aesthetics, it should be noted that the desire for rigor is not alien to seventeenth-century poetics.[68] One thinks, in particular, of German baroque poets such as Martin Opitz, whose 1624 *Book of German Poetry* codifies a poetic style that emphasizes rhyme schemes, rigid verse forms, and the purity of language; the point is to undertake a virtuosic display of assembling words under heavy constraints.[69] Leibniz's *Dissertation* contains extensive discussions of poetic forms that have a combinatoric aspect, including what Julius Caesar Scaliger had called *protean verses*—verses whose parts can be arranged in multiple orders—and palindromes.[70] For instance, Leibniz calculates that the following line by Scaliger can be rearranged in sixty-four ways:

Perfide sperasti divos te fallere, Proteu.
Deceitfully you hoped to cheat the Gods, Proteus.[71]

Such practices were, indeed, widespread in the period, reflecting the general valuing of "art" in the sense of artifice. Leibniz was working within a culture that valued the following of rules to a greater extent and in a wider range of fields than it would be valued in later periods; at least some of the aspects of his work that look to us like mathematization would have been, at the time, matter of course.

This cultural context is important to keep in mind while evaluating Leibniz's significance in the history of computation. The universal characteristic is often viewed as akin to either symbolic logic or a programming language; in both interpretations, Leibniz is taken to privilege formal relations of symbols over their meanings. But his goals are wider than those of either modern logic or computer science. The point was not just to provide rules for what to do with symbols but also to establish a universal sign system that would grant demonstrative certainty to a wide range of domains, including law, music, metaphysics, and theology. In a 1678 letter to Walter Von Tschirnhaus, Leibniz describes the characteristic as a generalization of the algebraic methods of Cardano and Viète to work with "other formulas which have nothing in common with magnitude."[72] He assures his skeptical friend that this method will not lose its grounding in reality: "No one should fear that the contemplation of characters will lead us away from the things themselves; on the contrary, it leads us into the interior of things."[73] This confident statement raises the question of how the characteristic numbers are supposed to relate to the ideas and things that the symbols were supposed to express.

In this regard, the comparison with Wilkins is illuminating. Although Leibniz is clearly following a different plan from the English bishop, the two are alike in what they are conspicuously not concerned about—namely, the role of words in establishing the meanings of symbols. At the beginning of the *Essay*, Wilkins claims that previous authors of real-character schemes "did generally mistake in their first foundations" because they attempted "the framing of such a *Character* from a *Dictionary of Words*, according to some particular Language, without reference to the *nature of things*."[74] Wilkins's character is not supposed to encode the English language; it is supposed to encode bare ideas that exist in the mind prior to any language. But the classification tables are nonetheless built from English words, without which they would be meaningless. That the use of words in the construction of the system did not seem to bother Wilkins at all is characteristic of the pre-Lockean attitudes toward language that made the real-character idea possible. As far as Wilkins is concerned, conceptual clarity stems not from how the symbols are explained but from the fact that the system as a whole is arranged according to "the rules of art" and thus reflective of a scientific understanding of the world.

As far as the issue of meaning goes, Leibniz's position is more akin to that of Wilkins than to modern formalism. Rather than exclude issues of communicational meaning from his symbolic system, he presumed that

THE MATTER OUT OF WHICH THOUGHT IS FORMED   ›   67

properly designed signifiers would disclose their meanings promptly to everyone. In the letter to Tschirnhaus, he makes a (retrospectively) questionable claim similar to one Wilkins makes in the *Essay*: that the system will be as easy to learn as numerals. On the basis of the universal characteristic, Leibniz writes, "a spoken and written language can also be developed . . . which can be learned in a few days and will be adequate to express everything that occurs in everyday practice, and of astonishing value in criticism and discovery, after the model of the numeral characters."[75] His confidence that this language would be so much easier to learn than, say, English stems from a belief that linguistic confusion results primarily from the fact that "the characters we use are badly arranged."[76] With a properly designed characteristic in place, we would "need merely to see the characters in order to have adequate notions brought to our mind freely and without effort."[77] The assumption is that learning a universal language does not involve learning a new set of ideas, a whole way of thinking about the world. The "notions" are already there in the mind, waiting to be perceived; all one need do is learn the symbols.

Leibniz never gave up on the idea of characteristic numbers, but he later grew somewhat more measured in his claims. For instance, he switched his linguistic efforts toward developing a simplified version of Latin; his work on logical calculi thus diverged from his attempts at a universal language.[78] His ambitions for what the characteristic could do also contracted. One of his most mature, albeit often misunderstood, statements on these issues appears in the 1685 "Art of Discovery." In this essay, Leibniz discusses the possibility of an art that would "accomplish in other matters something similar to what Algebra does with numbers."[79] He contrasts two approaches to this goal. The first is a system of signs so perfect that it contains the seeds of all possible knowledge. "If words were constructed," he writes, "according to a device that I see possible, but which those who have built universal languages have not discovered, we could arrive at the desired result by means of words themselves, a feat which would be of incredible utility for human life."[80] It is important to stress, however, that Leibniz is describing this idea here so as to concede to the Cartesian objection against it. The dream, Leibniz concludes, is theoretically possible, but it could not be achieved without a complete and correct theory of everything—that is, without Descartes's "vraie philosophie."

Leibniz's alternative to this unattainable dream is less like a computing machine than a method of checking one's work. "In the meantime," he continues, "there is another less elegant road already open to us, whereas the other would have to be built completely new."[81] This other approach involves "making use as mathematicians do, of characters, which are appropriate to fix our ideas, and of adding to them a numerical proof."[82] This statement might be taken to indicate something like symbolic logic, which, although it does not (and, as we now know, cannot) provide a purely mechanical procedure for answering any question, does allow proofs to be expressed in a form that allows their correctness to be checked mechanically. But Leibniz's explanation suggests something else. The certainty of mathematics, he states earlier in the essay, stems from the fact that "we can continually submit to trials or tests not only the conclusion but also, at any moment, each step made from the premises, by reducing the whole to numbers."[83] It is because of the lack of such "tests," he continues, that other fields are not as certain as mathematics. It is in this context that we find the famous "let us calculate" passage. This context suggests that what Leibniz has in mind is something like what programmers call a "sanity check": to test that 3 is a root of $y^3 = 8y + 3$, one plugs it in and verifies that 3^3 does, in fact, equal $8 \cdot 3 + 3$.

This sort of test was, indeed, a common element of the computational practices of the period. As I mentioned in chapter 1, early algorism texts often included special procedures, sometimes called "proofs" or "proves," designed to check the results of calculations. One "proof" of addition, for instance, was the procedure called *casting out nines*. One adds up the individual digits of the addends, subtracting nine whenever the total exceeds nine, and does the same for the sum. If these numbers are unequal, then the sum must be wrong. Leibniz had referenced this procedure explicitly in the 1677 version of his "Let us calculate" statement:

> Moreover, we should be able to convince the world what we should have found or concluded, since it would be easy to verify the calculation either by doing it over or by trying tests similar to that of casting out nines in arithmetic. And if someone would doubt my results, I should say to him: "Let us calculate, Sir," and thus by taking to pen and ink, we should soon settle the question.[84]

If this is what he means by "numerical proof," then what he is proposing is less a logical calculus in the modern sense, in which the certainty of conclusions stems from formal inference rules, than a pragmatic measure meant to reveal errors. As Leibniz undoubtedly knew, the casting out nines procedure does not always work; it fails to detect errors when the result is off by a multiple of nine. The point, then, is not to ensure correctness by replacing reasoning with a mechanical process; it is "to examine arguments," as he writes in the 1677 essay, and thus reveal their flaws.[85]

A similarly pragmatic attitude may be discerned in Leibniz's remarks about calculating machines. In an oft-quoted manuscript written the same year as "The Art of Discovery," he describes the purpose of the machine he built: "it is unworthy of excellent men to lose hours like slaves in the labor of calculation, which could be safely relegated to anyone else if the machine were used."[86] While this reference to slavery has often been taken to indicate that he means to liberate people from work by automating calculation, this is a misreading of the passage. The point of the machine is to ensure that calculations can be "safely" assigned to a subordinate who, presumably, cannot be trusted to do them correctly with a pen and paper. As in the numerical "tests," the aim is to compensate for the human tendency to err; "it is known," Leibniz writes earlier in the text, "from the failures [of those] who attempted the quadrature of the circle that arithmetic is the surest custodian of geometrical exactness."[87] By making calculation easier and more reliable, the machine would encourage astronomers and other "excellent men" to check their results through numerical tests and thus improve the reliability of knowledge.

If Leibniz was moving away from some of his more extreme claims about the characteristic in the 1680s, he retained his faith in the power of symbols. In the years following the writing of "The Art of Discovery," he developed some of his most sophisticated attempts at the rational calculus he had promised in the *Dissertation*. The best known of these appear in a pair of manuscripts about what is sometimes called the *plus–minus calculus*.[88] The idea is to generalize arithmetical addition into a "real addition," as he called it in the second manuscript, that deals with collections of entities rather than abstract quantities.[89] As he explains, $2 + 2 = 4$ in common arithmetic, but this only holds if the two pairs being combined are assumed to be different; if one has two coins and adds the same two coins, one still has only two coins.[90] Real addition is supposed to represent how addition works with collections that may overlap like this. Whereas his previous efforts had worked either through placing letters together (*ra* means *rational animal*) or through prime numbers, this system employed operators similar to the algebraic + and , creating an apparent relation between logic and algebra.

In the first version of the system, Leibniz uses $A + B = L$ to indicate that A is contained in L, with B being the remainder. He also defines a complementary relation $A - B$, which expresses whatever is left over when B is taken away from A. Leibniz's system works somewhat like numerical algebra, although it is not exactly the same: if A and B contain something in common, then $A + B - B \neq A$, since the subtraction "destroy[s]" something that is contained in A.[91] In the second manuscript, he leaves out subtraction and focuses on real addition, which he now distinguishes from numerical addition by drawing a circle around the plus sign. He suggests two distinct interpretations of this operation. In most cases, he takes it to express the combination of attributes, so that $R \oplus A$ might signify what is both rational and animal. The operation can also, he notes in passing, express the relation by which multiple species compose a genus, so that (to invent an example) $F \oplus B$ might signify a class containing both fish and birds.[92] He works out the consequences of this system based on a set of axioms, including $A \oplus A = A$, which would later become a central law of Boolean logic.

The legacy of Leibniz's logical calculi has long been filtered through a series of rereadings and revisions that took place in much later periods. While some commentators have intimated that George Boole built his system on foundations laid by Leibniz, the claim of a direct line of influence is not supported by historical evidence; Boole was probably unaware of Leibniz's relevant writings when he developed his logical system.[93] When Boole finally did read Leibniz's logical work in the 1850s, he later recalled, he concluded that the Leipzig philosopher had stated his views "in language which to those imbued with later and juster views of the functions of Logic must appear extravagent."[94] Around 1900, Bertrand Russell attempted to make a logical formalist of Leibniz, arguing that the universal characteristic "was evidently akin to the modern science of Symbolic Logic" as practiced by Boole; Louis Couturat, C. I. Lewis, and Ernst Cassirer expanded on this view.[95] But these readings had to work around aspects of Leibniz's thought that violate the disciplinary bounds of modern logic.

A particular difficulty regards how Leibniz addresses the objection (often raised against classical deductive logic) that the results were mere restatements of definitions and were thus ultimately arbitrary. In an earlier manuscript, Leibniz responds that, "even if certain propositions are arbitrarily assumed, as the definitions of terms, yet there arises from these a truth which is far from arbitrary."[96] As an example, he gives the numeral system, which consists of arbitrarily defined symbols but which nonetheless produces conclusions of absolute truth. This argument anticipates formalism in its focus on the relations of symbols; it resembles how programming languages treat names as arbitrary in the sense of interchangeable. Yet formalism does not, in its modern form, license the sort of connection Leibniz wanted to make between such symbols and the material world. Leibniz did not present the plus–minus calculus as a mere way of working out the consequences of axioms: the symbols were supposed to express truths about the natures of things that would enable one to draw conclusions about them through calculation alone, thus doing for other subjects what the Indian calculating techniques did for quantities.

Leibniz's confidence that this was possible makes sense only if we consider the seventeenth-century crucible from which his thinking flowed. Advocates of universal characters, including Wilkins and Leibniz alike, presumed the existence of a universal set of ideas that exist in the mind before one learns any language. Leibniz states this position explicitly in the 1686 *Discourse on Metaphysics*: "there is nothing we could ever learn of which we do not already have in our mind the idea, which is like the matter out of which the thought is formed."[97] The assumption that the human mind comes already outfitted with ideas entailed that, once an adequate symbolism was developed for a given area of knowledge, the symbols would readily become comprehensible to anyone intelligent enough. This rationale for trusting in the transparency of symbols is incompatible with any serious sort of cultural or linguistic relativism. As with Wilkins's real character and Oughtred's version of specious arithmetic, Leibniz's characteristic employs symbols as arbitrary signifiers, but if one admits that ideas, too, can be arbitrary, put together in ways that depend on the contingent circumstances of one's upbringing, the dream promptly withdraws from grasp.

Among Leibniz's early followers, the most fruitful parts of his work toward the characteristic were the attempts at a logical calculus. One of Leibniz's most prominent acolytes (after the tireless Christian Wolff) was the Swiss logician Johann Heinrich Lambert, who attempted, starting in the 1750s, to discover what "was concealed in the Leibnizian characteristic and in the *ars combinatoria*."[98] Lambert developed a logic system based on a representation somewhat like Venn diagrams, which was meant, as he presented it, to do for qualities what the algebraic notation had done for quantities. The goal was to produce a scientific sign system that represents the world with such veracity that, he writes, "the theory of things and the theory of signs become interchangeable."[99] Lambert presents his attempt at such a scientific symbolism in his book *Disquisitio*, published in 1767.[100] What he developed, however, did not live up to the expectations he had created; early readers of the *Disquisitio* were disappointed to find that his logical calculus merely used letters to represent qualities rather than, as he had promised, symbols that naturally corresponded to the essences of things.[101] In spite of Lambert's efforts, the dream of a system that can produce true knowledge about all manner of subjects through symbols alone remained unrealized.

If the characteristic turned out to be vaporware, Leibniz was a massive success in another area: mathematics. His version of the infinitesimal calculus draws much of its power from his carefully designed symbols, such as dx and $\int dx$; while the Leibnizian notation has its detractors, it is almost universally viewed as superior to that of Newton. As the next section shows, Leibniz's work on calculus contributed to a reframing of the discourse surrounding symbolic methods around the end of the seventeenth century. His notational innovations offered proof that, in a way, advocates of real characters were right: symbols really could do things that words could not. But Leibniz's work also raised new conceptual problems. It was easy enough to explain what, say, the + sign meant, but explaining the meaning of dx was not so easy. Critics thus questioned whether Leibniz's methods were merely instrumental procedures that fell short of the standards of demonstrative knowledge set by Euclid and Archimedes. If, in his work on the universal characteristic, Leibniz maintained confidence that symbols would remain firmly affixed to universal "notions," his work on calculus troubled that faith, raising the possibility that symbols could produce true results even though they meant nothing at all.

AN ALGORITHM, SO TO SPEAK

In his comprehensive history of mathematical notations, Florian Cajori states that only two mathematicians ever managed to create more than two symbols that caught on: Leibniz and Leonhard Euler.[102] Although Leibniz did not take symbolism to the extreme that Pierre Hérigone had a few decades before—he did not attempt to replace words altogether—he did give symbols a prominent role in his mathematical practice. Central to his view of notation is the idea of *blind thought*. In the *Dissertation*, Leibniz states that it is possible to think of many things at once in a single intellectual act, "as, for example, when by reading numerals on a piece of paper we grasp in a kind of *blind thought* some very large number that all the years of Methuselah would not suffice to count up explicitly."[103] The point is that one need not have a clear mental image of a million things in order to reason correctly about that quantity—one need only understand how the symbol 1000000 is constructed so as to "express," as Leibniz liked to put it, the number.[104]

Although Leibniz applied his idea of blind thought to some practices that we would now see as algorithmic, it would be a mistake to identify it with algorithmic thinking in its modern form. Blind thought did not mean blind rule following; it was a form of contemplation, not an unintelligent, mechanical process.[105] Leibniz used symbols not just for computing results but also for visualizing mathematical data in a tabular form so as to make patterns visible.[106] One may find a clear example of this approach in one of the texts that has led to outsize claims about Leibniz's role in the development of computers. Leibniz was, as is well known, interested in binary notation. In an article published in the journal *Memoires de l'Académie Royale des Sciences* in 1703, he introduces the binary numeral system and shows how the basic operations of arithmetic can be performed with only 1 and 0. He argues that binary arithmetic explains the meaning of the *I Ching* and suggests that it may have played some role in the formation of the Chinese characters as well, speculating that these characters might reveal "something considerable about numbers and ideas."[107] He ends the article with a suggestive comment linking these ideas to the universal characteristic, in which "every reasoning that one can draw from notions could be drawn from their characters by a manner of calculation."[108]

The binary system that Leibniz discusses in this article is indeed the same one that forms the basis of modern computation. But the reasons this system interested Leibniz are not the reasons it is important to computer science. Binary arithmetic became central to modern computation primarily because it forms a convenient basis for electrical computing machines. For mechanical calculators using rotating drums, higher bases were better, which is why Leibniz used base ten rather than binary for his calculating machine. Leibniz's interest in binary stemmed not from its utility in mechanization, but from its visual qualities: making patterns in numbers easier to see enables the easy perception of mathematical laws that cannot readily be discerned in decimal numbers. In a table of binary numbers, for instance, one "sees in a single glance the reason for a *celebrated property of the double geometric progression*"—namely, the fact that any number can be composed by summing powers of two.[109] The point is not to develop a new method of computation, for the purpose of which, as he notes in the article, it would be best to stick with the base ten system. The point, rather, is to think about what different types of symbol can tell us about mathematical entities.

Leibniz had the greatest success with symbolic methods in his work on calculus. As readers who have studied the field will know, calculus focuses on two operations that we now call *integration* and *differentiation*. The basic problem of integration (which was usually called *quadrature* in the seventeenth century) is to determine the area of a shape with a curved boundary, such as a circle. In Euclidean geometry, no one had ever been able to "square the circle," meaning to construct a square with the same area as a given circle. Mathematicians had long recognized that it was possible to approximate this area by computing the areas of polygons that come close to filling it; the more sides the polygon has, the more accurate the approximation can be (figure 2.4). The Persian mathematician Jamshīd al-Kāshī had used such techniques to approximate the value of pi in the early 1400s.[110]

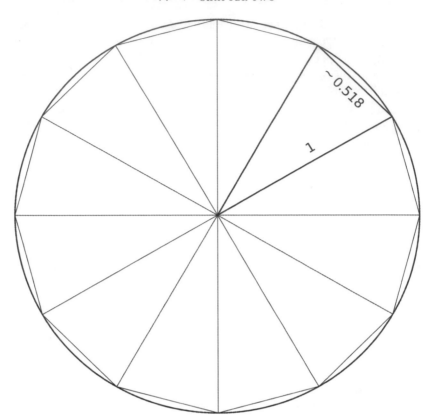

Figure 2.4. Estimating pi using a twelve-sided polygon. Using some geometric tricks, Jamshīd al-Kāshī was able to determine the lengths of the polygon's sides, which, added together, form an approximation of the perimeter of the circle. Using a hexagon produces the estimate $\pi \approx 3$, whereas the twelve-sided polygon produces $\pi \approx 3.106$. The smaller the sides, the more accurate the estimate.

With the methods of calculus, one can compute, at least in many cases, the value that this approximation approaches as the complexity of the polygon grows toward infinity, thus exactly determining the area of a shape. In the time of Newton and Leibniz, this method was often explained in terms of infinitely small magnitudes—a claim that raised philosophical problems that would occupy mathematicians through the eighteenth century. Differentiation, which is the inverse of integration, is about finding the rate of change of a function (or, equivalently, the slope of a curve) at a given point. Differentiation can be used to find the minimum and maximum points of curves, which was a problem that particularly interested Leibniz.

Although Newton famously accused Leibniz of plagiarism, the consensus now is that the two developed their versions of calculus independently. Newton's version, which he called the "method of fluxions," involved the intuition of flowing numbers; it is a notoriously difficult theory to understand. Leibniz's version is closer to how calculus is taught today, especially in terms of its notation. He first published this work in the 1684 article, which focuses specifically on differentiation.[111] To differentiate an equation is to transform it into another equation that indicates the rate of change at a given point. Although there are some important differences, Leibniz's rules for differentiation are recognizably similar to the ones used in modern calculus: x^a, for instance, becomes ax^{a-1}.[112] Leibniz published his method of integration (which he preferred to call *summation*) in another article two years later; it is here that he introduced the notation $\int dx$ for integrals. The integral sign is supposed to resemble an *s* for *sum*; the notation was originally printed, using the long form of *s* that was common at the time, as "ʃdx."[113]

While they both employed symbolic notations, Leibniz and Newton had different views on the theoretical significance of symbolism. Newton was generally opposed to methods that depended heavily on symbols, preferring geometric demonstration; in a 1715 *Philosophical Transactions* article about his dispute with Leibniz, Newton declared in the third person that "Mr. Newton doth not place his Method in Forms of Symbols, nor confine himself to any particular Sort of Symbols for Fluents and Fluxions."[114] Leibniz, on the other hand, developed new symbolic notations with abandon, and his 1684 article presented abstract rules of operation, which he called an "algorithm," shorn of any detailed proof of the procedure's validity. Modern scholars have taken varying views on whether Leibniz managed to develop a rigorous rationale for his calculus; certainly, he did spend a significant amount of time thinking about theoretical matters.[115] But even if he did have a rigorous theory to back up the method, his published work did not contain a clear statement of it, nor did his correspondence with other mathematicians. He left his followers, instead, to puzzle over the conceptual foundations of the new "algorithm."

Leibniz's tendency to avoid foundational questions in his published work has led some scholars to view him as an instrumentalist.[116] But even if Leibniz's followers were content to plow ahead without fretting over theory, programmatic instrumentalism was not an available option when they were pressed to justify their work. Simply pointing out that the "algorithm" worked was hardly satisfactory since, as long as there was no acceptable proof of its validity, it remained possible that it would fail in particular cases that had as yet gone untested. (As we will see in chapter 5, such problems would, centuries later, become a major motivation for the development of a theory of algorithms in computer science.) Some of Leibniz's early critics, such as Michel Rolle, attacked him from this angle.[117] Further, the idea of the infinitesimal, which played an important explanatory role in Leibnizian calculus textbooks such as those of Guillaume de l'Hôpital and Christian Wolff, seemed to contain a logical impossibility.[118] Leibniz's failure to state (at least publicly) a clear rationale for his use of infinitesimals led critics to question whether his methods were mere tricks that had no place in serious science.

The best-known attack on this front came from the Irish philosopher George Berkeley, who criticized the theories of both Newton and Leibniz in his 1734 pamphlet *The Analyst: Or, a Discourse Addressed to an Infidel Mathematician*. Berkeley makes a number of arguments in this text, among them the accusation that the calculus falls short of the status of a science because it merely provides methods for solving problems without placing them on solid conceptual ground. Berkeley asks, "Whether there be not a way of arriving at Truth, although the Principles are not scientific, nor the Reasoning just? And whether such a way ought to be called a Knack or a Science?"[119] In Berkeley's judgment, the calculus was a mere "knack" because it only provided procedures for solving problems without (in his view) founding those procedures on principles. This line of critique did more than try to show that Leibniz's "algorithm" could produce wrong results. For Berkeley, even if the results were always right, the procedure could still be unscientific if its conceptual basis were flawed; in particular, it might contain two errors that canceled each other out.

A number of scholars have pointed to an apparent inconsistency between Berkeley's arguments against calculus and his earlier writings on arithmetic and algebra, in which he took an unapologetically instrumentalist stance.[120] In the early notebooks known as the *Philosophical Commentaries*, probably written in 1707–8, Berkeley articulated a protoformalist view of arithmetic and algebra in which "Numbers are nothing but Names, meer Words."[121] Berkeley's account of numbers is, at least in some aspects, reminiscent of Leibniz's idea of blind thought. Arithmetic and algebra, Berkeley argues, are "sciences purely Verbal, & entirely useless but for Practice in Societys of Men. No speculative knowledge, no comparing of Ideas in them."[122] The value of these sciences, Berkeley suggests, consists precisely in these names, which tell us something that cannot immediately be perceived in things themselves: "I am better inform'd & shall know more by telling me there are 10000 men than by shewing me them all drawn up."[123] Earlier he links the certainty of such verbal sciences to the fact that symbols (unlike ideas in Berkeley's view) may be chosen at will:

"The reason why we can demonstrate So well about signs is that they are perfectly arbitrary & in our power, made at pleasure."[124]

To an extent, the apparent disjunction between this argument and the polemics of *The Analyst* stemmed from his political motives in attacking Newton and Leibniz. But that Berkeley's instrumentalist theory of algebra should translate into a general acceptance of symbol manipulation in all mathematical fields would not have been nearly as obvious in his time as it might be in ours. Newton and Leibniz both treated calculus as primarily a matter of geometry, which was, as the model demonstrative science, sacrosanct at the time in a way that algebra was not. As Douglas Jesseph has stressed, Berkeley likewise viewed calculus as a part of geometry, which he treated very differently from arithmetic and algebra.[125] Newton himself stated, in the English edition of *Universal Arithmetick*, that "Multiplications, Divisions, and such sort of Computations, are newly received into Geometry, and that unwarily, and contrary to the first Design of this Science."[126] Applying symbolic methods to geometry was a violation of disciplinary boundaries, and there was no widely accepted epistemological framework in which this violation could be justified.

The problem is manifest in Leibniz's decision to describe his procedure for differentiation as an "algorithm." This usage appears, among other places, in the 1684 article in which he first introduces the procedure: "From this rule, known as an algorithm, so to speak, of this calculus, which I call differential, all other differential equations may be found by means of a general calculus."[127] Although Leibniz does call his method an algorithm (*Algorithmo*) here, he qualifies this characterization with "so to speak" (*ut ita dicam*), suggesting that the usage is not quite standard. From his other writings, it is apparent that he specifically meant *algorithm* to imply an analogy with arithmetic. In a 1692 journal article written in French, Leibniz describes the calculus as "a new algorithm [*algorithme*], that is to say a new manner of adding, subtracting, multiplying, dividing, extracting [roots], proper to incomparable quantities, that is to say those that are infinitely large or infinitely small in comparison to others."[128] For Leibniz, *algorithm* was not yet a general term for any sort of clearly defined procedure, carrying instead a reference to the five (as he counted them here) basic operations of arithmetic.

By this use of the word *algorithm*, Leibniz contributed to the expansion of the term's meaning from a particular set of computational methods to a general style of procedure that could be applied in other areas. This process had already begun in the sixteenth century, when Michael Stifel described an "algorithm" for manipulating square roots. By Leibniz's time, some writers had begun to use *algorithm* to refer to algebraic procedures as well, and Leibniz pushed it further by extending the term to calculus.[129] This extension took root among Newtonians as well. In 1702, the popularizer John Harris referred to the rules of Newton's differential calculus as "the Algorithm or Arithmetick of Fluxions."[130] The term *algorithm*, then, was already, long before the computer age, coming to refer to practices beyond the original algorism; computation was, in spite of Newton's blandishment, encroaching on geometry.

One ought to question, however, just what sort of arithmetic Leibniz meant to single out when he called his method an "algorithm, so to speak." His differentiation procedure is less akin to common arithmetic than to specious arithmetic—that is, to the seventeenth-century version of symbolic algebra. Rather than producing numerical results, specious arithmetic provides rules for the transformation of symbolic formulae, such as $B(A + E) = BA + BE$ or $\frac{AE}{A} = E$. Leibniz's "algorithm" similarly formulates its rules as equations, as in his multiplication rule $\overline{dvx} = xdv + vdx$. While Leibniz did indeed inspire an expansion in the use of *algorithm*, his followers applied the term mainly to rules of this algebraic sort. For instance, in the late eighteenth century, the German mathematician Abel Bürja described an "algorithm of logarithms," meaning "a manner of representing them algebraically, with the forms of calculation that result from it"—that is, algebraic rules such as $\log a - \log b = \log (\frac{a}{b})$.[131] The point is not that the practice follows a strictly defined sequence of steps, as in modern algorithms, but that it involves operations expressed in symbols; extending the reach of "algorithms" thus went hand in hand with the development of new notations.

Algebra was already facing theoretical difficulties in the decades following Leibniz's 1684 article, and the calculus made the problem worse. With each expansion of the meaning of *algorithm*—from arithmetic to algebra to calculus—the conceptual soundness of the practices it denoted became more debatable. Arithmetic was more or less uncontroversial, and the qualms about negative and imaginary numbers in algebra only surfaced now and then; but questions about the foundation of calculus came to pervade the discourse. The intensity of this debate stemmed in part from the counterintuitive nature of infinitesimals, but it was also stoked by an increased awareness of the importance of defining symbols. As the next section shows, Leibniz was forced to defend his use of symbols in his engagement with the philosopher John Locke. Locke's work heralded a new cluster of epistemologies for which symbols did not offer a solution to the uncertainty of language, as they had for Oughtred and Leibniz; from a Lockean perspective, the symbols simply were a language, and their meanings were just as open to questioning as those of words.

FROM ART TO NATURE

If Leibniz's dispute with Newton was partially motivated by personal factors, his dispute with Locke was partially mitigated by them. In his 1689 *Essay Concerning Human Understanding*, Locke famously asserted that all knowledge derives from sensory observations rather than from innate principles or ideas. After reading the *Essay* in 1695, Leibniz repeatedly attempted to engage Locke in a debate, but he could not persuade the English philosopher to take him seriously. Starting in 1700, Leibniz began writing a book-length response to Locke that would eventually become *New Essays on Human Understanding*. Locke's death in 1704 led Leibniz to scrap his plans out of respect for the dead, and the manuscript went unpublished for decades. By the time it finally appeared in print in 1765, Locke had already gained the upper hand. Locke's argument dealt a crushing blow to the early modern belief in universal harmony of which Leibniz was a late holdout, a turn that would play a large role in setting the agenda for the way symbolic methods would function in the eighteenth century.

The nature and merits of Leibniz's defense of innatism have been a subject of much debate among scholars. The dispute had to do in part with religion; Locke was widely perceived to sympathize with a dissenting sect known as the Socinians, to whom Leibniz was strongly opposed.[132] Leibniz's innatism has also been likened to Immanuel Kant's argument, later in the eighteenth century, that certain categories are necessary to all rational thought.[133] But Leibniz's domain of innate ideas was rather more populous than Kant's transcendental realm. Leibniz wrote some time around 1680 that the ease with which children can understand "mathematics, morals, jurisprudence, and metaphysical matters" proves that "the seeds of all these concepts were already in the child."[134] This list of disciplines, not incidentally, includes just those areas in which Leibniz most fervently hoped to apply the *calculus ratiocinator*. By denying the existence of prelinguistic "notions" that may be presumed to exist in all rational minds, Locke's arguments undermined the way Ward, Wilkins, and (I am arguing) Leibniz had justified the view that symbols could circumvent the uncertainty of words. At stake was thus the possibility of the universal characteristic—and perhaps even the status of symbolic methods in mathematics.

In the 1689 *Essay*, Locke famously considered whether ethics could be made as certain as mathematics, in the course of which he gives an account of how, in his view, symbolic computation works. In Locke's view, the process of "casting up a long sum, either in *addition, multiplication,* or *division*" is nothing but "a progression of the mind, taking a view of its own ideas, and considering their agreement or disagreement."[135] Notation plays an important role, since "without setting down the several parts by marks, whose precise significations are known, and by marks, that last and remain in view, when the memory had let them go, it would be almost impossible to carry so many different ideas in mind."[136] Yet these symbols are mere memory aids that are only as good as the meanings assigned to them. This account of the function of mathematical symbols goes directly against Leibniz's case, in the letter to Tschirnhaus, that characters can lead us "into the interior of things." In Leibniz's view, the uncertainty of ethical arguments stems primarily from the fact that the characters we use to express them are not as well "arranged" as mathematical notations. For Locke, by contrast, simply improving the notation would not suffice to make ethics demonstrative like mathematics. The problem lies in the formation of the ideas, which cannot be "fixed" by means of characters alone.

Locke's account of the cognitive function of signs surfaced a difficulty that the real-character view tended to gloss over. Words, Locke argues, refer not to things themselves but only to what he called *nominal essences*—arbitrarily constructed bundles of qualities such as *smooth red sphere*. Nominalism had a long history in Scholastic thought, but Locke differs from earlier nominalists in the extent to which he treats linguistic differences as irreducible. Locke states that words are "voluntary signs," meaning that everyone has "a liberty, to make words stand for what ideas he pleases."[137] One does not, however, have the liberty to control how other people understand words. As a result, to suppose that one can get around the dependence of signification on the existence of commensurable ideas in the minds of the speaker and hearer—to act as if words can refer either to ideas in other people's heads or to things themselves, unmediated by mental conceptions—is to push language beyond what it can legitimately do. This account of the "abuse of words" constitutes Locke's explanation for the linguistic confusion that Bacon and his followers had long lamented.[138] The problem, as Locke has it, is not just the corrupt nature of the languages people speak at present. The problem is that language itself, meaning any form of signification whether spoken or written, is fundamentally incapable of establishing the sort of transparent communication that the Baconians desired.

Leibniz was probably not one of Locke's immediate targets in this argument, but Wilkins probably was. As Hannah Dawson has stressed, Locke's insistence that signs are "voluntary" was directed against the confidence of seventeenth-century semioticians like Wilkins in the unproblematic nature of concepts.[139] Locke argues this point in part by giving examples of cultures whose ways of thinking differ radically from those of his own country. For instance, he lists a variety of societies that supposedly exposed infants in the wilderness, killed people once they reached certain ages, and fatted and ate their own children.[140] These protoanthropological sketches, which Locke borrowed from unreliable travel narratives, are supposed to prove that moral ideas develop very differently across cultures and thus cannot have been divinely etched on the mind at birth. Locke was no moral relativist—he was, for all his heterodoxy, an ardent Christian—but his philosophy did have dire implications for the idea that a real character could establish universal agreement on moral issues. The possibility of communication, for Locke, depends on the existence of an accord about what signs mean, and simply arranging symbols in a table can provide no guarantee of such an accord.

Wilkins's project had already collapsed under its own weight by the time Locke published this argument, but Leibniz's universal characteristic was still alive, and its projector set out to show in *New Essays* that Locke was undervaluing the role of signs in reasoning. In opposition to Locke, Leibniz maintained that mathematical ideas were innate, present in the mind from birth and waiting to be revealed. Citing Julius Scaliger, he describes these innate ideas as "living fires or flashes of light hidden inside us but made visible by the stimulation of the senses, as sparks can be struck from a steel."[141] One might question whether this dispute was much ado about nothing—does it really matter whether ideas come into the mind through sensory input or are merely "sparked" by it? But the answer makes a difference in regard to symbolic methods. Blind thought had no place in Locke's conception of rational activity; one had to assemble clearly formed ideas to attach to the symbols or they would amount to mere nonsense. Leibniz's innatism enabled him to treat symbols, instead, as a means of revealing new truths whose conceptual bases had not yet been discovered.

Mathematics arose in this debate in part because Locke had used it as an example of the difficulties surrounding innatism. If one accepts that the mind has an *"implicit knowledge"* of mathematical principles, Locke had argued, then "all mathematical demonstrations, as well as first principles, must be received as native impressions on the mind: which I fear they will scarce allow them to be, who find it harder to demonstrate a proposition, than assent to it, when demonstrated."[142] How, that is, can mathematical knowledge be innate if discovering it can be arduously difficult? While Leibniz concedes that mathematics can be hard to learn, he nonetheless counters that it is built from ideas that exist in the mind from birth: "It would indeed be wrong to think that we can easily read these eternal laws of reason in the soul, as the Praetor's edict can be read on his notice-board, without effort or inquiry; but it is enough that they can be discovered within us by dint of attention: the senses provide the occasion, and successful experiments also serve to corroborate reason, somewhat as checks in arithmetic help us to avoid errors of calculation in long chains of reasoning."[143] The role Leibniz here assigns to the senses is much like that of the universal characteristic as he describes it in "The Art of Discovery": sparking thought and testing conclusions. The senses are not the source of all our knowledge, but they provide a necessary means of "rectifying" our reasoning so that we can better discover what is within us.

These two positions entail different attitudes toward the role of signs in thought. For Locke, signs are arbitrary in both the signifier and the signified: one is free to choose both what to call trees and how to define a tree. Symbols are thus, just like words, no better than the ideas by which they are defined. For Leibniz, by contrast, the "sensible traces" we use to record our ideas play an essential role in the discovery of the ideas themselves: "we cannot have abstract thoughts which have no need of something sensible, even if it be merely symbols such as the shapes of letters, or sounds; though there is no necessary connection between such arbitrary symbols and such thoughts."[144] If sensory input were not necessary for thought, Leibniz continues, "the pre-established harmony between body and soul" would not obtain.[145] This line of thought suggests that symbols can reveal something about the true natures of things, since they take part in the divinely instituted harmony between the soul and the world. It also suggests, at a more mundane level, that the problem of communication is less intractable than Locke claimed. Even if the symbols themselves are arbitrary, the ideas to which they refer are not, and as a result it would not take a Herculean effort to ensure that they mean the same thing to everyone.

This rationale for the use of symbols depends, ultimately, on an idiosyncratic variant of the early modern belief that the human mind is a mirror of the world. In a Leibnizian universe, every being perceives every other being; the only reason we do not know everything is that most of these perceptions are to some degree confused. "It can even be said," Leibniz writes in the preface to *New Essays*, "that by virtue of these minute perceptions the present is big with the future and burdened with the past, that all things harmonize . . . and that eyes as piercing as God's could read in the lowliest substance the universe's whole sequence of events."[146] The doctrine of harmony enabled Leibniz to recognize the historical contingency of natural languages without giving up on the belief that symbols—be they real characters or sounds—could convey truths about mathematics, morals, metaphysics, and other domains in ways that left no room for disagreement. If Locke pointed toward a form of cultural relativism that recognized the dependence of moral ideas on the customs of particular countries, Leibniz gestured the opposite way: by harmonizing with ideas rather than signifying them by convention, the universal characteristic would short-circuit culture.

By the time Leibniz's *New Essays on Human Understanding* was finally published in 1765, his opponent had secured the high ground. Locke's *Essay* was initially condemned by Oxford for heterodoxy, and it faced immediate criticism from numerous quarters.[147] But Lockean thought would go on to become one of the dominant philosophical strains in the eighteenth century, whereas the tide turned against Leibniz. The 1755 Lisbon earthquake, which Voltaire famously depicted in *Candide* and in his "Poem on the Lisbon Disaster," was only one of a number of factors that placed Leibniz's optimistic metaphysics under question. Voltaire's poem satirizes Leibniz's doctrine that "all is necessary," which would have entailed that the infants and children who died in the earthquake must have in some way deserved their fate.[148] At a less existential level, ideas about language had changed by the mid-eighteenth century so as to erode the distinction between characters and languages. Wilkins's real character had failed to catch on, and starting in the 1680s, opinion turned against the idea that symbols, no matter how "artificially" designed, could suffice to ensure that people were understanding things correctly; clearing up the "pipes" of communication, it was increasingly recognized, would take more than a new form of writing.

Locke's work offered an appealingly humble alternative to such high-flown schemes. Rather than trusting in the power of symbols to overcome cultural differences, Locke emphasized definitions—a pragmatic measure for addressing local conceptual problems by clarifying how ideas were assembled. He also encouraged cultivating good habits so that rationality would become second nature. In *Some Thoughts Concerning Education* (1693), he argues against the use of explicit rules in pedagogy: "Children are *not* to be *taught by Rules* which will be always slipping out of their Memories. What you think necessary for them to do, settle in them by an indispensible Practice, as often as the Occasion returns; and if it be possible, make Occasions. This will beget *Habits* in them, which being once establish'd, operate of themselves easily and naturally, without the Assistance of the Memory."[149] Locke is not talking about computational rules here—he means rules like "No pudding before dinner"—but his argument would have significant influence on mathematical pedagogy for well over a century. Explicit directions, for Locke, are an ineffective means of shaping human behavior; it is better, as he later wrote more in the context of logic, to "settle the habit of doing without reflecting on the rule."[150]

Whether through Locke's direct influence or not, eighteenth-century mathematicians generally followed him in emphasizing definitions as a means of establishing clarity, as well as in avoiding explicit rules in favor of theoretical explanations that were supposed to make procedures intuitively obvious. Some authors, especially in Britain, grew suspicious of symbols.[151] In his preface to a widely reprinted eighteenth-century edition of Euclid's *Elements*, John Kiell declares, "The Elements of all Sciences ought to be handled after the most simple Method, and not to be involved in Symbols, Notes, or obscure Principles."[152] The turn also affected British popularizations of calculus, which (while they made liberal use of symbols) tended to follow Newton in emphasizing geometry over algebra. In his 1742 *Treatise of Fluxions*, an early Newtonian calculus textbook, Colin MacLaurin states that algebra "may have been employed to cover, under a complication of symbols, abstruse doctrines, that could not bear the light so well in a plain geometrical form."[153] The key to avoiding this obscurity, MacLaurin continues, is "defining clearly the import and use of the symbols, and proceeding with care afterwards."[154] The contrast with Oughtred's statement of a century before is stark. Symbols are no longer treated as a way of placing mathematical ideas directly on the page without the interference of words; instead, they are suspected to be meaningless until proven otherwise.

Continental mathematicians were more sympathetic to Leibniz than MacLaurin was, and their ideas of clarity were more under the sway of Descartes. The development of symbolic methods continued unabated. Yet the French and German traditions also increased their focus on clarifying the meanings of symbols. Jean Le Rond d'Alembert, who wrote most of the entries on mathematics for the great monument of the French Enlightenment, the *Encyclopédie*, paid great attention to defining such tricky concepts as negative numbers and differentials. He defines *differential* (*différentiel*), for instance, as "a quantity infinitely small, or less than all assignable magnitude."[155] His article on *negative* (*négatif*) begins by noting that some mathematicians define negatives as numbers "smaller than zero," but he thinks this definition is wrong; negatives, he explains in great if somewhat knotty detail, must be understood as positive numbers placed in a relation of opposition to certain other numbers, as two curves may extend in opposite directions from a given point.[156] Like MacLaurin, d'Alembert gives a privileged role to geometric interpretability even when employing symbolic notation. D'Alembert's conception of mathematical knowledge had no place for methods that were not backed up with some conceptual basis, abstract as that basis may be.

This intellectual shift did not defeat Leibniz's reputation for good. The idea of extending "calculation" into a universal method would gain a new life in the nineteenth-century work of Charles Babbage and Ada Lovelace, as would Wilkins's real character: in his memoir, Babbage recounts a youthful attempt to create a comparable system.[157] Leibniz's writings on logic would return to prominence later in the nineteenth century as Boole, Frege, and Russell looked for a kindred spirit in the past. As a result, some histories of computation proceed directly from Leibniz to the nineteenth-century work of Babbage, Boole, or Giuseppe Peano.[158] But to skip ahead to this revival would be to miss something important. As the next chapter shows, the idea of universal computation was not absent in the century between Leibniz and Babbage; it simply took on a different form. The culmination of this eighteenth-century version of universal computation appeared in the work of Nicolas de Condorcet, who sketched out something like a *calculus ratiocinator* within an epistemology more in line with Locke's views than with Leibniz's. For Condorcet, the purpose of the system was not to reveal ideas that were already present in the minds of all intelligent people; it was to replace people's existing conceptions of the world with better ones.

CHAPTER THREE

Symbols and the Enlightened Mind

"When I use a word," Humpty Dumpty said in rather a scornful tone,
"it means just what I choose it to mean—neither more nor less."

—LEWIS CARROLL, *Through the Looking-Glass*

THE SCHISM

In the winter of 1794, Marie Jean Antoine Nicolas Caritat, Marquis de Condorcet, knew that he would not live long. Best known as a mathematician and political thinker, Condorcet was an enthusiastic participant in the French Revolution of 1789. A titled aristocrat in the Old Regime, he shocked an audience in 1791 with a forthright declaration of support for the republic.[1] Just over a year after the National Assembly was dissolved, however, his support for the moderate Girondin faction made him into an outlaw. On October 3, 1793, the Jacobins released a warrant for his arrest, and he was forced to flee from his home.[2] During his eight months in hiding, he passed the time by writing political texts that show little hint of the direness of his situation. In these manuscripts, which include some of his best-known work, he sketched out a utopian plan for humanity that stood in sharp contrast to the realities of revolutionary France.

In one of the fragments he wrote while a fugitive, later published as *Sketch for a Historical Picture of the Progress of the Human Mind*, Condorcet suggests two means by which the improvement of the human race can be assured: first, the adoption of "technical methods," by which he means "the art of arranging a large number of subjects in a system so that we may straightaway grasp their relations, quickly perceive their combinations and readily form new combinations out of them"; and second, a "universal language" that "expresses by signs either real objects themselves, or well-defined collections composed of simple and general ideas, which are found to be the same or may arise in a similar form in the minds of all men, or the general relations holding between these ideas, the operations of the human mind, or the operations peculiar to the individual sciences or the procedures of the arts."[3] Such a language, he writes, would not have "the disadvantages of a scientific idiom different from the vernacular"; it could be learned by all, as schoolchildren learn the language of algebra, providing universal access to the best knowledge available and ensuring that there could be no disagreement about either the meaning of terms or the validity of arguments.[4]

Condorcet's work has long stood as a metonym for the desire to extend mathematical reasoning to all domains of knowledge. In *The Order of Things*, Michel Foucault cites Condorcet as one of the French classical thinkers who attempted "to mathematize empirical knowledge" in domains outside of physics and astronomy.[5] Similarly, Umberto Eco, in *The Search for the Perfect Language*, takes Condorcet's approach to creating a universal language as proof that "the search for perfect languages was definitively turning in the direction of a logico-mathematical calculus," and Roger Chartier holds Condorcet's project up as an example of a desire for "formalizing cognitive operations and logical reasoning."[6] Indeed, Condorcet's biographer Keith Michael Baker argues that some of his remarks anticipate twentieth-century developments in computation.[7] But Condorcet treated the communicational side of symbolic methods very differently from later practitioners of logic and computation. Unlike a modern programming language, Condorcet's system was meant not just to produce results but also to provide a set of tables that classify all the things in the world. Condorcet's project thus exemplifies a distinctly eighteenth-century approach to symbolism, one that keeps the human mind at the center even as reasoning becomes a mechanical process.

Condorcet did not live to see his remarks about the universal language published. On March 27, 1794, he was arrested while attempting to flee the house where he was hiding, and two days later he died in his cell of unknown causes. His papers, however, survived, and the *Sketch* was published the following year. Along with the manuscript for the *Sketch*, he left behind an unfinished plan for the universal language.[8] Over the course of about ninety handwritten pages, he shows how the symbolic language of algebra can be made to subsist on its own, without the need for sentences linking the pieces together (figure 3.1). The method is based on boxlike symbols indicating which equations are taken as given, supposed as hypotheses, proven, and so forth; he also introduces symbols for different types of number as well as for functions, series, approximations, and con-

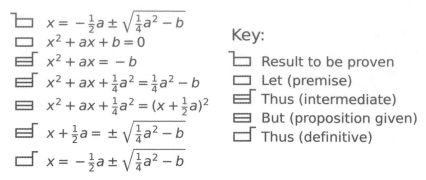

Figure 3.1. Nicolas de Condorcet performs a mathematical proof without using any words. My reconstruction follows the 2004 edition (Condorcet, *Tableau historique des progrès de l'esprit humain*, 960) but corrects the omission of *a* in the second-to-last line.

ditional expressions. Although he begins with algebra, his goal is to extend this symbolization to all sciences. The manuscript includes rudimentary notations for geometry, mechanics, astronomy, and natural philosophy as well as a proposal for a "hieroglyphic" language that represents certain ideas pictographically.[9] Just before the manuscript cuts off, Condorcet promises to explain how his system can be extended to metaphysics, linguistics, morals, and politics.[10]

This project appeared amid a broad revival of the utopian hopes that had attended symbols in the seventeenth century. In a 1785 letter, Condorcet had cited G. W. Leibniz as a precedent for the idea, although he probably did not know the details of Leibniz's attempts at a universal characteristic.[11] As another potential source, the editors of the 2004 edition of Condorcet's manuscript suggest Pierre Hérigone's *Cvrsvs mathematicvs*—the early-seventeenth-century book that presents mathematical proofs through symbols designed to be readable in both Latin and French.[12] Condorcet was one of a number of thinkers who revisited such ideas in the last decade of the eighteenth century. A few years later, in 1797, Joseph de Maimieux published what he called the *Pasigraphie*: a system of symbols that were supposed to share the translinguistic character of mathematical notations.[13] Condorcet, however, was after more than just international communication. He also wanted to preserve knowledge against catastrophe by developing a form of writing that future generations could read even if all existing languages were forgotten.

Given what Condorcet came up with, it is not hard to see why his system has struck some observers as protocomputational. His universal algebra hints at many of the elements of programming languages, including loops, if–then conditionals, and the binding of variables. For instance, using the struck-through letters $N\!\!\!/$ to mean *integer*, he writes the following to indicate the value $+a$ if n is even and $-a$ otherwise:[14]

$$m = N\!\!\!/ \quad \begin{matrix} +a, n = 2m \\ -a, n = 2m + 1 \end{matrix}$$

There is much here that anticipates later developments; indeed, the use of m to refer to an arbitrary integer seems to anticipate the "there exists" quantifier of modern symbolic logic. But Condorcet's work points in a different direction from the historical current that leads from early modern algebra, via Leibniz and George Boole, to the computer.[15] In contrast to Leibniz and Boole, Condorcet makes his logical notations entirely distinct from algebraic operators, and he gives other sciences such as geometry and astronomy their own symbols, including special lines indicating to which science each symbol belongs. (This is why the symbol for *integer* appears to be crossed out.) Instead of trying to develop a single method that could apply to all sciences, he is out to gather multiple sciences together; instead of presenting formal rules for how to use the symbols, he focuses on teaching people what the symbols mean, leaving the operations to be guided by reason.

In the natural sciences, this system works by the ordering device that is, for Foucault, distinctive of the Enlightenment period: the taxonomy. For example, one could have the number 145702342 designate a plant, with 145 representing the class, 70 the genus, 23 the species, and 42 a particular variety.[16] This numerical classification, which constitutes Condorcet's "technical method," would provide both a notation for scientific writing and a lookup system that could work with a dictionary or filing bureau. He goes to great length to show how one can explain the meanings of signs without using any preexisting language—for instance, through pictures, analogies, and experimental instructions. This re-creation of language was premised on what Tristram Wolff has called *linguistic voluntarism*—the idea that, since the meanings of signs are arbitrary, those meanings can and should be altered to better suit the natures of things.[17] New terms, Condorcet explains, will be established through explicit definitions, not tacit agreement.[18] As he states in the discussion of geometry, his goal is to sketch out a system complete enough that new symbols may be defined using nothing but the other symbols already introduced, "independently of all anterior convention."[19]

Even before Condorcet begins to move beyond mathematics, however, he begins to show some anxiety about the possibility of language finding its way back in. Before the universal algebra can be applied in a particular case, it is necessary to establish the meanings of the symbols, and it was not apparent that this could always be done without some recourse to words. Condorcet considers this a flaw in his scheme, although not a fatal one:

> We observe first that if, in a rare circumstance, it were impossible to make understood an absolutely new theory, to designate an object which had not yet been considered, to develop an operation of which one has not yet formed any idea, without having recourse to some verbal explications, the universal language would not merit less the name, would not be less useful. It would happen then, but in an opposite sense, what happens in spoken language, when sometimes one is obliged to show the object itself or its representation, because of a lack of having the expressions to describe it. One would need one language to supplement [*suppléer*] the other. But one might believe that this defect will not be encountered but very rarely in the language of universal algebra.[20]

To make theories understood, to designate objects, to form ideas of operations—these are all matters of mediating between the symbols and a person's mental conceptions of the world. When an adjustment has to be made to the alignment between symbols and ideas, "verbal explications" must intervene. This is an objection that Joseph Marie de Gérando would judge, a few years later, to be fatal to the idea of a philosophical language: one would have to explain the meanings of the newly minted words in an existing, presumably imperfect language, thus tainting the new one.[21] Yet Condorcet is confident that it will not be a problem in the majority of cases. One can mostly avoid the taint of language, he thinks, by taking care always to proceed "from known to unknown" and by expressing new ideas as "generalizations" or "restrictions" of existing ones.[22] In this way, the algebraic system can be made as self-contained as possible, and words can, for the most part, be held at bay outside the walls.

The urgency with which Condorcet wanted to get rid of natural language manifests much the same distrust of words that had motivated his seventeenth-century precursors. But his apparent anxiety about the possibility that, despite his best efforts, language would seep back in—the fact that he sees the occasional need for French definitions as a "defect" of his plan—is distinctive of the eighteenth century. For Leibniz and John Wilkins in the 1660s, the need to use words in establishing the meanings of symbols had not arisen as an epistemological problem. The two had different views as to how the system should be constructed, but there was never much doubt in their time that it would be possible to fix the meanings of the characters with certainty, independently of any verbal language.[23] By the 1790s, however, it had become clear that the idols of the market existed not just in words, but in the minds of people who spoke them. A new attention to the histories of European vernaculars in the mid-eighteenth century had led to the widespread belief that different languages embedded different ways of thinking, which implied that, contrary to Leibniz's view, the errors of vernaculars could still affect mathematical reasoning even when that reasoning was expressed entirely through symbols. To truly dispense with the idols that lurk in one's mind, one would need not just new symbols but also new ideas.

The stakes of this issue were heightened by the conceptual nature of eighteenth-century mathematics. As Amir Alexander has argued, mathematicians in the period viewed their work as a means of understanding quantitative and spatial relations in the material world.[24] As I show in this chapter, the emphasis on grounding methods in physical reality led to a divide between two ways of relating symbols to language. In the midcentury, Jean-Jacques Rousseau praised the virtues of what he called "natural man"; for Rousseau, people were naturally virtuous but have been corrupted over time by civilization, which creates artificial wants that turn people against each other.[25] On one interpretation, this primitivism suggests the need for reform. Our language has become corrupt through centuries of abuse, the thinking goes, so we must start afresh by building a new language from the ground up, being sure to avoid tainting it with traces of received conceptions and attitudes. Such reasoning led, as de Gérando put it, to the "seductive hope" that reforming language would resolve disputes and eliminate political discord.[26] Condorcet picked up on this hope in his universal language scheme; the symbols were supposed to put an end to the violence that had engulfed France by settling, once and for all, the meanings of such divisive words as *liberté, égalité,* and *fraternité.*

Yet Rousseau's argument also admits another interpretation that might be called conservative. The earliest languages, the thinking goes, were the best—they were the clearest, the most vivid, the least corrupt—and so we ought to stay as true as we can to the original meanings of words. This interpretation speaks against linguistic reform, suggesting that any linguistic change ought to be lamented and, if possible, suppressed. These two positions in eighteenth-century linguistic thought translated into divergent visions for what symbolic methods could do. If Condorcet wanted to make people more rational by replacing words with newly minted symbols, another school of thought located rationality in language as it already existed and tried to keep symbols as true to the ordinary meanings of words as possible. Both sides, however, shared a set of assumptions about the nature of mathematics. For a time, mathematics was supposed to be based on reason, not on arbitrary axioms and methods, and equations had to have meanings that were clearly understood. As a result, the question of how the symbols were defined—and thus, their relation to words—could not be ignored.

This chapter examines the variety of ways people addressed this question in the eighteenth century. For radicals like Condorcet, algebra offered a model for how all communication should work, an example of a language derived from reason rather than from corrupt social practices. For others, the symbols' apparent independence from established conventions was dangerous. As an example of the latter position, I consider the English mathematician Francis Maseres, who is best remembered for his vehement opposition to the use of negative numbers. Joan L. Richards and others have argued that the French and English mathematical traditions had different attitudes toward symbolic methods in the eighteenth century.[27] But the difference between Condorcet and Maseres was not just a matter of national cultures; it also had to do with conflicting ideas about language and politics. The eighteenth-century disputes about algebra parallel, in some striking ways, the twenty-first-century discourse about the political implications of algorithms, raising questions such as whether strict rules can serve a liberatory purpose. Yet these disputes took place within Enlightenment epistemologies that draw no sharp line between computational procedures and human thought even when those procedures were literally mechanized. To illustrate the stakes of the issue, I begin with an instance where mathematics became expressly political: the division of land.

ALGEBRA AND THE ORIGIN OF LANGUAGE

The Hindi–Arabic algorism had long found a major application in surveying. The word *geometry* means, etymologically, the measurement of land, and numerical calculations were necessary for determining the areas of variously shaped plots based on measurements of their boundaries. Surveying was important to taxation, since the state had to know exactly how much land a person held to determine how much the person owed. This practice gained a new political resonance during the French Revolution. France had long been divided into thirty-four provinces, but shortly after the revolution began, the National Constituent Assembly replaced these provinces with other divisions called *départements*. One faction campaigned for the establishment of *départements* that were perfectly square and thus reminiscent of a checkerboard. In this proposal, calculation was no longer just a way of measuring plots of land whose divisions have already been established; now it would play a role in the reconstruction of the country's administrative structure. Whereas this plan was not enacted in France, a version was put in force in the United States, where it resulted in the perfectly square counties that still exist in much of the West.

This geometric model of land division exemplifies a late-eighteenth-century tendency that John Heilbron calls "the quantifying spirit."[28] Across a range of contexts, mathematics was thought to have a persuasive power that could break down the structures of monarchical and aristocratic authority with the force of undeniable truth.[29] Such reform faced powerful opposition. In the 1790 book *Reflections on the Revolution in France*, Edmund Burke takes the checkerboard scheme as characteristic of the oppressive nature of revolutionary reform. In using such a "geometrical distribution and arithmetical arrangement," Burke writes, the Jacobins "treat France exactly like a country of conquest."[30] Since the checkerboard plan pays no attention to the traditions and emotional attachments that defined places for residents, it eliminates the possibility of feeling a personal connection to one's region. For Burke, this is an outrage; but for the reformers who advocated such methods, it was precisely the point—the new divisions would supersede the aristocratic attachments that stood in the way of equal representation.

On the basis of remarks like Burke's, one might suppose that mathematics stood on one side of a political divide—that reformers were attempting to impose mathematical rationality on society, whereas counterrevolutionaries wanted to resist such efforts as a way of maintaining traditional hierarchies and values. But the reality is more complex than this. The political factions of the time were not just tussling over how to use mathematics; they also had different ideas of what mathematics was. In the seventeenth century, as we saw in chapter 1, Thomas Hobbes had objected that algebraic symbols constitute a "narrow" language that is alienated from the way most people in a country speak. In the eighteenth century, the apparent disconnect between symbols and ordinary language became a point of contention, at times promising a radical break from established practices and at others threatening to unmoor mathematics from reality.

Positions regarding the politics of language arose in even the most abstract, theoretical forms of mathematics in the eighteenth century. A striking example appears in the work of the mid-eighteenth century's greatest mathematician: Leonhard Euler. One of Euler's interests was the problem of computing the sums of infinite series. A well-known infinite series is the sum $\frac{1}{2} + \frac{1}{4} + \frac{1}{8} + \frac{1}{16} + \cdots$, or, as we might alternatively express it, $\sum_{i=1}^{\infty} \frac{1}{2^i}$. As we add up these diminishing numbers, the total gets closer and closer to 1, although it never quite reaches it; the sequence is thus said to *converge* to the value of 1, which is usually seen as the sum of this series. Some series, however, do not converge on any value. For instance, the series $1 - 1 + 1 - 1 + 1 - 1 + \cdots$, sometimes known as Grandi's series, alternates between 1 and 0 forever as we repeatedly add and subtract units. This series is thus called *divergent*.

Even though this series never converges, we might still wonder if the series as a whole—including not just the first n terms, but also the whole infinity of terms expressed by the "$+ \cdots$"—might still be assigned a sum. We could plausibly argue that the sum is zero by grouping the numbers in pairs:

$$(1 - 1) + (1 - 1) + (1 - 1) + \cdots = 0 + 0 + 0 + \cdots$$

By grouping them differently, however, we could argue with equal plausibility that the sum is one:

$$1 - (1 - 1) - (1 - 1) - (1 - 1) + \cdots = 1 - 0 - 0 - 0 - \cdots$$

In 1703, Guido Grandi had proposed that the true sum of this series was $1/2$, but this result was not universally accepted.[31] Euler constructed what he thought was a definitive proof of Grandi's result, but he could only do so by making a theoretical gambit.

Euler's argument, which was published in 1760 though probably developed in the 1740s, involves an expansion technique previously applied to this problem by Leibniz.[32] This argument is often stated in terms of Isaac Newton's generalized binomial theorem, but I will explain it in a more elementary way. We may begin by observing the following identity:

$$\frac{1}{1+a} = \frac{1+a-a}{1+a} = 1 - a\frac{1}{1+a}$$

By applying this identity recursively to itself, we can transform the expression on the left into polynomials of ever higher orders:

$$1 - a\frac{1}{1+a} = 1 - a\left(1 - a\frac{1}{1+a}\right) = 1 - a + a^2\frac{1}{1+a}$$

$$= 1 - a + a^2\left(1 - a\frac{1}{1+a}\right) = 1 - a + a^2 - a^3\frac{1}{1+a}$$

$$= 1 - a + a^2 - a^3\left(1 - a\frac{1}{1+a}\right) = 1 - a + a^2 - a^3 + a^4\frac{1}{1+a}$$

And so forth. From this endless expansion, Euler concludes (in a move that would not be accepted in modern mathematics) that $\frac{1}{1+a}$ is equivalent to an infinite series:

$$1 - a + a^2 - a^3 + a^4 - a^5 + \cdots$$

Let $a = 1$; the fraction evaluates to $\frac{1}{2}$ and the expansion reduces to Grandi's alternating series. Euler took this as a proof of Grandi's claim. This result came, however, at the cost of abandoning the usual definition of the word *sum*. Mathematicians, Euler writes, traditionally define the sum of an infinite series as "that quantity to which it is brought closer as more terms of the series are taken."[33] In this definition, divergent series clearly do not have sums. However, we could also define the *sum* as "that quantity which generates the series."[34] Since this new definition coincides (according to Euler) with the old one in the case of convergent series, it has no drawback; and since it enables a wider array of problems to be solved, it is superior.

Although Euler's explicit goal was to put an end to the disputes over divergent series, his argument remained controversial for centuries, and it later became an emblem of the perceived sloppiness of eighteenth-century mathematics. In 1826, the mathematician Niels Henrik Abel called divergent series "the Work of the Devil."[35] For our purposes, what matters is less whether Euler was right or wrong by modern standards than the role language played in his argument. Those who deny that divergent series have sums, he argues, "have stumbled into a mere battle of words"; these difficulties will vanish if "we change the accepted notion of sum."[36] Euler's confidence in this rhetorical approach rests on much the same assumption that animated Condorcet's utopian scheme: that one can settle disagreements by more clearly defining words. Whether this was an acceptable form of argument came down to the issue of to what extent definitions are arbitrary and to what extent that arbitrariness licensed scholars to deviate from received usages.

Such issues formed a major topic of discussion in the linguistic thought of the time. In his 1746 *Essay on the Origin of Human Knowledge*, the Abbé de Condillac presents a speculative account of the invention of language that serves at once as a historical reconstruction and as a model for how sciences should remake their languages anew.[37] Later writers debated a number of aspects of this account. Johann Peter Süßmilch argued that language could not have been invented, since there would be no way to define words without some preexisting system of communication; he thus concludes that language was a divine gift to humankind.[38] Johann Gottfried Herder disagreed, enthusing over the human ingenuity that produced language.[39] Thinkers also debated whether languages improved or declined over time, with Rousseau making a case that the languages of advanced civilizations lacked the vibrancy and emotion of primitive ones. A further question was whether commoners created language or whether, as the eccentric Lord

Monboddo argued of the Greek language, it was designed by a council of the learned.[40] These origin stories contained such a profusion of evidence-free speculation that the field of linguistics has all but disavowed them. But in regard to uncontroversially artificial languages such as algebra, the question of origins made a difference. If language was an object of intentional design, then there was no problem in proposing a new definition of *sum*. If, on the other hand, it arose naturally from the activities of commoners, such a move would create an impassible barrier between algebra and the vernacular tongue.

The Marquis de Condorcet's narrative of progress incorporates just such an origin story, and he makes its implications for mathematical symbols plain. In Condorcet's view, the first languages were the work "of the whole society," not of any particular genius.[41] Yet language later became an agent of social division with the invention of writing. The first writing systems, according to Condorcet, were hieroglyphs, whose meanings were kept secret by the priesthood as a means of deluding the masses.[42] The transition from hieroglyphs to alphabetical writing was progress because, by unifying written and spoken language, the alphabet removed this social barrier. Yet through "a strange revolution," Condorcet hypothesizes, hieroglyphs will one day return.[43] Condorcet envisions the future development of another type of writing, one "reserved exclusively for the sciences," that could facilitate "the precise and calculated operations of the understanding."[44] Whereas in ancient times, such a scheme would only "have helped to prolong ignorance," specialized forms of writing can become agents of Enlightenment once they lie "in the hands of philosophy."[45] This three-stage sequence makes the possibility of a universal algebra contingent on social progress. For Condorcet, symbols can only be rational if they arise in a rational society.

Condorcet's confidence that his algebraic hieroglyphs would unite rather than divide depends on a belief that the natural reasoning faculties of human beings are basically reliable. Condorcet was, in eighteenth-century terms, a sensationalist: he maintained that all ideas derive from the senses.[46] His *Sketch* opens with a taxonomy of mental faculties that emphasizes the analysis of sensory data:

> Man is born with the ability to receive sensations; to perceive them and to distinguish between the various simple sensations of which they are composed; to remember, recognise and combine them; to compare these combinations; to apprehend what they have in common and the ways in which they differ; to attach signs to them all in order to rec-

ognise them more easily and to allow for the ready production of new combinations.[47]

The progression from sensations to ideas to signs, which Condorcet later attributes to John Locke, suggests that learning to understand a language is equivalent to learning a particular way of understanding the world.[48] The analysis of sensations is not wholly arbitrary; the understanding is developed "through the action of external objects."[49] The shared nature of these objects grounds the intersubjectivity of language and places the perfect communication Condorcet wanted to achieve within reach.

This sensationalism is typical of the discussions of linguistic origins. The Abbé de Condillac's work loomed large in this discourse. In his 1749 book *A Treatise on Systems*, he sums up his position with the gnomic, often-quoted statement that "a well-conducted science is merely a well-formed language."[50] A language, in Condillac's view, provides a "method of analysis," its words dividing the world into comprehensible chunks and indicating their relations. As a result of the fundamentally linguistic nature of science, he argues, "any science should be within the reach of an intelligent mind, since every well-formed language is comprehensible."[51] Condillac disarmingly turns the rubric of the well-formed language on himself. "If you do not understand me," he writes, "it is because I do not know how to write; and if you happen sometimes not to understand me, that is because I sometimes write badly."[52] This statement bears some resemblance to Leibniz's claim that a universal character would rapidly become comprehensible once it was properly "ordered." But Condillac expressly rejected Leibniz's doctrine of innate ideas, instead viewing the way we divide up and classify the objects of the senses as fundamentally arbitrary. The quality of a language, then, was measured by how sharply its concepts were cut.

The exemplary well-formed language, at least in the Abbé de Condillac's later writings, is algebra. In *Logic*, published a few weeks after his death in 1780, he argues that exact proofs are possible in algebraic notation because it is tainted by neither "vulgar words that have no determinate sense" nor "foreign or barbarous words that are poorly understood."[53] The success of symbolic algebra, he continues, proves that "scientific progress depends solely on the progress of languages."[54] The Abbé de Condillac and the Marquis de Condorcet had disparate views on a number of matters, including the equivalence of algebra to conceptual analysis. Condillac reportedly "detested" Condorcet and blamed the marquis for serving him the tainted hot chocolate that, he believed, caused the "putrid bilious fever" from which he eventually died.[55] But it is hard to deny that they were thinking within the same broad terms. Both were under the sway of Locke, both took the senses as the origin of knowledge, and both regarded algebra as a model for how a well-constructed language can clarify our understanding of the world.

Such thinking provided a way of understanding the function of symbolism that was distinctive of the eighteenth century. In this period, algebraists viewed their science as a highly abstract representation of the properties of material things. As Denis Diderot wrote, "Every abstraction is merely a symbol devoid of particularized meaning. Every abstract science is simply juggling with symbols. The exact picture was dropped when the symbol was separated from the physical object, and it is only when the symbol and the physical object are brought together again that the science once again becomes a matter of real things."[56] This insistence on grounding mathematics in the sensible world at once stoked the hope of a universal language modeled on algebra and raised theoretical questions. What was it that gave algebra its superior clarity? And was it really so clear after all? These issues were up for dispute in the eighteenth century, and the answers had political stakes—they affected the legitimacy of the sort of rationalizing reform that Condorcet was trying to undertake.

A PARTICULAR KIND OF LANGUAGE

As much as algebra exemplified perspicuous expression in the eighteenth century, its advocates were haunted by a lingering suspicion that it did not actually make sense. In *Elements of Philosophy* (1759), Jean Le Rond d'Alembert asks why algebra, in spite of the certainty of its principles and inferences, "is not yet entirely exempt from obscurity in certain regards."[57] As a specific example, he offers negative numbers, of which, he writes, he does not know a single work that provides a clear theory.[58] D'Alembert is referring to one of the most notorious conceptual readjustments caused by the widespread adoption of symbolic notation. Antoine Arnauld, one of the authors of the well-known Port-Royal Logic, argued in a 1667 geometry text that, intuitively, the proportion of a larger number to a smaller one should be larger than the reverse.[59] Yet, he points out, this is not the case with negative numbers, at least in symbolic algebra: $\frac{1}{-1} = \frac{-1}{1}$. This was far from the only conceptual conundrum regarding negatives; in his article on divergent series, Euler considers the suggestion that the value of $\frac{1}{-1}$ should be higher than infinity because it involves dividing by something less than zero.[60]

What to make of such paradoxes was a matter of dispute among the French philosophes. The adoption of symbolic methods fundamentally altered the definition of quantity, leading to an epistemological gap between the symbols and classical theories of number. As an explanation of this gap, d'Alembert concludes that algebra is "a kind of language which has, like the others, its metaphysics."[61] Just as blindly following the rules of grammar does not lead to an understanding of language, "the vulgar only celebrate the result" of algebraic methods, whereas the truly enlightened can also "see the germ that produced it."[62] To perceive the meanings of algebraic proofs, one must learn to think algebraically; otherwise, the symbols will only lead to confusion. This argument assigned algebra a narrow role within knowledge more generally. For d'Alembert, algebraic analysis must especially be distinguished from logical analysis—that is, the division of concepts into their component parts—since conflating the two would obscure algebra's basis in notions of quantity.

The Abbé de Condillac took an emphatic stand against this thinking. In *Logic*, he states, "I do not agree with mathematicians who claim that algebra is a kind of language. I say it is a language and cannot be anything else."[63] As W. R. Albury points out, two of the mathematicians he means are d'Alembert and Alexis-Claude Clairaut, who, in his 1746 book *Elements of Algebra*, calls algebraic notation a "particular kind of language" that uses simple signs to make it easier to see one's operations at a glance.[64] The distinction between algebra being "a language" and being "a particular kind of language" might seem pedantic, but something serious is at stake. Against the objection that his description of "reasoning" only captures "the way we reason in mathematics, where reasoning is carried out with equations," Condillac asserts that "*equations, propositions*, and *judgments* are at bottom the same thing, and that consequently we reason the same way in all the sciences."[65] As a result, the difference between words and algebraic symbols is superficial: "We should not suppose that the sciences are exact—or that we prove rigorously—only when we use x's, a's, and b's."[66] This argument marks a reversal in the relation of symbols to language. In the seventeenth century, Seth Ward was vehemently insisting that symbols were not a language; a century later, Condillac was insisting with equal vehemence that they were a language. At stake in this insistence was whether the symbols really did have the ability to convey ideas clearly or whether, to the contrary, they confused things.

In practice, the algebra of the period did not always correspond as neatly to language as the Abbé de Condillac asserted. Starting in the early 1770s, the Italian French mathematician Joseph-Louis Lagrange had been attempting to systematize and generalize the equation-solving methods of his precursors.[67] In 1795, Lagrange delivered a series of lectures at the École Normale in Paris, in the course of which he attempted to clear up the difficulties that had surrounded cubic equations since Cardano came across the troublesome $\sqrt{-1}$ in the 1540s. Even though such expressions are not "susceptible of being numerically evaluated," Lagrange suggests, they can still "be employed in the operations of algebra" as a means of solving problems.[68] "It would," he states elsewhere, "be the height of injustice to algebra to accuse it of not yielding results which were possessed of all the generality of which the question was susceptible. The sole requisite is to be able to read the peculiar hand-writing of algebra, and we shall then be able to see in it everything which by its nature it can be made to contain."[69] He notes, in particular, that mathematicians have overlooked the fact that cube roots have three values, some real and some not. Lagrange's rationale for accepting such values depends on a distinction between algebraic quantities, which are defined by how they relate to each other through operations such as $\sqrt{}$, and ordinary conceptions of number.[70] Algebra, as d'Alembert might say, had a metaphysics of its own, different from how people think in other sciences.

In spite of such difficulties, the quantifiers of late eighteenth century mostly shared the Abbé de Condillac's optimism about the universal comprehensibility of mathematical arguments. Although he tended to side with d'Alembert on the nature of analysis, the Marquis de Condorcet rhetorically equates mathematics with reason in one of his most influential works of "social mathematics," *Essay on the Application of Analysis to the Probability of Majority Decisions* (1785). In this book, he develops a means of quantifying the reliability of decision-making bodies such as juries based on their size and the degree of "enlightenment" of their members. Recognizing the difficulty that the book's thickets of symbols will present for readers, Condorcet also explains his argument nontechnically in a "Preliminary Discourse" more than half the length of the main text of the book. This way, he writes, "readers who are not Geometers, will only need, in order to judge the work, to admit as true that which is given to be proven by calculation."[71] Even though the innumerate reader will have to take Condorcet's word that the proofs are correct, Condorcet does not expect much surprise as to the results: "Almost everywhere one will find results consistent with what the simplest reason would have dictated; but it is so easy to obscure reason by sophisms and by vain subtleties, that I would consider myself happy when I have done nothing but support by the authority of a mathematical demonstration one single useful truth."[72] The assumption is that, even if one cannot follow the intricacies of the computations, their conclusions will remain intuitively true; the purpose of the proofs is not to produce new results but rather to guard against the corrupt forces that lead people into error.

A similar thinking may be discerned in the Marquis de Condorcet's universal language project. As Baker points out, Condorcet's goal in this project was to extend the certainty of mathematical analysis to areas of inquiry that were ordinarily the domain of other, less reliable forms of analysis.[73] Doing so led him, as Baker points out, to the possibility of mechanizing logical inference.[74] Yet there is an important difference between Condorcet's thinking and the instrumental approach to mechanization that characterizes modern algorithmic thinking. Mechanical implements for reasoning, the marquis writes in a late fragment, "would appear ridiculous" until the "tables" of classification on which they are based are proven worthy by experience.[75] His universal algebra fed on a faith that these tables could be developed methodically enough that there would be no problem bringing people to an accord about what the categories meant. If this premise turned out to be false, then the system would not fit with his leveling politics at all; instead, his mathematical hieroglyphs would, like the secret language of the Egyptian priesthood, become an impenetrable mystery to all but a few.

The strongest arguments along these lines appeared in the German states. A few decades before Condorcet wrote his manuscript, the philologist Johann David Michaelis had made a detailed, well-argued case against philosophical language schemes in the French edition of his *Dissertation on the Influence of Opinions on Language and of Language on Opinions*. This essay won a Prussian Academy prize in 1759, and it was well received internationally, garnering high praise from d'Alembert.[76] The French translation of the essay includes an extensive section arguing against the idea of a "learned language," with apparent reference to Leibniz's universal characteristic.[77] In an argument that would go on to have significant influence in the German context and elsewhere, he concludes that, in spite of all the problems language causes, we are to a large extent stuck with it—the costs of doing away with words, as Leibniz had attempted to do, outweigh the benefit.

One of Michaelis's arguments regards an issue Condorcet raised as well: the potentially divisive quality of a learned language. A universal characteristic, Michaelis argues, would throw up a barrier between the learned and the uneducated; it could be used to delude the masses by concealing the way new inventions work and thus casting them as "false miracles."[78] In learned writing, "an author treats as sovereign master the technical language he makes use of. He says, this is the meaning I fix to this term, this is the definition I give of it: we then are all obliged to understand him, as he has declared he will be understood, and as little can we contest that right with him, as prescribe to the Algebraist what lines he shall call *a* and what *b*."[79] In contrast to this "sacred tyranny," in vernaculars "all is Democratic: words cannot be deprived of their received meaning but by the consent of the people, and by a contrary usage, that is introduced bit by bit."[80] Being foreign, as it were, to all nations, a learned language could never be absorbed as deeply as we absorb the languages we hear from the cradle.

Michaelis's essay suggests a different response to the flaws of language that would gain a great deal of influence by the end of the century: that imaginative literature, in appealing to the feelings as well as to reason, was better suited to influence the linguistic practices of a people than explicitly prescriptive projects such as Leibniz's characteristic or Condorcet's algebra.[81] This line of thought would gain dominance in Germany around the end of the century with the emergence of German classicism. A key figure in this movement was J. G. Herder, who draws a distinction between "the culture of the men of learning and the culture of the people":

> So it is that algebra is an occult science, for few in Europe understand it, though learning it is prohibited to none. Now we have indeed, in a useless and damaging manner, in many respects confused the spheres of learned and popular culture, thereby extending the range of the latter almost to that of the former; the ancient founders of states, who thought in more human terms, thought more wisely on this subject also. They rooted the culture of the people in sound morals and useful arts; they deemed the people neither qualified nor likely to benefit by grand theories, even in philosophy or religion.[82]

Like Michaelis, Herder distinguishes algebra from natural languages, which develop organically in communities and have the ability to resist intentional efforts to alter them. In the Herderian view, to suppose that one could entirely expunge the influence of one's mother tongue from one's way of thinking, as Condorcet wished to do, was to overlook the critical role that membership in a group plays in the development of human consciousness.

The arguments of Michaelis and Herder pointed out a practical problem with schemes like that of Condorcet: they would require not just designing a system of symbols but also training people to think in a new way. How one addressed this problem depended on what social function one expected science to serve. One option was simply to accept the disconnect between the learned and popular spheres; this was essentially the view of Leibniz, who had no problem restricting knowledge of science to a select group of people. Condorcet, however, was not comfortable with such a division. One of the most ancient hindrances to progress, he writes, is "the separation of the human race into two parts: the one destined to teach, the other for believing."[83] This separation, according to him, was widened by the alienation of hieroglyphs from spoken language. Condorcet's algebraic hieroglyphs were not supposed to return us to this ancient situation. The goal was to enlighten the whole society. Achieving this end would require not only the development of the language itself but also an infrastructure to support it—educational institutions that could teach people to read the symbols and a free press that could disseminate knowledge widely. Bringing the universal algebra into being would mean installing scientific ideas in every mind.

This revolutionary program was not, however, the only alternative to Leibniz's elitism. If Condorcet wanted to replace language with something more scientific, one could also go the other way, striving to keep learned discourses grounded in the language of the people.[84] This was (at least at times) the position of Herder, who wrote that, by pursuing Francis Bacon's program of expunging "idols" from language, "one has stripped oneself of the aid of all the centuries of one's fathers, and stands there naked."[85] Whereas Condorcet viewed the vernacular as a repository of errors and prejudices, Herder regarded it as a "treasure" that we inherit from our ancestors.[86] Similar thinking motivated Burke's defense of traditional land divisions. Condorcet was, from this standpoint, attempting to enact a sort of cultural imperialism by replacing practices he judged "barbaric" with those he deemed "enlightened" and, by doing so, trying to force a particular set of values on the world. These arguments might be identified with what Isaiah Berlin called the Counter Enlightenment, a movement that valued cultural traditions and looked askance at the program of rationalization that Condorcet was trying to enact.[87]

One should not, however, assume too simplistically that "the Enlightenment" and "mathematics" were always aligned in the eighteenth century. Even Herder, for all that he preferred feeling over cold rationality, propounded a series of pseudoalgebraic "laws" for history and ethics based on the mathematical work of Johann Heinrich Lambert.[88] There were also mathematicians in the period who valued tradition and linguistic continuity and whose use of symbols reflected these values. A notorious example is Francis Maseres. An ardent Whig, Maseres supported some liberal causes, such as the abolition of slavery, but he sided strongly with Burke in the wake of the French Revolution; in 1790, he complained of the "highly democratical spirit" that had taken root in France.[89] His mathematical work is characterized by a heroic effort to keep algebra aligned with what he took to be the common sense of English gentlemen. By doing so, he put up resistance against the break from ordinary language that Condorcet and other radicals were intentionally trying to create by means of symbols.

MERE NONSENSE AND UNINTELLIGIBLE JARGON

Maseres is now a marginal figure in intellectual history, but he was more typical of his time than is generally recognized. Maseres was not an academic success, losing a bid for the Lucasian Chair of Mathematics at Cambridge, but he published several mathematical books that garnered praise from various quarters.[90] From 1766 to 1769, he served as the attorney general of Quebec, and from 1773 on he held the post of cursitor baron of the Exchequer, a position involved in the collection of royal revenues. Maseres was part of a school of mathematical thought, also including Robert Simson and William Frend, that viewed the conceptual problems surrounding negative numbers as proof that they were simply meaningless.[91] This school of thought illustrates what an algebra grounded in concepts looked like for those who opposed the Enlightenment project of remaking language.

Maseres's first major mathematical work was his 1758 *Dissertation on the Use of the Negative Sign in Algebra*. This book, he writes, is "an attempt to treat the Science of Algebra with the same perspicuity and accuracy of reasoning that has usually been thought necessary in books of Geometry, but which, through causes not very easy to be fixed upon with certainty, has been almost universally neglected in books of Algebra."[92] Different as Maseres and Condorcet were, they shared a wariness of the potentially divisive nature of symbolic methods as well as a belief that the best hope for amelioration lies in the senses. An algebraist, as Maseres has it, may easily "fancy he has a meaning where, in reality, he has none" because symbolic arguments address themselves only to the understanding, not (as in geometry) to both the understanding and the senses.[93] Like Condorcet, Maseres expresses the concern that inscrutable symbols could be used to delude the underinformed. It is, Maseres states, easier to lie with words or symbols than with diagrams; "wherein the senses are not concerned, men are much more easily deceived."[94]

Grounding algebra in the senses means, for Maseres, eliminating the idea of a negative number. He begins the book with the declaration that "the clearest idea that can . . . be formed of a negative quantity is that of a quantity that is subtracted from another greater than itself."[95] Under this definition, a quantity considered in isolation has no sign, and so rules for computation with negatives, as in $-5 \times -5 = +25$, are "mere nonsense and unintelligible jargon."[96] A dissatisfaction with such rules was not unusual for the time. As we have seen, there were extensive debates in the period about what should happen when one performs mathematical operations with negative quantities; dividing by a negative seemed to raise conceptual conundrums, and the rules for logarithms of negatives remained unsettled. Maseres's theory aims to resolve all this confusion by preventing the issue from arising in the first place. There are no negative quantities; there is merely subtraction, and all of algebra is supposed to follow from the easily intelligible idea of a positive number.

This conceptual approach exemplifies an emphasis on integrating mathematics with other forms of knowledge that Joan L. Richards has identified as a characteristic of English mathematics.[97] Maseres sought, in particular, to maintain continuity between symbolic methods and the ordinary meanings of words. The primary sense of *multiplication*, he writes, "is evidently repetition, or the taking a quantity over and over again a certain number of times"; the word later came to be applied to fractions, but only, he claims, with some degree of distortion.[98] This pseudophilological thinking is not especially different from the etymological musings with which d'Alembert often begins his encyclopedia articles. Nor, indeed, is Maseres's explanation of negatives wholly distinct from d'Alembert's: both thought that negatives only made sense relative to some other quantity. But unlike the French mathematician, Maseres is not seeking to construct a new "metaphysics" that is more exact than "vulgar" notions of quantity. Instead, he is setting out to prove the validity of algebraic methods based on the true notions that already existed (he thinks) in the everyday language of his country.

From a modern perspective, Maseres's book hardly seems to clarify things. His insistence on avoiding negatives results in a proliferation of different rules for different cases, making algebra far more complex than it needs (from our perspective) to be. For instance, if $a > b$, one must employ a different procedure for computing $c + a - b$ and $c - a + b$. In the first case, one may perform the operations in order; but in the latter case, one must reverse the order and add b to c before subtracting a.[99] Things become even more complex when Maseres moves on to solving equations. As in the sixteenth-century texts of Robert Recorde and Gerolamo Cardano, Maseres divides quadratic equations into multiple types, such as $xx + px = r$, $xx - px = r$, and $px - xx = r$.[100] Since coefficients cannot be negative, these equations cannot be converted into one another; hence Maseres has to give different rules for how to solve each one.

On account of this seemingly needless complication, Maseres is often cast as a villain in histories of algebra, as a reactionary holding back progress.[101] What he was getting at becomes clearer in a relatively neglected part of his work. In 1783, he published a book that one might place along-side Michael Dary's pamphlet on compound interest in the genre of computational manuals. This book, titled *The Principles of the Doctrine of Life-Annuities*, presents a series of methods for computing the values of life annuities, which are investments that pay out in installments from a given start date to the end of a person's life. Maseres published it as part of an attempt to revive a failed parliamentary bill that would have enabled the poor to purchase annuities from church wardens, thus reducing the amount of state aid needed by poor people who were too old to work.[102] Maseres's procedures are prosaic but unmistakably algorithmic:

> Find the present value of one pound certain, to be received at the end of the given number of years, according to the given rate of interest, by the help of Mr. Smart's second table of compound interest, or otherwise. Then find in the given table of the probabilities of the duration of human life at the several different ages of it, the number of persons living at the age of the purchaser. Then add to the age of the purchaser the number of years at the end of which the sum of one pound is made payable to him, and find in the said table the number of persons living at the said greater age.[103]

And so forth. The book also presents the same procedure symbolically and through examples so as to make it, Maseres writes, "as clear and familiar as possible."[104]

The dryness of this passage can easily lead one to overlook the fact that it is about predicting death.[105] Since an annuity pays out for a person's entire life, the eventual cost to the issuer depends on how long the buyer lives, and so the issuer could minimize risk by adjusting the price based on information about life expectancy (figure 3.2). While mortality tables had long been compiled for governmental purposes, their use to set rates was new: the first life insurance company to use them was founded in 1762.[106] How best to use such statistics was a central problem of what Lorraine Daston has called "the classical theory of probability."[107] This school of thought, of which Condorcet was a major exponent, differed from modern probability theory in that it drew no clear distinction between objective probabilities and subjective degrees of belief. One of the later adherents of this school, Pierre-Simon Laplace, famously stated that probability theory was nothing but "good sense reduced to a calculus"; the goal was to formalize the human mind's innate ability to reason about matters of chance.[108]

Like Condorcet, Maseres presented his theory as grounded on and legitimized by common sense. But Maseres's approach left more room for

V. ·   T A B L E   I.

Reprefenting the probabilities of the duration of human life at the feveral ages therein mentioned from the time of birth to the age of an hundred years: grounded on the regifters of certain affignable annuities for lives granted by the government of Holland, which had been kept there for 125 years, and in which the ages of the feveral annuitants dying during that period had been truly entered.

By Mr. K E R S S E B O O M.

Age.	Perfons living.	Age.	Perfons living	Age.	Perfons living.	Age.	Perfons living.
0	1400	26	760	52	482	78	130
1	1125	27	747	53	470	79	115
2	1075	28	735	54	458	80	·100
3	1030	29	723	55	446	81	87
4	993	30	711	56	434	82	75
5	964	31	699	57	421	83	64
6	947	32	687	58	408	84	55
7	930	33	675	59	395	85	45
8	913	34	665	60	382	86	36
9	904	35	655	61	369	87	28
10	895	36	645	62	356	88	21
11	886	37	635	63	343	89	15
12	878	38	625	64	329	90	10
13	870	39	615	65	315	91	7
14	863	40	605	66	301	92	5
15	856	41	596	67	287	93	3
16	849	42	587	68	273	94	2
17	842	43	578	69	259	95	1
18	835	44	569	70	245	96	0.6
19	826	45	560	71	231	97	0.5
20	817	46	550	72	217	98	0.4
21	808	47	540	73	203	99	0.2
22	800	48	530	74	189	100	0.0
23	792	49	518	75	175		
24	783	50	507	76	160		
25	772	51	495	77	145		

Figure 3.2. A table of death. From Francis Maseres, *The Principles of the Doctrine of Life-Annuities* (London, 1783), 6.

individual agency than the theories of his French contemporaries. While he maintains that his numbers are solid, he also declines to take the calculations as the final word: one must use "judgment and discretion" in determining whether exceptions should be made in particular cases.[109] Likewise, in a pamphlet on the annuity bill, he suggests allowing for deviations from the prescribed interest rate "if the Churchwardens or Overseers of the poor shall think fit."[110] Mathematically, this stance translated into an insistence on keeping technical definitions in line with the ways people (in his view) ordinarily think. True to his word, he scrupulously avoids negatives in the book on annuities; even in tables, he always presents differences as positive regardless of their direction. He also undertakes extensive comparisons of mortality tables and computational methods, which he judges based on both their conceptual soundness and what he takes to be the reasonableness of their results.

To twenty-first-century critics of algorithmic power, this may seem a refreshingly humanistic approach. Much like some currents of critical algorithm studies, Maseres pushes against blind trust in rules. Yet his commonsense approach has its limitations. The political thrust became more overt in the work of Maseres's most incendiary follower, William Frend. Frend was notorious for writing a 1793 pamphlet entitled *Peace and Union Recommended to the Associated Bodies of Republicans and Anti-Republicans*, which contained criticisms of the clergy that got him expelled from Cambridge. His 1796 book *Principles of Algebra*, which includes a lengthy appendix by Maseres, attempts a comprehensive treatment of algebra on commonsense grounds. In *Peace and Union*, he stresses the importance of writing laws that the governed can understand; his algebra was similarly supposed to exclude ideas (e.g., negatives) that were incomprehensible and that thus had to be taken on trust.[111] The anti-Catholic overtones of this argument are apparent: Frend likens negatives to the Athanasian Creed, declaring with Archbishop Tillotson, "I wish we were fairly rid of it!"[112]

Frend's algebra text underscores the centrality of language to the eighteenth-century debates about the nature of number. His goal, he tells us, is to explain algebra in a language suitable for "English boys and girls"—that is, direct, nonmetaphorical, idiomatic English that avoids "foreign" or technical terms such as *quadratic* and *square number*.[113] "People err much," he writes, "in supposing that a word is of little consequence, if it is explained. If that word has a very different meaning in other respects, the learner will confound frequently the different meanings, and pass through life without having a clear idea upon the subject."[114] That is, the associations already attached to the word *square* will inevitably shape how one understands square numbers even if the word is given a wholly new technical definition. In the context of eighteenth-century algebra, this position about the influence of words on thought had epistemological, not just pedagogical, ramifications. Demanding adherence to received usage would, for instance, preclude argumentation along the lines of Euler's redefinition of *sum*. For Euler and Condorcet, mathematicians could freely define words however they pleased, and explicit definitions could thus settle controversy. For Frend, such definitions were to be distrusted since they could never wholly displace the conventional meanings of words. As Michaelis had declared language a democracy, governed by the people rather than by the learned, Frend sought true notions in the everyday vocabulary of his country.

And yet Maseres and Frend were pushing more for incremental reforms than for a fully democratic society. Making laws more tolerant, Frend argues in *Peace and Union*, helps "preserve a steadier attachment to the established authority."[115] Frend may have meant this argument as a rhetorical wedge, but it describes just the approach Maseres took while serving as attorney general of Quebec. Quebec was a predominantly Catholic and Francophone colony that Britain had captured from France in 1759, during the Seven Years' War. On the advice of Burke and others, the British established religious tolerance in Quebec as a pragmatic measure because outlawing Catholicism would have led to its clandestine practice.[116] Maseres took a similarly tolerant tack in dealing with land rights. In 1767, he was involved in an effort to establish property laws that would allow for some degree of continuity in the established practices of the country so as to avoid alienating the French Canadian populace. This approach contrasts with the "geometrical" scheme that Burke would soon criticize. As a way to avoid "unsettling men's ancient and accustomed rights and natural expectations founded thereon," states the draft law that Maseres prepared, French customs regarding land rights are "deemed and taken to have continued without interruption from the time of the conquest of this country."[117] Continuity and tradition serve, in this instance, as a protection against the confusion and discontent that sudden change would cause and thus, ultimately, as a way of shoring up British rule.

Maseres's preference for "judgement and discretion" over strict rules, then, does not necessarily place power in the hands of the people. Fitting a system to the existing practices of a nation can instead solidify the authority of those who write the rules. In his book on annuities, Maseres treats rules as necessary primarily to protect commoners from themselves: the law must secure the poor who buy annuities "against their own folly and weakness, by making it impossible for them to sell their annuities for a small part of their true value, over a pot of ale and without a proper degree of deliberation."[118] The judgment Maseres trusted, this argument suggests, would seem to be primarily that of gentlemen like himself, not that of the lower classes. This stance would have had a particular resonance given his position on the Court of the Exchequer. The Exchequer operated with little oversight from Parliament at the time, and reformers were arguing that positions like the one Maseres held were needlessly high-paying sinecures given out as political favors.[119] In the eyes of such critics, the Exchequer was an opaque institution that operated by its own self-serving traditions—in short, just the sort of entrenched interest the French Revolution was threatening to uproot.

The difference between Condorcet and Maseres—the former keen to eliminate language, the latter keen to keep it in sight—may be ascribed in part to national context. In the eighteenth century, Continental mathematicians adopted Leibniz's notation for calculus, whereas the English largely followed Newton in his aversion to methods that depended too strongly on symbols. The received story is that this aversion held English mathematicians back until, in the 1810s, Charles Babbage and others introduced French mathematical ideas to Cambridge through their translation of Silvestre-François Lacroix's textbook on calculus.[120] But the differentiation of national traditions is not as clear-cut as often supposed. Sensationalist epistemology not unlike that of Condorcet was present in England in the late eighteenth century. Among the champions of such thinking was Condorcet's English ally Charles Mahon, third Earl of Stanhope, for whom mathematical reason served as a weapon against the self-sustaining privilege the Exchequer represented. The example of Lord Stanhope shows that the divergent epistemologies of the period are not wholly reducible to national culture—they also had a political meaning.

LOGIC MACHINES IN THE AGE OF REASON

The Marquis de Condorcet wrote about mechanizing logical reasoning; Stanhope actually did it. An anonymous 1818 obituary of the earl in *Annals of Philosophy* explains: "It has been asserted, upon grave authority, that his Lordship conceived the possibility of forming a reasoning machine, by which the results of certain combinations of ideas, or of elementary propositions, might be ascertained with as much ease and accuracy as those of figures."[121] Stanhope did not build a *reasoning machine* as we might understand the term today. He did, however, construct what Martin Gardner calls a *logic machine*: "a device, electrical or mechanical, designed specifically for solving problems in formal logic."[122] Logic was an entirely separate discipline from algebra at the time, and so these devices cannot be placed in the universal-algebra tradition of Leibniz and Condorcet. Yet Stanhope's machines, which he began building around 1801 based on a theory he had been working on for decades, express much the same desire to bring people into agreement by replacing language with something clearer. The "Demonstrator" will display reasoning, Stanhope wrote, "by means of a symbol purely mechanical, and without using any of those symbols which are called words."[123] The machine would thus cut through the sophistry of verbal argumentation to reveal truths that no one could deny.

Stanhope's reasons for building "reasoning machines" can only be understood in light of his radically democratic politics. He was a close friend of Condorcet, and the two corresponded on a range of mathematical and political matters. The "Jacobin Earl," as his contemporaries called him, also had personal ties to François-Alexandre-Frédéric de La Rochefoucauld and other figures involved in the French Revolution, and he continued to support them even after France declared war on Britain.[124] Predictably, this position did not win him many allies in the House of Lords, and Stanhope's own children bristled at his leveling views. His son ran away from home at the age of twenty, and his daughter, Lady Hester Stanhope, cut ties with him, declaring herself "an aristocrat" and denouncing the "dirty Jacobins" with which he associated.[125] Through this whole series of disappointments, the earl devoted his time to practical efforts in science and engineering. Like Thomas Paine, whose work he often championed, he advocated scientific reason as a way of challenging the authority of the aristocracy.[126] The Demonstrator was one of his numerous inventions, which also included a series of mechanical calculators, a means of fireproofing buildings, and a new type of printing press that was widely adopted in the early nineteenth century.

Philosophically, Stanhope's logic system was aligned with the later work of the Abbé de Condillac.[127] Characteristically of the period, Stanhope shows a concern with making logic "productive"—that is, enabling it to produce new knowledge about the world rather than merely restating premises in different forms. Stanhope's adult life, stretching from the 1770s to the 1810s, roughly corresponds to the peak of what Wilbur Samuel Howell has called "the new logic."[128] While Leibnizians such as Lambert and Gottfried Ploucquet were experimenting with mathematical approaches to logic, another school of logical thought, including Condillac as well as Thomas Reid and George Campbell, shunned formalism in favor of practical education. This school was generally skeptical of the utility of Aristotelian syllogisms; in *The Philosophy of Rhetoric* (1776), Campbell declares that no one "will ever be made a reasoner, who stands in need of them."[129] In contrast to the old rule-based system, the "new logic" was meant to train people to think clearly, methodically, and without prejudice in real-life situations.

Stanhope's logical theory was in part an attempt to rescue syllogisms from this criticism by showing that they could produce more than mere tautologies. The basic idea is to introduce numerical quantities into cate-

DEMONSTRATOR,

INVENTED BY

CHARLES EARL STANHOPE.

The right-hand edge of the gray points out, on this upper scale,
the extent of the gray, in the logic of certainty.

The lower edge of
the gray points out,
on this side scale, the
extent of the gray,
in the logic of
probability.

The area of
the square opening,
within the black
frame, represents
the holon, in
all cases.

The right-hand side of the square opening points out, on this
lower scale, the extent of the red, in all cases.

The right-hand edge of the gray points out, on the same
lower scale, the extent of the consequence,
(or dark red,) if any, in the
logic of certainty.

Rule for the Logic of Certainty.

To the gray, add the red, and deduct the holon: the remainder, (or dark red,)
if any, will be the extent of the consequence.

Rule for the Logic of Probability.

The proportion, between the area of the dark red and the area of the holon,
is the probability which results from the gray and the red.

PRINTED BY EARL STANHOPE, CHEVENING, KENT.

Figure 3.3. Text pasted to the front of one version of the Demonstrator devised by
Charles Mahon, the third Earl of Stanhope. The black square is a window into which two
panels could be slid from the side of the device for comparison. From Robert Harley,
"The Stanhope Demonstrator," *Mind* 4, no. 14 (1879): 203.

gorical logic. One cannot logically infer anything from the premises "Some
people are poor" and "Some people are intelligent," but if one has "Sixty
percent of people are poor" and "Sixty percent of people are intelligent,"
one can infer that at least twenty percent of the poor are intelligent. Stanhope's device performs such inferences visually (figure 3.3). It represents
a whole category using a square aperture, which Stanhope referred to as
the *holos* or *holon*; two sliders are inserted through slots so as to represent
other categories that make up some proportion of this whole. If the sliders
are forced to overlap, a conclusion can be deduced about their relation. To

take an example from Robert Harley, suppose we have the premises "No boaster deserves respect" and "Some heroes are boasters."[130] We use the *holon* to represent boasters, since this is the term that is common to the two premises. One slider represents people who do not deserve respect; since this includes all boasters, the slider must cover the whole aperture. The other slider represents heroes; since there may be other boasters who are not heroes, this slider covers only some of the aperture. Because the sliders must overlap, we can conclude that "Some heroes do not deserve respect."

This idea may not seem to have much to do with symbolic computation as we now understand it. Certainly, Stanhope's system lacks the grounding in algebra that characterizes the logical systems of Leibniz and Boole. But the Demonstrator raised epistemological questions about symbols not unlike the ones those thinkers faced. The 1818 obituary was not sanguine about the idea that a logic demonstrator could achieve true certainty:

> It is scarcely necessary to observe that, independent of other difficulties, no *mechanical* process for reasoning can ever be employed until mankind have agreed upon certain general principles as decidedly as upon the value of certain numbers, and until all doubt has been removed respecting the import of words, or the combinations of them. A machine for resolving political queries would give very different answers, according as it was constructed under the superintendance of an advocate for reform, or an admirer of the infallible wisdom of our ancestors.[131]

This critique of Stanhope's project is arguably a precursor of what has become a critical commonplace in the twenty-first century: machines reflect the political biases of their creators. Stanhope, like advocates of algorithmic fairness, is looking for technical solutions for problems that properly belong to the domain of rhetoric. This stance of neutrality, the obituary writer counters, is sleight of hand concealing an agenda.

But Stanhope's neutrality is not the neutrality twenty-first-century technocrats assign to algorithms. As the economic historians Shilov and Silantiev observe, using the Demonstrator requires about as much work as it would take to think through the syllogism oneself.[132] This difficulty is precisely the point. Stanhope viewed the mechanization of logic primarily as a way to "strengthen the human mind" and provide "an anti-dote to self-conceit"; as a contemporaneous biographer puts it, the machine is meant "not only to detect false reasoning, however sophistically combined, but to shew the various links of the chain by which these false conclusions have been deduced."[133] This insistence on revealing chains of reasoning, which we might liken to the "trials or tests" Leibniz describes in "The Art of Discovery," had a political thrust in Stanhope's context. In *An Enquiry Concerning Political Justice* (1793), William Godwin insisted that one must know the reasons behind things: "Wherever truth stands in the mind un-accompanied by the evidence on which it depends, it cannot properly be said to be apprehended at all."[134] Although Godwin is here referring to the situation in which "government assumes to deliver us from the trouble of thinking for ourselves," his argument could just as well apply to the design of a reasoning machine.[135] The machine is not supposed to think for you; its purpose is to make you think.

This interpretation still leaves a difficulty. If the results are based on the application of natural reason to sensory data that are accessible to every-one, why is the machine needed at all? Why not just, as Godwin writes, think for oneself? A clue as to what Stanhope was trying to achieve may be found in his political career. As a member of British Parliament, he pushed for encoding rational principles in enforceable laws as a way of preventing corruption. In 1786, for instance, he published a pamphlet criticizing the plan of the prime minister, William Pitt the Younger, for reducing the na-tional debt. Pitt's plan involved creating a commission that could redeem any annuities they "deem it expedient" to redeem.[136] (One of the commis-sioners, it is worth noting, was to be Maseres's boss, the chancellor of the Exchequer.) Pitt's plan, Stanhope argued in his pamphlet, would enable the commissioners to make fortunes "by *gambling in the public funds*" — that is, it would enable them to buy up stocks personally, knowing that their values will soon be increased by an infusion of government money.[137]

This argument did not go over especially well in Parliament. In the en-suing debate, the Earl of Bathurst took Stanhope's reference to gambling as a slight against the commissioners' honor.[138] But Stanhope was concerned, as he stated in the debate, that Pitt's plan depended on the goodwill of fu-ture administrators, who may not be so trustworthy. His alternative plan would render the fund "UNALIENABLE" so that ministers can never divert it for personal or political gain.[139] Notoriously, Stanhope published this plan in a style that mimics the structure of a mathematical proof, beginning with "AXIOM I" and ending with the QED-like pronouncement, "And *this* is the *proposition*, which I had proposed *to prove*"; he even, according to a later account, employed one of his calculating machines to ensure the cor-rectness of his numerical results.[140] As it happened, Stanhope's proof did not persuade; Pitt's bill passed.

At issue in this debate was, once again, whose judgment was trusted. Pitt's plan left decisions about how to handle funds to the discretion of ap-pointed individuals, on whose honor the plan relied. Stanhope's response is characteristic of the Rousseauian picture of human nature that informed so much political radicalism in the late eighteenth century. People are naturally rational and naturally good, but in practice they have been corrupted by social forces, and so they cannot be relied on to reason justly. Whereas Maseres perceived this corruption in commoners, Stanhope specifically distrusted aristocrats since their social attachments and political commitments would give them an impetus to distort the truth.[141] In this regard, the Demonstrator would have served much the same function that the rule of law served for Thomas Paine—it would ensure that no one, not even those in power, stood above the laws of reason. If there is an aspect of Stanhope's work that anticipates modern algorithmic thinking, it is this preference for intentionally planned systems over informal agreements and individual discretion. Like some forces within the French Revolution, Stanhope saw a positive value in rules, which promised a way of rooting out entrenched interests by ensuring that decisions were made consistently.[142]

And yet the systems Stanhope wanted to construct were not exactly algorithms in the modern sense. Unlike Alan Turing's "effective procedures," the Demonstrator was not to be judged solely by how efficiently it could produce the correct output; instead, it was supposed to enable people to work out their differences rationally. To fulfill this political purpose, the system could not depend on a technical cant that required specialized training to understand. Stanhope shared this desire for a common ground with Condorcet. This desire may also be discerned in Maseres, who wanted to root algebra in everyday language as a way to ensure that it is comprehensible. The difference between the two factions—Condorcet and Stanhope on one side and Maseres and Frend on the other—regarded how this common ground would be established and who would be included. Maseres located reason in the existing common sense of gentlemen and trusted in the existing language of his nation. Stanhope was out to create a new, logical form of communication, and he wanted to include everyone.

This leveling program was not destined to be realized. If Maseres was too sanguine about received practices, Stanhope and Condorcet were too optimistic about the ease of replacing them. Neither side won. By 1801, when he began working on the Demonstrators, Stanhope was clinging to an epistemology that was already on its way out; in the ensuing century, mathematicians and logicians withdrew their ambitions of social reform, and the gap between mathematics and common sense became insurmountable. The dream of a universal algebra did not, to be sure, disappear, but it took on a new form in which aligning the mechanical operations of computation with actual human thought no longer seemed so important, and in which the perceived perfidy of words no longer seemed like such a problem. The clashes between disparate views of language gave way to a compromise: symbols and words could work together, with the symbols providing rigor and the words providing the meaning that formed an essential component of the cultivation of human thought. To understand how this compromise was possible, it is necessary to consider another intellectual current that existed in a complex relation to the radical and traditionalist factions in the 1790s: Romanticism.

OUT OF THE LIGHT'S DOMINION

The Enlightenment had a problem with culture. When they discussed the conceptions people receive from past generations, the philosophes largely cast them in negative terms, as "errors" and "prejudices" that had to be expunged in the interest of spreading light. According to Condorcet, governments have repeatedly erred in the practice "of turning prejudices and vices to good account rather than trying to dispel or repress them."[143] This practice, he elaborates, stems from the "mistake of identifying the natural man with the product of the existing state of civilization, with, that is, man corrupted by prejudices, artificial passions and social customs."[144] While Condorcet sometimes used the word *prejudice*—in French, *préjugé*—to refer to biases against groups of people, it had a broader meaning in the eighteenth century.[145] In the *Encyclopédie*, Louis de Jaucourt defines *prejudice* as a "false judgement that the soul carries of the nature of things, after an insufficient exercise of the intellectual faculties."[146] It was often assumed that these false judgments were transmitted through language, which could therefore serve as scapegoat for broader concerns about the perpetuation of unwarranted preconceptions, scholastic errors, and mistaken folk beliefs. "Hardly are we born," wrote the polymath Pierre Louis Maupertuis, "but we hear repeating an infinity of words that express rather the prejudices of those who surround us than the first ideas that are born in the spirit."[147] Condorcet's version of Enlightenment meant breaking the chain of transmission by which prejudices persist across generations by replacing language with something better.

This antipathy toward culture had a flip side. Since the goal was to start anew, received practices and, in particular, language appeared as a threat to science. Condorcet hoped that his universal algebra could be understood even if all languages were lost; like a Svalbard Seed Vault for knowledge, it would thus provide an insurance against lapses into darkness and chaos like the one that, no doubt, he feared France was facing during the Reign of Terror.[148] His approach to achieving this universal intelligibility is modeled on Lockean sensationalism, which, he writes, "has forever imposed a barrier between mankind and the errors of its infancy; a barrier that should save it from relapsing into its former errors under the influence of new prejudices."[149] But this "barrier" protecting the present from the past depended on empirical and thus potentially falsifiable claims about the nature of the human mind. What if people were not as naturally rational as Condorcet thought? What if he mistook for natural reason what was, in reality, only more prejudice—only the arbitrary values and assumptions of European society?

The eighteenth century's most heroic effort to fend off such objections seemed not to enter Condorcet's perception. It was in 1784, around a decade before Condorcet wrote the *Sketch*, that Immanuel Kant had declared his time an "age of enlightenment."[150] Kant wanted, just like Condorcet, to free people from traditions and dogmas; he construed enlightenment as the courage to think for oneself rather than relying on the guidance of others. Yet Kant makes no appearance in Condorcet's narrative of progress. This oversight may have stemmed from Condorcet's preferences or the limits of his reading. Yet Kant's critical philosophy does point in another direction from Lockean sensationalism, and its ultimate effect was to undermine the possibility of—and, it is important to note, the necessity of—permanently overthrowing culture in the way Condorcet wanted to do with his universal algebra.

Kant's work has long occupied an ambiguous position in intellectual history. Self-consciously a proponent of Enlightenment as he was, he also influenced Romantic thinkers who are commonly seen as repudiating Enlightenment thought. This ambiguity has led Clifford Siskin to declare that the Enlightenment movement had to come to an end to succeed in changing the world.[151] In *Critique of Pure Reason* (1781), Kant attempts to show that all rational beings must think in terms of twelve universal categories that determine the conditions in which empirical knowledge is possible. Although these categories include conceptions of space relevant to geometry, they certainly do not include the rules of algebra. Yet Kant provided a new set of terms by which mathematicians could redraw the disciplinary boundary around symbolic methods. Kant's followers appropriated his ideas to support a position that went against his own explicit views—the idea that algebra is a matter of "pure reason," stemming from thought alone and not (contra Maseres, d'Alembert, and Condorcet) having any dependence on the knowledge we receive through the senses.

Although Kant's critique is largely about determining the bounds of what pure reason can do, it also provided a way of shoring up the autonomy of the thinking subject. Like many other philosophers of the period, Kant insisted that it was best to think through philosophical matters oneself; if a student memorizes a philosopher's system without going through the reasoning, Kant writes, "he has grasped and retained, that is, he has learnt well and has become a plaster cast of a living person."[152] In this suspicion of rote learning, Kant is not too different from Locke, who wrote that a person who memorizes conclusions "makes his understanding only the warehouse of other men's lumber."[153] Yet this ideal of individual autonomy faced practical difficulties in empirical fields. The experimental science promoted by the Royal Society, for instance, depended on trust: no one could witness every experiment firsthand, and so people had to credit the word of others. Such socially transmitted forms of knowledge sat uneasily with the Enlightenment credo "dare to know" because they implied that authority played a necessary role in how the individual learns.

Kant responded to this difficulty, in effect, by restricting autonomy to certain types of knowledge. In the second part of the book, Kant distinguishes what he calls "rational" and "historical" knowledge. Knowledge is "historical," he writes, "if he who possesses it knows only so much of it as has been given to him from outside, whether through immediate experience or through narration, or also through instruction (of general knowledge)"; knowledge is rational, on the other hand, if it stems solely from reason.[154] In subjects that are properly historical—language, zoology, human history—one often has no choice but to rely on information received from other people. Rational knowledge is fundamentally different for Kant; it has nothing to do with the specific content of sensory experience, and thus it is possible to obtain independence from others in the realm of reason without throwing out everything one knows.

In support of this way of setting the bounds of reason, Kant enlisted a now familiar pair of terms: *objective* and *subjective*. Prior to Kant, these terms had been part of the vocabulary of Scholastic logic, in which they had roughly the opposite of their current meanings. As the *Oxford English Dictionary* puts it, *objective* meant "Existing as an object of thought or consciousness as opposed to having a real existence," whereas *subjective* meant "Relating to the subject as that in which properties or attributes inhere; inherent; relating to the essence or reality of a thing; real, essential."[155] After Kant, their meanings reversed: *objective* came to refer to the aspects of knowledge that were common to all people, whereas *subjective* meant those specific to the individual. There is some ambiguity in how these terms came to be used—*subjective* could also mean interior to the mind, which does not self-evidently exclude certainty or universality—but they provided a powerful way of delineating disciplinary knowledge from the practicalities of education.[156] Students may understand a philosophical system in different ways and to different degrees, but this variation was a mere "subjective" matter that did not affect the objective rationality of the system itself.

While Kant's realm of pure reason is highly limited, his followers wasted no time in outfitting it with a wide range of logical, mathematical, and poetic practices. Kantian thought inspired logicians to divide standards of validity from empirical facts about human thought.[157] A similar division took place in algebra, which came (at least in its theoretical forms) to be seen as an abstract science detached from empirical reality. In the final years of the eighteenth century, this refiguring led to another revival of the idea of a universal algebra, this time in a very different guise from Condorcet's scheme. In the untitled series of fragments written in 1798–99 that their twenty-first-century translator David W. Wood has dubbed *Notes for a Romantic Encyclopaedia*, the German poet, mining engineer, and philosopher Novalis (born Friedrich von Hardenberg) revived Leibniz's idea of a universal characteristic on the ground Kant had tilled.[158] Novalis's fundamental idea is to generalize algebra so as to apply to qualities as well as quantities.[159] His ambitions were (as they always have been in such projects) encyclopedic: he considered applying algebraic methods to fields as diverse as music, metaphysics, and poetry. "All sciences," he declares, "should become *mathematics*."[160]

Novalis quotes approvingly from Condorcet's *Sketch* in his notes, but stylistically the two might as well have come from different worlds.[161] If Condorcet took pains to ground algebra in the senses through clear definitions, Novalis viewed it as a pure dance of symbols. In the remarkable 1798 text titled "Monologue," he suggests that even language itself needs no reference to the world to be meaningful:

> If one could only make people understand that it is the same with language as with mathematical formulae. These constitute a world of their own. They play only with themselves, express nothing but their own marvelous nature, and just for this reason they are so expressive—just for this reason the strange play of relations between things is mirrored in them.[162]

Such statements have led the philosopher Paul Redding to place Novalis in the protocomputationalist tradition of Leibniz.[163] Yet Novalis was far from a doctrinaire Leibnizian. Following Kant's lead, he drew a hard line between mathematical and empirical knowledge; from this perspective, Leibniz was mixing the two together incoherently. Rather than produce knowledge about the world of the senses, Novalis's universal algebra would turn seemingly empirical sciences into matters of pure abstraction.

From a strictly Kantian perspective, this conflation of pure reason with algebra was not quite right. Like other idealists such as Johann Gottlieb

Fichte, Novalis disobeyed Kant's stipulations about the limits of pure reason. As a result, the Scottish logician William Hamilton would later accuse the idealists of being Leibnizians in Kantian clothing.[164] Legitimately or not, Novalis was venturing boldly toward something like algorithmic thinking. "Pure mathematics," he wrote in a 1799–1800 fragment, "has nothing to do with quantity. It is the bare study of names—become *mechanical*, in relationships of orderly conceptual operations. It must merely be *arbitrarily–dogmatically* instrumental."[165] The word *instrumental* here indicates less an embrace of instrumentalism than a comparison with instrumental music. Like music with no words, algebra is not about anything in particular, gaining its meaning instead from the "operations" by which its elements are related. If Condorcet saw the symbolic "hieroglyphs" of algebra as a barrier between present and past, Novalis, drawing on Kant, repositioned this barrier to lie between rational and historical knowledge as they exist here and now. Algebra is pure thought, having no dependence on any conceptual content, and it is thus potentially applicable to anything.

Novalis was an unusual figure, and his pronouncements about algebra were largely met with puzzlement. But his remarks were not out of line with the direction the field was taking. As Novalis declared algebra a pure dance of symbols, nineteenth-century algebraists would redefine their science to deal only with the operations expressed in symbolic equations, not with notions of quantity. This idea had roots in Lagrange's insistence on the generality of algebra, although later mathematicians found some aspects of Lagrange's work lacking in rigor. In his 1797 *Theory of Analytical Functions* (which Novalis cites in his notes), Lagrange presents an attempt to ground the calculus on what he calls "the algorithm of functions"—a usage that likely contributed to the wider adoption of the term *algorithm* in the nineteenth century.[166] Lagrange's "algorithm" is based on an approximation method called the Taylor series, named after the early eighteenth-century mathematician Brook Taylor. Given a function f that is infinitely differentiable for value a, we can often approximate the function as $f(x) \approx f(a) + (f'(a)/1!)(x - a) + (f''(a)/2!)(x - a)^2 + (f'''(a)/3!)(x - a)^3 + \cdots$, where f' is the first derivative of the function, f'' the second, and so forth. Instead of working with Leibniz's infinitely small dx, Lagrange suggests, one could simply define the derivatives as the functions that appear in the Taylor series.

Lagrange's "algorithm of functions" had numerous admirers, among them Charles Babbage, who promoted Lagrangian ideas at Cambridge and extended Lagrange's method of generalization to other areas such as the analysis of games.[167] Around 1820, however, Augustin-Louis Cauchy and Bernard Bolzano would undermine Lagrange's foundational program: it is possible to construct counterexamples in which the Taylor series does not converge to the original function, which came (fairly or not) to be seen as a fatal flaw in Lagrange's theory.[168] This judgment resulted from much the same epistemological shift that led Abel to reject Euler's work on divergent series. Although standards of mathematical proof remained in flux through much of the nineteenth century, the turn away from d'Alembert's "metaphysical" approach was general. By the late 1800s, mathematicians had stopped fretting over how to define such tricky terms as *negative* and *infinitesimal*; conceptual explanations came to be seen as secondary to formal definitions and rules.[169] Condorcet's project of making symbols comprehensible by themselves, without any use of language, thus lost its urgency. The continued need for verbal explanations could now be filed away as a merely "subjective" issue relevant mainly to pedagogy, persuasion, and the cultivation of a mathematical community, but not internal to algebra itself.

The outcome, to be sure, was not that all mathematicians became Kantians. Mathematicians had varying evaluations of Kant, with many rejecting his ideas or ignoring him altogether; even those who admired him drew more on the general spirit of his philosophy than on his specific opinions about mathematics.[170] What happened was less a philosophical shift than a change in priorities. Due in part to the perceived excesses of the French Revolution, the reforms of the 1790s gave way to calls for balance; science came to be seen not as a means of making people more rational by replacing received modes of thinking but rather as a method or body of knowledge that must be balanced with the study of literature and the cultivation of sentiment. Some fruits of Enlightenment reform lasted, most notably the metric system; weights and measures would only move further toward standardization in the nineteenth century. Matters affiliated with the emerging category of subjectivity, however, came to be treated with a new softness. Advancing science no longer meant expunging old ways of thinking from the mind altogether, as Condorcet's universal algebra was supposed to do; genius had to work its mysterious magic, thought given room to breathe.

If Condorcet's predictions came true anywhere, it was in the sphere of economics. In the *Sketch*, he envisions the formation of "a great people whose language is universally known and whose commercial relations embrace the whole area of the globe."[171] His proclamations about commerce herald the rise of international trade and the establishment of a worldwide system of monetary equivalences. Yet globalization did not establish the brotherhood he desired. With the benefit of hindsight, Condorcet's political writings plainly expose the fundamental contradiction of Enlightenment universalism. Condorcet was an opponent of slavery (although this certainly does not mean he was devoid of racism), and he supported gender equality.[172] How can these commitments be reconciled with a respect for other cultures—how can advocates of Condorcet's political program foster equality worldwide without forcing their ideas on everyone? Condorcet comes down wholly on one side of this question, embracing a form of equality that presumes the superiority of a particular set of values. He condemns colonial oppression of non-Western peoples, which stemmed from a "murderous contempt for men of another colour or creed," but he envisions a future time when, "no longer presenting ourselves as always either tyrants or corrupters, we shall become for them useful instruments or generous liberators."[173]

To say that this position has unsavory implications would be an understatement. Conquerors have time and again deluded themselves that they would be greeted as liberators. But it would be naïve to suppose that one can avoid the trap into which Condorcet fell simply by detaching mathematics from politics. The conflict between the desire to spread liberty and the need to respect existing cultures was debated openly in the eighteenth century, and advocates of symbolic methods took explicit stances as to how their systems fell out in regard to these issues. After mathematics lost its link to the empirical, it became possible to declare a system universal without facing up to the stark choices universalism entails. Such an apolitical approach may be discerned in Boole, whose intentions were altogether quietistic—not to change culture but to find a "harmony" between cultures as they already exist. Boolean logic is arguably the greatest practical success of the universal algebra movement, and (although one could dispute how much credit should go to Boole himself) it really did change the world; but this success came at the cost of abandoning all hope that mere symbols could create the new beginning of which Condorcet dreamed while in hiding from the Jacobins during the Terror.

CHAPTER FOUR

Language without Things

We must regard it as more important for Ralph to acquire a just &
intimate knowledge of Nature & her exact <u>laws</u>, than of the <u>convention-</u>
<u>alities</u> of <u>man</u> & of <u>civilization</u>. <u>Language</u> is not <u>knowledge</u>, tho' a very
valuable acquisition for its own proper objects & purposes.

—ADA LOVELACE to Lady Byron, 1850

THE TRUCE

In discussing the algebra of the nineteenth century, historians of mathematics write of a "symbolic turn." In the eighteenth century, mathematicians such as Jean Le Rond d'Alembert and the Marquis de Condorcet had thought of symbols as a "particular kind of language" for analyzing the world; the validity of symbolic methods thus came down to whether one could attach the symbols to clear ideas. By 1900, symbols had come to be viewed, instead, in terms of form. Mathematical rigor came not from the clarity of verbal definitions but from the construction of axiomatic systems judged primarily by their internal consistency; symbolic algebra could be open to multiple interpretations or, perhaps, used without any interpretation at all.[1] To connect this development to computing machines is not anachronistic: at the forefront of the symbolic turn was Charles Babbage, promoter of French mathematical thought, designer of the never-completed steam-powered computers called analytical engines, and forthright apologist of the factory system.[2]

The symbolic turn in algebra coincided with a seemingly contrary movement in linguistic thought. In the previous century, Enlightenment philosophers such as the Abbé de Condillac had held all languages, including both French and algebra, to the standard of "natural reason"; this way of thinking had given rise to the revolutionary program of replacing existing languages with something more precise. In the nineteenth century, philologists such as Wilhelm von Humboldt treated language, instead, as an organic product of the "mental life" of a people.[3] Romantic writers such as Samuel Taylor Coleridge pushed to collapse the antithesis of words and things implicit in Francis Bacon's critique of language; while Coleridge thought it was possible for individual writers to influence language, he saw this process as a negotiation with tradition rather than a reconstruction from first principles.[4] The debates about who made language were pushed to the fringes, as philologists came to agree that language was a collective creation rather than an invention of "the learned or the priesthood," as Franz Bopp put it.[5] Far from a hindrance to scientific progress, language was now seen as something valuable, an important element of the culture of a nation that had to be preserved and cherished.[6]

This chapter argues that this renewed appreciation for vernacular languages played a role in one of the period's most important advances in algorithmic thinking: Boolean logic. Developed in the 1840s, George Boole's system of symbolic logic is a precursor of the *and* and *or* operators used in search engines and in many other aspects of computer systems; it enabled the construction of compound expressions with a nested structure based on algebra, such as, in modern notation, (mathematics *or* algebra) *and* logic. By characterizing his work as "a step toward a philosophical language," Boole echoed the claims of G. W. Leibniz, Condorcet, and many others before him.[7] But unlike the schemes of his precursors, Boole's logic system does not promise a complete, self-sufficient replacement for existing languages; instead, it provides abstract symbolic forms that must be interpreted within the languages of other disciplines to become meaningful. This approach, I hope to show, reflected a dualistic view of culture promoted by Romantic thinkers such as William Wordsworth, who sought to balance the rigor of scientific methods with the poetic insight and moral feeling that only natural language could provide. Although it was overtly anti-mechanistic, this dualism ultimately granted mechanical systems of symbol manipulation a greater degree of autonomy from human thought than the Cartesian and Lockean epistemologies of the Enlightenment had allowed them, thus enabling *algorithms*—a word that was then coming close to its modern sense—to take a newly central role in the production of knowledge.

Born in 1815 to a working-class Lincolnshire family, Boole had little formal education. He was, however, a voracious autodidact, teaching himself modern and classical languages as well as algebra and calculus, and he established himself as a respected mathematician; in 1849, he was appointed the inaugural professor of mathematics at Queens College in Cork, Ireland, without ever having earned a degree. In his books *The Mathematical Analysis of Logic* (1847) and *An Investigation of the Laws of Thought* (1854), as well as in a brief 1848 journal article, he presents a method for representing logical propositions with algebraic symbols.[8] This system built on nineteenth-century developments in algebra, which he sums up at the beginning of *Mathematical Analysis*: "They who are acquainted with the present state of the theory of Symbolical Algebra, are aware, that the validity of the process of analysis does not depend upon the interpretation of the symbols which are employed, but solely upon the laws of their combination."[9] Starting in the 1810s, a new generation of algebraists, including George Peacock and Boole's mentor Duncan F. Gregory, redefined rigor based on the following of formal rules concerning the arrangements of symbols.[10] Algebraic symbols, as Peacock puts it, are "governed by laws which must likewise govern, and to a certain extent determine, their interpretation, and not conversely."[11] This reversal opened the possibility that algebraic operators can have multiple interpretations that may or may not relate to quantity.

Boole's logical theory proceeds from the insight that this newly flexible form of algebra could be applied to the analysis of logical propositions.[12] In Boole's system, addition represents the combination of two categories, so that $h + z$, for instance, might mean the class of things that are either horses or zebras. In contrast to the *or* operator of modern Boolean logic, Boole's addition operator only applies to categories that do not overlap; this is a necessary consequence of his insistence on keeping his system aligned with ordinary algebra.[13] Multiplication represents the intersection of categories, so that br might represent brown rabbits; and subtraction represents the exclusion of a subcategory, so that $s - c$ might represent snakes that are not cobras. This system uses 1 to represent everything in the universe of discourse and 0 to represent nothing; thus $1 - s$ might represent everything that is not a snake. Boole also presents a second interpretation of his algebra that deals with truth values rather than categories; in this interpretation, 1 represents *always true* and 0 *never true*.[14]

Both interpretations of Boole's system maintain a close analogy with numerical algebra, differing only in the addition of one law: $x(1 - x) = 0$ or, equivalently, $xx = x$. This law, which he called the *law of duality*, represents the axiom that no category overlaps with its opposite—nothing can be both a snake and not a snake. Adding this law causes the logical equations to behave somewhat like linear differential equations, which enables Boole to develop a general procedure for solving them. To adapt one of his examples, suppose we are given the premise that "Every poet is a man of genius," and we need to know what this tells us about "men of genius."[15] We express the premise as $p(1 - g) = 0$, which means, more literally interpreted, that the category of entities that are poets (p) and not men of genius ($1 - g$) is empty; one then solves for g using Boole's method.[16] The result, $g = p + v(1 - p)$, states that the category "men of genius" contains all poets along with some indefinite number (represented by the special symbol v) of other beings who are not poets.

Much like the previous schemes of Leibniz and Condorcet, this system reflected a desire to resolve disputes and bring people with disparate backgrounds and beliefs into accord. Boole, in short, wanted everyone to get along. This desire for unity reflected his personal background—a religiously unorthodox Englishman who taught at a Protestant university in predominantly Catholic Ireland, he was always eager to ease tensions and avoid becoming embroiled in conflict. In the introduction to *Mathematical Analysis*, he expresses the characteristically early Victorian "conviction, that with the advance of our knowledge of all true science, an ever-increasing harmony will be found to prevail among its separate branches."[17] Like Francis Maseres and William Frend, Boole had Unitarian sympathies; he believed that all religions and philosophical systems expressed a single truth that was directly accessible to all human minds. Boole's logic system was based on similar thinking, offering a symbolic representation of the common logical relations that (he thought) already existed in every language.

It is hard to deny the Leibnizian resonance of this thinking. Boole first found out about Leibniz's logical work in 1855, after he had already published his system.[18] His wife later reported that he reacted with "childlike delight" upon discovering that the Leipzig philosopher had anticipated his law of duality.[19] Like Leibniz, Boole was deeply invested in the power of symbols, and in *The Laws of Thought* he echoes Leibniz's doctrine of harmony by speculating (although Boole adds significant caveats) that "the constitution of things without may correspond to that of the mind within."[20] But Boolean logic is fundamentally different from what Leibniz was trying to create. Boole's system would include no encyclopedic catalog of things; instead, it expressed formal relations that did not depend on the empirical content of propositions. Making its results meaningful was a matter of interpretation, in the undertaking of which the mother tongue still reigned. The two positions on the politics of language that clashed in the eighteenth century, reform and traditionalism, thus reached a truce: one could maintain arbitrary control over mathematical symbols while still deferring to received usage with regard to the words used to explain them.

This chapter offers an account of the conditions that made this reconfiguration possible. I begin with a general illustration of the intellectual climate of Boole's time and then examine some of his specific sources. A major influence on Boole was Immanuel Kant; like some other mathematicians in the period, Boole (mis)interpreted Kant's idea of pure reason to show that mathematical knowledge existed apart from the material world.[21] Boole's work also reflects the rise of what was then a relatively new value category: culture. Science, in the early Victorian period, was no longer about replacing established ways of thinking with more rational ones, as Condorcet and so many other thinkers had attempted to do during the radical ferment of the 1790s. The trend, instead, was toward seeking a balance: scientific methods must work together with people's organically developing languages and modes of thought to produce results that were both rigorous and meaningful. This nineteenth-century intellectual turn, I hope to show, was a crucial moment in establishing the subject–object divide that undergirds the modern idea of algorithm. I begin with a phenomenon that illustrates the shift: mental calculation.

THE GENIUS AND THE CALCULATING BOY

In 1818, two students at the Westminster School in London announced a calculating contest.[22] One competitor was the thirteen-year-old Zerah Colburn, the son of a farmer from Cabot, Vermont. Colburn was remarkable in at least two ways: he was born with an extra finger on each hand, and, starting at the age of five, he had developed an astounding ability to multiply, extract the roots of, and factor large numbers in his head (figure 4.1). Colburn's challenger was William Rowan Hamilton, an Irish boy about a year younger. Hamilton would go on to become one of the most important mathematicians of the nineteenth century. But Colburn was the victor. Using skills he developed almost entirely on his own, Colburn could determine the cube roots of numbers as large as 268,336,125 in a matter of seconds, with no need for paper.[23]

Public reactions to such calculating prodigies register shifts in general attitudes toward computation. As Lorraine Daston has shown, people in the eighteenth century viewed computational prodigies as the epitome of intelligence.[24] A 1796 magazine article, for instance, praises the prodigy Jedediah Buxton for the "astonishing strength of mind" he showed through such feats as counting the number of words spoken by each actor in a performance of *Richard III*; the author regrets that Buxton did not have the ambition to apply his talents to great ends.[25] Another eighteenth-century prodigy was Thomas Fuller, who was born in West Africa and enslaved in a Virginia plantation. As he demonstrated to Benjamin Rush, Fuller could mentally compute the in-

Zerah Colburn,
Aged 8 years,
Remarkable for solving Arithmetical questions.
Copied by permission from the original.
Pub. April 7 -1813, by R.S Kirby, 11, London House Yard, S.! Pauls

Figure 4.1. A collectible engraving of Zerah Colburn. [London] (11 London House Yard, St. Pauls): R. S. Kirby, 7 April 1813. Public domain image from Wellcome Library, no. 231i.

terval between any two points in time down to the second, even accounting for leap years.[26] Jacques-Pierre Brissot, head of a French abolitionist society cofounded with Condorcet and others, took this instance as a proof that Africans were just as intelligent as their enslavers and would be able to excel in all sciences if they were permitted liberty and education.[27] This elevation of prodigies harmonized with the spirit of the Enlightenment, which emphasized the mind's innate capacity for rational thought. In *New Essays*, Leibniz takes the existence of mathematical prodigies as proof that arithmetical knowledge is innate.[28] Even for those who sided with John Locke against innate knowledge, prodigies seemed to indicate that the human mind had reliable natural reasoning abilities that did not require external aid.

Zerah Colburn was initially treated with much the same reverence. Francis Baily, who spent a great deal of time with the boy, wrote in 1812 that he could make a great contribution to mathematics, "for the elucidation of which his mind appears to be peculiarly formed by nature."[29] A group of "friends of science" from Dartmouth College proposed a plan to give Colburn a liberal education so that he could put his skills to good use.[30] The hope was not just that he would grow up a genius but also that he had already discovered something important: he could rapidly factor large numbers, a task for which no general method was known. In his 1812 article, Baily observes that the idea of "expressing the powers and roots of quantities by means of *indices*"—that is, the notations x^n and $\sqrt[n]{x}$—led to the development of a general "algorithm of powers" that enabled the invention of logarithms.[31] Baily hopes that, "when his mind is more cultivated," Colburn may be able to explain how he does it and thus contribute to the development of more new "algorithms."[32]

It was not to be. Colburn ended up less a mathematician than a sideshow attraction, his father parading him around the world and charging twenty-two cents to see "the Calculating Boy." Colburn later took to acting and finally returned to Vermont to become a language teacher. He eventually came to doubt the practical value of his abilities. In his 1833 memoir, he writes in the third person: "Were it his opinion that a full account of his remarkable gift, and the methods by which he effected his calculations, would be of any service to the mathematical world, he should have published it long ago."[33] As a boy, he recalls, he told a Boston woman that he could not share his gift with others: "God put it into my head, and I cannot put it into yours."[34] His fate, sadly, was to be remembered less as a great mathematician than as a psychological curiosity. In an 1884 article entitled "The Mathematical Failure," Coleman E. Bishop wrote that Colburn "had mathematics in 'the natural way.'"[35] This was not a good thing for Bishop, who takes the boy's example as proof that arithmetic stifles the general development of the mind: "the more he used it the stupider he grew."[36] This was an uncharitable conclusion—Colburn seems to have been of at least normal intelligence—but it indicates how attitudes were turning away from the old Hobbesian equation of reasoning with computation.

In regard to what a mathematician was supposed to be in the nineteenth century, William Rowan Hamilton is more exemplary than Colburn. Although he could not beat the "Calculating Boy," Hamilton (who is not to be confused with the logician William Hamilton) was himself a precocious child. He was initially most interested in languages, and as a young man he considered devoting himself to poetry.[37] It was his 1818 contest with Colburn, as he later recounted, that first inspired him to study mathematics.[38] The two met again when Colburn's acting troupe visited Dublin in 1820, and they discussed the method the American had invented for factoring numbers. The method, it turned out, involved knowing all the two digit numbers whose products end in a given two digits; Colburn had effectively memorized a table of 820 rows.[39] As Hamilton determined, this method worked well when one of the factors was below two hundred, but in more difficult cases it required guesswork.[40] This reflectivity about methods set Hamilton apart from Colburn, who (as he admitted in his memoir) had difficulty with problems that could not be solved through procedures he found obvious.[41] For Hamilton, mathematics was about discovering procedures, not performing them.

Hamilton is best remembered for developing an algebraic system known as the quaternion, which he described in a series of articles published between 1844 and 1850. Quaternions are a generalization of complex numbers, which are numbers with real and imaginary components, as in $4 + 6i$, where i indicates the square root of -1. Since they contain two components, complex numbers may be thought of as points in a two-dimensional space. A quaternion expands the number of dimensions to four by adding two more imaginary components: $\mathbf{Q} = w + ix + jy + kz$. Hamilton's key insight involved working out consistent rules for quaternion algebra. The procedure for adding quaternions is obvious, but multiplying them produces nine distinct combinations of the imaginaries, whose values need to be determined. Hamilton proposes the following rules:

$$i^2 = j^2 = k^2 = -1$$
$$ij = k; \; jk = i; \; ki = j$$
$$ji = -k; \; kj = -i; \; ik = -j$$

The way Hamilton justifies these rules indicates just how much the field of algebra had changed since a century before, when Euler and d'Alembert had disputed the definition of negative. The first objection he anticipates is that quaternions break the commutative law: $ij \neq ji$.[42] He argues that this sacrifice is necessary because it makes the system more convenient in other ways, and that his equations should be judged by whether they "conduct to results of sufficient consistency and elegance."[43] The problem is not to find a conceptual justification for the rules, but to set up the rules in such a way that the system has as many desirable properties as possible.

By the 1840s, this new approach to algebraic rules was already well established. The centuries-long debates about negative and imaginary numbers had sputtered out: the clarity of verbal definitions no longer seemed as important as determining the rules by which symbols were to be used. An important step in working out these rules is attributable to the German mathematician Martin Ohm. (The ohm, a unit of measurement for electrical resistance, is named after Martin's brother Georg, a mathematician and later physicist who went on to specialize in the study of electric current.) Beginning in 1816, Martin Ohm attempted to eliminate the paradoxes that continued to plague arithmetic, developing the first set of consistent rules for a^b that account for imaginaries. In his 1842 book *The Spirit of Mathematical Analysis and its Relation to a Logical System* (published in an English translation the following year), Ohm explains his thinking in somewhat Kantian terms.[44] His solution to the confusions that surrounded imaginary numbers is based, he writes, on the realization that mathematical expressions "do *not* represent *magnitudes* (quantities), but *mental acts* (in systematic language: 'symbolized operations'), which stand in certain relations to one another, that are enunciated in 'equations.'"[45] All numbers, real or imaginary, are "*nothing but forms* per se," and "the whole of mathematical analysis is solely employed in the *transformation* of given forms."[46]

By reframing calculation in terms of "operations" rather than conceptual definitions, Ohm breaks with the sensationalist view that mathematical expressions must refer to ideas derived from our understanding of or intuitions about space and time. Addition and subtraction become matters of "mere *form*," consisting of "nothing more than the *construction* of these *forms a + b, and a − b*" which is, "objectively considered, the mere *writing* of them *down*."[47] The correctness of a mathematical inference, in this view, can be determined solely through an examination of the physical marks a person makes; "subjective" considerations, such as what one thinks the symbols mean, are secondary. Ohm represents his version of formalism as a rejection of the prevailing wisdom, claiming that, upon the first publication of his *Attempt at a Perfectly Consequential System of Mathematics* in 1822, several other mathematicians declared his ideas "insane."[48] By the

1840s, symbolic methods were winning out over the conceptual ones that had reigned from the time of René Descartes to that of Condorcet.

This embrace of symbols was especially abrupt in the British Isles. The Cambridge Analytical Society, founded in 1811 by the then undergraduates Charles Babbage and Edward Ffrench Bromhead (not a misspelling), promoted symbolic methods in Britain.[49] The stated purpose of the society was to advocate the use of Leibniz's notation for calculus, which had (at least in their view) been unfairly overlooked in England because of a nationalistic preference for Isaac Newton.[50] In 1816, Babbage (along with two other members of the society, John Herschel and George Peacock) published an English translation of a calculus textbook by the French mathematician Sylvestre-François Lacroix as a way of promoting the symbolic approach of Continental mathematicians. The Analytical Society devoted a great deal of attention to the linguistic aspects of algebra, although its members were more measured in their ambitions than Condorcet. In an 1821 talk titled *On the Influence of Signs in Mathematical Reasoning*, Babbage extensively quotes from the French thinker Joseph Marie de Gérando, whose criticisms of philosophical language schemes I mentioned in chapter 3. For Babbage, algebraic symbols are fundamentally different from words because their definitions are simple, including no extraneous meanings to distract from the matters at hand.

The Cambridge Analytical Society was enamored (in a way that was politically dangerous during the Napoleonic Wars) with the late eighteenth-century French tradition that viewed algebra as a form of "analysis." Peacock, however, would later take a major step beyond this tradition. In his 1830 *Treatise on Algebra*, Peacock argues that symbols such as + and − need not be constrained by the words used to explain them: "The imposition of the names of Addition and Subtraction upon such operations, and even their immediate derivation from a science in which their meaning and application are perfectly understood and strictly limited, can exercise no influence upon the results of a science, which regards the combinations of signs and symbols *only*, according to determinate laws, which are altogether independent of the specific values of the symbols themselves."[51] Peacock's theory is meant in part to resolve the difficulty surrounding negative numbers by showing that, even though algebra was originally developed on the basis of arithmetic, it need not be limited by this starting point.

This shift coincided with the first uses of *algorithm* in something like its modern sense. Although the word had long been used to refer to the method of differential calculus, it was now coming to serve as a general term for procedures performed using symbols. The *Oxford English Diction-*

ary dates the first instance of this sense in English to 1811; an undeniable instance appears in the 1836 version of Charles Hutton's *Course of Mathematics*, in which we are instructed to "apply *Horner's* algorithm," meaning a procedure for transforming a polynomial.[52] An even more striking use of the term appears in the work of the Polish mathematician–philosopher Józef Maria Hoëné-Wroński, who declared that the "algorithm of indefinite summation . . . may only be one of the divers constituent parts of a system of universal algorithms, which, in their collective strength, might embrace all the possible generations of quantities, and form thus *the absolute system of the science.*"[53] As one of Wroński's followers defined it, an algorithm is a "general form that indicates operations to execute for constructing a numeric quantity," such as $a + b = c$.[54] Although Wroński existed somewhat outside of the mathematical mainstream, this thinking was in line with a general trend. Like Ohm, Wroński gave operations primacy over conceptions of quantity; "algorithms" were coming to be seen not just as practical methods of computation but also as the means by which new types of number such as irrationals become accessible.

It should, however, be noted that the meaning of *algorithm* was still rather fluid in the nineteenth century. The 1845 *Encyclopædia Metropolitana* states (citing Michael Stifel's work from the 1540s) that the word *algorithm* is sometimes used "to denote any species of notation whatever for the purpose of expressing the assigned relations of numbers or quantities to each other," such as the a/b notation for fractions.[55] Note that this *algorithm* does not refer to a procedure: it refers to a notation. While some authors did distinguish algorithms from notations, the word was still strongly associated with algebraic symbolism. This association persisted through the end of the century, as in James Byrnie Shaw's 1895 book *Mathematics: The Science of Algorithms*, which uses the word to refer to algebra-like systems such as quaternions.[56] Something like algorithmic thinking was beginning to emerge in the mid-nineteenth century, but the idea of algorithm had not fully gained its independence from algebra.

There was also a fair amount of discomfort about the seemingly mechanical nature of the new methods. While it is now a cliché that algebra involves the "manipulation of symbols," this phrase was novel in the nineteenth century, and it often served as a way of mocking the perceived shallowness of modern algebra. For instance, in a paper read at the Royal Society of London, the Reverend James Booth castigates English mathematics teachers for focusing on "nimble dexterity in the manipulation of symbols" rather than on "the knowledge of principles and familiarity with methods of investigation."[57] John Venn uses the phrase in a similarly negative way in his 1866 book *Logic of Chance*, questioning whether Pierre-Simon

Laplace's theory of induction can provide anything more valuable than "formulæ for the manipulation of symbols."[58] These comments suggest a class-based fear of mathematics becoming a physical activity rather than an intellectual one—"manipulation" referred primarily to physical acts, especially with the hands—as well as an apprehension that mathematical theories were becoming disconnected from reality.

Avoiding this lapse into the "merely mechanical" was a matter of intellectual development and, hence, of pedagogy: rather than follow rules blindly, one had to grasp their rationales. Thus, while the Babbages and Hamiltons of the world were designing mechanical methods of symbol manipulation, the textbooks of the period generally continued the previous century's emphasis on conceptual underpinnings. Sarah Ricardo Porter's 1835 *Conversations on Arithmetic*, later revised under the title *Rational Arithmetic*, teaches computation on Lockean principles. Instead of explaining "the mere mechanical operations of numbers," the book aims to teach "the science, in combination with the cultivation of the reason."[59] The body of the book takes the form of a conversation between a young boy and his mother, with occasional interjections by his two siblings; this dialogue is supposed to lead the student to intuit the rules of algorism without stating them directly. Ada Lovelace's letters on pedagogy reveal a similar emphasis on grasping the principles behind procedures. In 1834, she wrote to her student Annabella Acheson about what is now called the Euclidean algorithm: "I wish particularly to know if you make out the greatest common measure—I mean the rationale of the process—to your satisfaction. It is important that you should thoroughly understand the principles of this operation."[60] Lady Lovelace, typically of upper-class educators in the period, treated mathematics not as a mechanical skill but as a form of mental "cultivation" whose primary purpose was to instill habits of rigorous thinking.[61]

As the example of Lovelace shows, the nineteenth-century idea of culture was not at all incompatible with what looks to us like algorithmic thinking. It was, after all, Lovelace herself who published (in 1843) what is sometimes characterized as the first true algorithm: a procedure for computing Bernoulli numbers intended as a program for Babbage's never-completed computing engine.[62] Mechanical as such "operations" may be, understanding the ideas behind them required thinking in a nonmechanistic way that made room for imagination and intuition. Such, indeed, was Zerah Colburn's conclusion. The mind, he writes, has always been "a machine of gigantic powers," yet it is not enough to let the machine run on its own; true progress requires the mind to be "raised, refined, and regulated by suitable culture."[63] This culture cannot be reduced to mechanical rules;

it must be allowed to grow organically. Never mind the whir of the analytical engine in the background.

A WHOLESOME SEPARATION

"A real mathematician," Boole reportedly said, "must be something more than a *mere* mathematician, he must be also something of a poet."[64] Boole did, indeed, write a large number of poems, and poetry played a role in his rationale for algebraizing logic. As Daniel Brown has shown in his study of Victorian scientist-poets, the juxtaposition of mathematics and poetry was less a paradox than a commonplace at the time.[65] I have already mentioned Novalis's claim that "*Algebra* is *poesy*" and William Rowan Hamilton's dalliance with becoming a poet. Ada Lovelace famously asked for "poetical science" in a letter to her mother, and Edgar Allan Poe's short story "The Purloined Letter" extols the superior intellect of a character who is both a poet and a mathematician.[66] Later in the century, Karl Weierstrass—a central figure in the rigorous formalization of calculus—remarked that "a mathematician who is not somewhat of a poet will never be a complete mathematician."[67] What was poetry, in particular, supposed to supply that "mere mathematics" could not?

In the case of Boole, it is helpful to consider what poetry he was likely reading. Boole's tastes in poetry hewed toward the formal; his favorite poet was Dante, and he effusively praised John Milton. He thought that the Augustans were unfairly maligned by the Romantics. Boole's own poetry, however, has a conversational quality that Seán Lucy has likened to the later work of William Wordsworth.[68] Wordsworth's influence was so widespread by the 1840s that it is hard to imagine that Boole was unfamiliar with him, and there are clear traces of Wordsworthian thinking in Boole's prose writings. Wordsworth, along with his sometime collaborator Samuel Taylor Coleridge, pushed for a balance between the analytical methods of science and the sense of moral direction that those methods left out, and Boole employed just this division when he described the limits of symbolic logic. Poetry expressed the emotional attachments that governed human behavior in actual life but that formal logic abstracted out; logic and poetry thus appeared as two halves of a whole.

A consideration of Wordsworth's views of science sheds light on the role poetry played for nineteenth-century mathematicians. The popular version of literary history has assimilated Wordsworth to the view of Romanticism as antagonistic toward science. This reading is reinforced by his 1798 poem "The Tables Turned," in which he states that "Our meddling intellect / Mis-shapes the beauteous forms of things; / —We murder to dis-sect."[69] On account of such statements, the Romantic movement is often viewed as a reaction against the "disenchantment" of the world produced by science. But *science* did not mean the same thing in the eighteenth century that it means today. Wordsworth states in the "Advertisement" to the book that "The Tables Turned" was a rebuke to a friend who was too at-tached to "modern books of moral philosophy."[70] The reference is to what Lorraine Daston has called "the moral sciences"—a dissection of morality underwritten by an excessive confidence that analytical methods can make people virtuous.[71] Wordsworth's quarrel is less with science per se than with such incursions of science into the realm of culture.

Like others in the first generation of British Romantics, Wordsworth initially supported the French Revolution, and he traveled to France in 1791 to participate in what he saw as a transformation of the world. He later recounted how the Revolution seemed an opportunity to put radical ideals into effect, "Not in Utopia, subterraneous Fields, / Or some secreted Island, heaven knows where! / But in the very world, which is the world / Of all of us,—the place where in the end / We find our happiness, or not at all!"[72] The reality fell short of these expectations. France descended into violence, and the projects of English reformers were disappointing. The utilitarians of the 1790s raised a cohort of second-generation intellectu-als who were traumatized by their parents' attempts to shape them into perfectly rational beings: William Godwin with his daughter Mary Shel-ley, James Mill with his son John Stuart Mill.[73] Lady Hester Stanhope, too, reported that her father (whom she called by the nickname "The 'Logi-cian'") was overbearing in attempting to force her to think logically.[74] The moral sciences had sought to improve people, but people had proven more complex than the theories had let on, and so the social function of science needed rethinking.

One of Wordsworth's most explicit statements on the nature of science appears in the preface first added to the 1800 edition of *Lyrical Ballads*, the collection he coauthored with Coleridge. In a footnote, Wordsworth rejects the "contradistinction of Poetry and Prose" in favor of "the more philosophical one of Poetry and Science," meaning, as he clarifies in the revised version of the preface, a distinction between language that deals with feeling and language that deals with facts.[75] The best poetic language, for Wordsworth, is that which has a deep connection to the "durable" ex-periences of the natural world; it thus provides "a more permanent and a far more philosophical language" than that of John Dryden and Alexander Pope, whose poetry is hobbled by "false refinement or arbitrary innova-tion."[76] The problem with such "arbitrary innovation," from Wordsworth's perspective, is that its results lack the emotional resonances of natural languages such as English, which are deeply intertwined with the lives of human communities and laden with centuries of history.

While Wordsworth's immediate target in the preface is the highly formal poetic diction of the Augustans, his use of the word *arbitrary* inculcates him in the broader eighteenth-century debates about the nature of language. Lockeans such as the Marquis de Condorcet had presumed that language could and should be altered to better reflect a rational understanding of the world. An important example for Wordsworth is Maria Edgeworth's 1798 book *Practical Education*, which includes a few chapters written by her father, Richard Lovell Edgeworth.[77] Inspired by Enlightenment philosophers such as the Abbé de Condillac and Dugald Stewart, Maria Edgeworth advocates using the "technical terms" of chemistry and other scientific fields in ordinary situations, thus enabling children to absorb their meanings deeply.[78] Her pedagogy involves a softer version of the sort of linguistic reform that Condorcet proposed: an attempt to outfit students with a "philosophical vocabulary" built from the ground up on clear definitions.[79] For Wordsworth, such analytical methods reach their limit when language has to deal with feeling. One might be able to conduct a scientific inquiry using an analytical language, but one could not write genuinely moving poetry with it.

Some scholars have taken Wordsworth's preface as an argument for disciplinary specialization, a proposal that poets and scientists should concern themselves with different aspects of human life.[80] If so, then matters of feeling would simply fall outside the scope of scientific disciplines such as logic and mathematics. But the distinction Wordsworth is constructing is not simply a matter of discipline or genre. If poetry's end is pleasure, as Wordsworth asserts, that does not mean it is irrelevant to science: "We have no knowledge, that is, no general principles drawn from the contemplation of particular facts, but what has been built up by pleasure, and exists in us by pleasure alone. The Man of Science, the Chemist and Mathematician, whatever difficulties and disgusts they may have had to struggle with, know this and feel this."[81] To produce genuine knowledge in any domain, one needs both the analytical rigor of science and the pleasure of poetry.

One ramification of this call for balance was a withdrawal of science from overt social action. In his autobiographical poem *The Prelude*, Wordsworth characterizes geometry as "an independent world / Created out of pure intelligence."[82] He treats this "independent world" as an escape from both morality and politics; having become disillusioned first with the French Revolution and then with Godwinian radicalism, Wordsworth

"Yielded up moral questions in despair, / And for my future studies, as the sole / Employment of the enquiring faculty, / Turned toward mathematics, and their clear / And solid evidence."[83] There is arguably a biographical parallel between Wordsworth and Condorcet here, as both sought comfort in mathematics during the political turmoil of the mid-1790s. But Condorcet certainly had not "Yielded up moral questions"—instead, he had viewed mathematics as a means of pinning morality down. Wordsworth narrates his disillusionment with such incursions of science into the moral realm in the later books of *The Prelude*, stating that he has learned "to keep / In wholesome separation the two natures— / The one that feels, the other that observes."[84]

Wordsworth's remarks about mathematics were, whether owing to his direct influence or not, indicative of the direction the field took in the nineteenth century. Pure mathematics was no longer grounded in physical reality; over the course of the century, it would come to be seen as a way of describing other worlds apart from our own. The field of logic, too, withdrew from ambitions of actually making society more logical and came to be concerned solely with speculative knowledge. In the 1826 *Elements of Logic*, Richard Whately argues that previous logicians have failed to be scientific because, in aiming to teach reasoning in general, they left the bounds of their discipline vague.[85] Whately addresses this vagueness by limiting logic to dealing with formal structures of syllogisms, not with the full complexity of actual human thought.[86] John Stuart Mill similarly separates logic from education in his 1843 *System of Logic*: logic deals only with "how to do the thing," not with "how to make ourselves capable of doing it."[87] Science was withdrawing from the sort of all-encompassing social reform that Condorcet had envisioned and coming to be concerned only with matters of fact; the cultivation of feeling could be left to the poets.

Boole's system differs from these earlier logics in its symbolic nature, but it follows them in this narrow conception of what logic should be. In the final paragraph of *An Investigation of the Laws of Thought*, Boole draws the bounds of his discipline in terms quite in line with Wordsworth's dichotomy between science and feeling:

> If the mind, in its capacity of formal reasoning, obeys, whether consciously or unconsciously, mathematical laws, it claims through its other capacities of sentiment and action, through its perceptions of beauty and of moral fitness, through its deep springs of emotion and affection, to hold relation to a different order of things. . . . And if we embrace in our survey the interests and duties of life, how little do any processes of mere ratiocination enable us to comprehend the weightier questions

which they present! As truly, therefore, as the cultivation of the mathematical or deductive faculty is a part of intellectual discipline, so truly
is it only a part.[88]

The pairing of "sentiment and action" here is especially telling. Boole's
anatomy of the mind effectively drains mathematics of the overt political
content that it had had for the French analysts, slotting it away as one discipline among many and distinguishing it, in particular, from the emotional
factors that motivate actual human behavior.

While this passage might be taken as a mere nod to traditional pieties,
Boole's published lectures suggest that he was sincere in his adherence
to Romantic ideas of culture. In an 1847 address, he criticizes those who
believe that the physical sciences are fit "to effect a species of intellectual
regeneration in society."[89] He recommends, instead, "cultivating the love
of Nature" by going out of the city into the countryside.[90] Appreciating
natural beauty, Boole writes, provides a corrective for those who have become "too artificial," too attached to "the conventional refinements of society."[91] Like Wordsworth, Boole breaks with Enlightenment conceptions
of nature that emphasize human reason; instead, he associates "Nature"
specifically with the wilderness and rustic life, which present an antidote
to the changing fashions of the city. In Boole's view, an aesthetic appreciation of landscapes is a better way of developing moral character than scientific study. Elsewhere Boole maintains, echoing an earlier argument by
Coleridge, that empirical methods can only produce "a mere collection of
facts" unless they are guided by a higher purpose.[92] Science is an important
part of culture for Boole, but it cannot substitute for the moral feelings that
poetry is best suited to regulate.

The idea that science had to be balanced with other forms of cultivation was widespread in the mid-nineteenth century. In a pair of essays
published in 1838 and 1840, John Stuart Mill cast the poet Coleridge and
the utilitarian philosopher Jeremy Bentham as two opposing forces in English society.[93] Mill argues that each of these thinkers sees what the other
fails to see. Bentham's blind spots have to do with culture: he can wrap his
head around neither the necessity of "self-education" in developing moral
character nor the differences in "national character" between countries.[94]
Coleridge, Mill tells us, has a better understanding of these cultural matters
as well as a better appreciation for what is valuable in the experiences of
the past, yet he fails to see the errors in established institutions. While Mill
expressly sides with the Benthamites' empiricism over the idealism of the
Coleridgeans, he asserts the value in both perspectives, concluding that
"these two sorts of men, who seem to be, and believe themselves to be,

enemies, are in reality allies," their powers forming "opposite poles of one great force of progression."[95]

This balancing act was, on the surface, reassuring: science would not meddle in the moral development of people, as it had so disastrously done in the 1790s. Human thought would be allowed to develop organically, with room left for varieties of feeling and poetic insight that could not be reduced to strict rules. Yet this dualistic idea of culture opened the way for reasoning to become far more mechanical within certain restricted domains. In *System of Logic*, Mill offers the following "aphorism": "Whenever the nature of the subject permits the reasoning process to be, without danger, carried on mechanically, the language should be constructed on as mechanical principles as possible; while in contrary case, it should be so constructed that there shall be the greatest possible obstacles to a merely mechanical use of it."[96] The "danger" that mechanization poses, Mill explains, is the potential that, in their eagerness to extend the power of algebra to other areas of knowledge, philosophers will lose sight of the sensible meanings of the symbols. In purely deductive forms of inquiry, however, mechanization is safe and, indeed, laudable. The category of culture saved universal algebra from its association with revolutionary politics. Bentham went down easier when sweetened with Coleridge.

Boole disagreed with Mill on many things, but he shared this keenness to separate practices that could legitimately be mechanized from ones that could not. In the preface to *The Mathematical Analysis of Logic*, he (mis) quotes Mill's "aphorism" in full.[97] For Boole, mechanizing logic safely means treating other forms of knowledge with respect: "It were perhaps the best security against the danger of an unreasoning reliance upon symbols, on the one hand, and a neglect of their just claims on the other, that each subject of applied mathematics should be treated in the spirit of the methods which were known at the time when the application was made, but in the best form which those methods have assumed."[98] Before one can employ symbolic methods, that is, one's mind must be "prepared to receive them"; symbols must be used "with a full understanding of their meaning" rather than as "mere unsuggestive characters, the use of which is suffered to rest upon authority."[99] Once this preparation is complete, the symbols can take over and logic can become purely mechanical.

The rhetoric of this passage might come off as disingenuous. What does it really mean to use applied mathematics "in the spirit of" another discipline? Certainly, to some extent, Boole was looking to rationalize a departure from received practices and fend off the accusation that he was replacing intelligent thought with a machinelike system. But his algebraic logic really did handle culture with a softer touch than previous attempts at a universal algebra did. Unlike Condorcet, Boole places no barrier between present and past; instead, his system is supposed to work together with the science to which it is applied, as that science has developed up to the present. The practical implications of this difference are strongest in the way he handled the meanings of the symbols. For once, language was a friend, not an enemy.

RIGOR AND SPIRIT

In 1878, Boole's wife, Mary Everest Boole, published some reminiscences about her late husband's character. She reports that his approach to teaching language was reverent: "I was never allowed to encourage our children in any babyish corruption of language. He would sit for a length of time with an infant on his knee, teaching it to pronounce its first words with perfect distinctness."[100] The care he took in teaching language was partly "to give the children habits of accuracy" and also because, she surmises, "language being a common property, he wished to discourage the idea of individuals having a private right to use it as they pleased."[101] Finally, it reflected his belief that "people who make any sort of profession of believing in the Bible ought to be very reverent in their use of words."[102] George Boole was, if this account is just, a linguistic involuntarist: he saw words as collective creations that existed outside the control of any individual. This reverence toward language gave him a different perspective on what symbols could do than those of Leibniz and Condorcet. For Boole, the task was not to expunge all traces of words from algebra but to infuse algebra with the meaning that can only come from words.

Instilling respect for language was a common part of the pedagogy of the time. In Sarah Porter's *Conversations on Arithmetic*, the pupil's older brother repeatedly suggests ways of making the English language more rational, and in each instance his mother warns him against such presumption. He calls the words *eleven* and *twelve* "barbarous," and she responds that "we must thank our Saxon ancestors" for them.[103] Later, he asks whether "carrying" a number is not "a very silly mode of expression"; his mother responds that "it is well not to make any innovation," and that what matters is that the term is clearly defined.[104] The mother also discourages the child from thinking about changing the system of arithmetic. At one point, he raises the possibility of numerations other than base ten. "It would be just as sensible," the mother says, "to transpose the characters of the alphabet at your arbitrary pleasure, and then to form words and sentences on the supposition."[105] Speculating about alternative systems may be pleasing to "eminent mathematicians," but she wishes her son to follow "the higher and more useful path," for which purpose it is best to follow the practices of others.[106]

This insistence on the sanctity of convention was a departure from the radical educational theories of the late eighteenth century. A few decades prior, Maria Edgeworth had emphasized building clear conceptions in one's own mind over following the practices of others; following the Abbé de Condillac, she averred that the best language was one that was constructed from the ground up. The reaction against such individualistic approaches to language occurred in scientific thought as well as in education. In *System of Logic*, Mill rejects Condillac's program of linguistic reform in favor of "the doctrine . . . of the Coleridge school, that the language of any people among whom culture is of an old date, is a sacred deposit, the property of all ages, and which no one age should consider itself empowered to alter."[107] In his 1845 book *Kosmos*, the naturalist Alexander von Humboldt praises the "animating power" of language, whose "mysterious influence still reveals itself most strikingly where it springs among free-minded communities, and attains its growth upon native soils."[108] This overtly nationalistic praise of the German language marks a break from the antipathy toward language that stretched from Francis Bacon to Lord Stanhope. Words no longer stood like a distorting lens between the observer and reality but rather provided an essential link to culture that gave life to scientific knowledge.

This rehabilitation of language occurred at just the same time—often involving the same people—as the symbolic turn in algebra. Words might be collective property, but algebraists could treat their as, xs, and even their +s as wholly arbitrary signs whose interpretations could be chosen at will. While it may be supposed that the symbols were regarded differently from words—that tradition was valued in regard to the English language but viewed as unimportant in regard to the value of a—matters were not so simple. The valuing of tradition could apply to notations, too; in an 1842 encyclopedia article for "Symbols and Notation," Boole's close friend Augustus De Morgan stresses the importance of maintaining continuity in mathematical symbolism, since excessive innovation renders older texts unreadable.[109] Conversely, arbitrary control could apply to words. In the modern age, wrote William Whewell in 1840, technical language is "constructed intentionally, with set purpose, with a regard to its connexion, and with a view of constructing a system."[110] In regard to both symbols and words, tradition and innovation, backed respectively by collective and scientific governance, were both, in their own ways, viewed as good things.

The coexistence of these two attitudes is best explained as a reconfiguration of the relation of science to the human mind. As it was adopted by mathematicians such as Boole and Ohm, the Kantian distinction between rational and historical knowledge enabled an increased separation of disciplinary criteria of rigor from the "cultivation" of thought. The aspects of language that required a connection with tradition were "subjective" matters that played a crucial role in the education and mental life of the scientist, but that did not bear on matters of truth. Even for mathematicians hostile to Kant, such as Bernard Bolzano, formal standards of rigor were increasingly privileged over the factors that went into the formation of human understanding. This division rendered the definitional concerns that had been central to mathematics in the eighteenth century—the problem of defining such tricky words as *sum, negative,* or *infinitesimal*—less important. It is a general truth that the less people think there is something wrong with language, the more easily they can ignore it. When words no longer appeared as a threat to scientific certainty, algebra was free to develop in new directions.

Boole discusses the linguistic implications of symbolic methods in the most detail in *The Laws of Thought*. After an introduction explaining his overall purpose in the book, he enters into a discussion of "signs and their laws."[111] He defines a "sign" as an "arbitrary mark" with a "fixed interpretation, and susceptible of combination with other signs in subjection to fixed laws dependent upon their mutual interpretation."[112] On account of this arbitrariness, he finds it "permissible" to replace a subset of words—those that express things or qualities of things—with single letters.[113] He then attempts to show that these signs follow laws of combination that are equivalent to the basic laws of algebra. The order in which such terms appear, he argues, makes no difference; "rivers that are estuaries" means the same thing as "estuaries that are rivers."[114] This equivalence translates to the commutative law, $xy = yx$. Likewise, "European men and women" means the same thing as "European men and European women," which translates to the distributive law, $z(x + y) = zx + zy$.[115] Boole also shows how *if, not,* and various other connecting words map onto algebraic expressions.

It is on account of these analogies between the symbols and words, as his other writings make clear, that he viewed his system as akin to a "philosophical language."[116] Any elements of language that are left out of the algebraization, he declares at the end of the chapter, either modify the signification of other words or else serve only "to express some emotion or state of feeling accompanying the utterance of a proposition, and thus do not belong to the province of the understanding, with which alone our present concern lies."[117] By "philosophical," this remark suggests, he means scientific in something like Wordsworth's sense—that is, concerned primarily with facts, not feelings. The logical algebra excludes these emotional aspects of language not necessarily because they are unimportant but because they are divided from logic by a disciplinary boundary.

While he illustrates his "laws of thought" with English phrases, Boole is decidedly not attempting to derive them through an empirical analysis of language. In an 1847 book that Boole cites elsewhere, the philologist Robert Gordon Latham had argued "that the structure of propositions in Language does not always coincide with the structure of propositions in Logic."[118] Boole similarly sets logic apart from the conventionalities of particular languages. He uses three quotations from John Milton's *Paradise Lost* to illustrate how a phrase can retain a logical structure even as it violates the usual rules of English word order:

> "*Offspring of heaven first-born.*"
> "The rising world of *waters dark and deep.*"
> "Bright effluence of *bright essence increate.*"[119]

Poetry, according to Boole, operates with a "lawful freedom" from the conventions of the language, a freedom that is "sanctioned by the intimate laws of thought."[120] Boole thus maintains that the law of thought $xy = yx$ "is actually developed in a law of Language, the product and the instrument of thought," even though, "for reasons of convenience," prose writers usually follow a uniform word order.[121]

This argument is worth puzzling over. How does the interpretability of some lines from Milton demonstrate the existence of a logical law? What kind of interpretation is Boole imagining? The field that dealt with these issues in his time was hermeneutics, the science of interpretation that originated in Germany and spread to England via the Romantics. Like Boole in his discussion of Milton, Romantic hermeneutics valued perceiving the spirit of a text over such "mechanical" matters as spelling and syntax.[122] An instructive example of this value hierarchy, including its potentially troubling implications, appears in Coleridge. In his "Essays on the Principles of Method," Coleridge imagines an illiterate person attempting to make sense of a Bible:

> Say that after long and dissatisfying toils, he begins to sort, first the paragraphs that appear to resemble each other, then the lines, the words—nay, that he has at length discovered that the whole is formed by the recurrence and interchanges of a limited number of cyphers, letters, marks, and points, which, however, in the very height and utmost perfection of his attainment, he makes twentyfold more numerous than

they are, by classing every different form of the same character, intentional or accidental, as a separate element. And the whole is without soul or substance, a talisman of superstition, a mockery of science.[123]

Coleridge goes on to imagine a "friendly missionary" arriving and teaching the man to read, after which point the "words become transparent, and he sees them as though he saw them not."[124] The point is that the analytical methods employed by Enlightenment thinkers cannot provide a self-sufficient means of reading a text. A form of writing understandable by the wholly uninstructed, as Condorcet's universal algebra was supposed to be, is impossible; all interpretation requires some link to tradition.

Hermeneutics has taken on various guises over the centuries, but in its original form it was primarily about interpreting the Bible and other religious texts. The interpretation Coleridge describes has specifically to do with scripture; the Bible becomes "transparent" not just by virtue of linguistic instruction but also because of the reader's ability to perceive the divine truth that shines through the text. Boole's rationale for his logic system was likewise religious.[125] He believed, according to his wife, Mary Everest Boole, "that there is direct contact between the Divine Magnetism and the nervous system of man"; in the last chapter of *The Laws of Thought*, he argues that his algebraic laws have appeared again and again in all religious and philosophical systems, which he takes as evidence of their authority.[126] In this regard, his choice of the Puritan poet Milton is appropriate. It is an old critical saw that Milton used intentionally difficult syntax so as to ensure that his readers chose a pious reading by their own free will; the meaning of *Paradise Lost* would be ratified by faith, not by the authority of the text itself.[127] What renders Boole's break from language "lawful" is the accordance of his system with the higher truth one perceives when one sees through the words of a text.

This, at least, would be Boole taken on his own terms. If we swap in a secular form of hermeneutics, the argument becomes more problematic. What if it is not the light of God that appears to shine through the text but simply the prejudices of a particular culture—or worse, ideology? The explicitly colonial situation Coleridge describes in his description of reading indicates the danger of claiming that a text is transparent. The written text is not merely acting as a conduit through which one being can communicate with another but also as an instrument by means of which a particular doctrine—in Coleridge's case, a Christian one—can be instilled. Boole's logic system could inspire a similar critique. To the extent that Boole's laws of thought lacked the divine certification he thought they had, his freedom from language was not truly "lawful." Instead, he was manipulating symbols through rules that were ultimately arbitrary, and his results therefore had no particular meaning that anyone was obliged to accept.

Such concerns arose frequently in discussions of symbolic methods. In a review of Peacock's *Treatise on Algebra* published in two installments in 1835, De Morgan writes that Peacock's work initially appeared like "symbols bewitched, and running about the world in search of a meaning."[128] This statement is often quoted as an example of the anxieties created by the symbolic turn; symbol manipulation, it suggests, is not truly rational, and losing oneself in mathematical abstraction is a way to solipsism or madness. Yet De Morgan ultimately came around to Peacock's position, and his review critiques the work from within the new epistemology that the symbolic turn constructed. De Morgan is not complaining that the symbols lacked clear conceptual definitions, as so many in the eighteenth century had done. His concern, instead, is to ensure that rules of their manipulation were grounded on sound principles.

De Morgan's review—which is mainly an essay about his own opinions on symbolic methods—attempts to settle the confusion regarding negative and imaginary numbers by clarifying the nature of algebra. Algebra, De Morgan argues, gains its generality from the use of "arbitrary conventions by which terms in common use are made to signify less than their vulgar meaning implies."[129] As a result, "certain formulæ may be chosen, not as consequences of any meaning given to the symbols, but as the definitions of the symbols themselves."[130] For example, the formula $a + b - b = a$ may be taken as a definition of subtraction that is more general than the arithmetical definition because it tells us less—specifically, it does not tell us whether a and b represent numbers or something else. Since the interpretation of the symbols + and − is arbitrary, the mathematician is "at liberty" to make those symbols stand for any two operations whose relationship corresponds to the rule.[131] This is just the sort of thinking that William Rowan Hamilton would soon employ in specifying the rules for multiplying quaternions and that, just a few years later, Boole would apply to logic; it points toward the development of what is now called *abstract algebra*—a field which, brought to its full development by Emmy Noether in the early twentieth century, studies the properties of generalized algebra-like systems that may or may not have a connection to numbers.[132]

This approach to definition suggests, contra the Lockean perspective, that algebraic symbols are fundamentally different from words. To ask whether formulae may have multiple interpretations, De Morgan writes, might seem like asking "whether two different languages might have all their words in common, but with different meanings, in such manner that by writing a treatise on astronomy in the first language, we write, *totidem verbis* [by all the same words], a treatise on music in the second."[133] This, he agrees, is absurd. But in algebra, the rules are determined not by "pure convention," as the meanings of words are, but by the universal truths of mathematics.[134] Defining an algebraic symbol is thus different from defining a word: "We say, *let $a^0 = 1$*; but the word is a great deal more like 'I think we must let $a^0 = 1$' than 'We are perfectly free to choose a meaning for a^0, and therefore let it be 1.'"[135] In hermeneutics, the definitions of words are arbitrary, but interpretation is guided by a higher logic by virtue of which the truth can shine through. In symbolic algebra, the situation is reversed: the definitions are guided by logic and truth, whereas interpretation is wholly a matter of choice.

A consequence of this reversal is that the apparent uncontrollability of language posed less of a problem for De Morgan than it had for eighteenth-century mathematicians such as Condorcet and Maseres. Words had not become unimportant, but De Morgan locates their relevance primarily in pedagogy, where they provide a route from familiar notions into the intricacies of advanced mathematics. Near the end of the review, De Morgan proposes "a treatise on the use of words in mathematics, and their connexion with symbols."[136] The new algebra, he writes, "will introduce a very new set of idioms" that may pose difficulties for students; an adept mathematician can navigate these difficulties by "helping himself to the meaning of the terms out of the algebraical context."[137] The conventional meaning of the word *sum*, that is, could aid in understanding $a + b$ even if those symbols technically represent a more abstract operation than common addition. This line of thinking is a direct reversal of William Frend's late eighteenth-century argument that the received meanings of words such as *square* might interfere with their technical definitions by calling to mind spurious associations. For De Morgan, the ordinary, nontechnical meanings of words provide a valuable "resource" that students may exploit in making sense of symbols, provided they have sufficient linguistic competency.[138]

As this passage suggests, the symbolic turn had not done away with words; it had merely altered their relation to symbols. Boole, too, treated words as a resource, a well from which meaning could be drawn. The role of words in Boole's logic system has often been overlooked. In the form used in programming languages, Boolean logic has only two values, 1 and 0 (or, if we prefer, *true* and *false*); using the rules for *and* and *or*, we can always evaluate a Boolean expression to one of these two values, as (0 *and* 1) *or* 1 has the value 1. Some popular accounts have implied that Boole's system worked like this, too.[139] But Boole's deduction procedure did not involve evaluating logical expressions numerically, nor did he restrict his system to the two values 0 and 1. If 1 means everything and 0 nothing, Boole's system also admits an indefinite number of other values that refer to specific categories of things, such as ducks and capybaras. These values do not correspond to any numbers; if one interprets x as *capybara,* then the value of the symbol simply is the meaning of the word. The rigor of the system is supposed to come not from the clarity of these meanings but from the fact that the meanings are irrelevant to the formal rules by which the logic operates. We may or may not have a clear idea of what a capybara is, but either way, we can be sure that Boole's laws of operation are solid.

The formal turn lifted a weight from the shoulders of language. No longer was it necessary to debate whether the meaning of *negative* was clear enough or whether the meaning of *sum* could be altered, except perhaps for the purpose of teaching. In the hands of Boole, the distinction between form and interpretation made it possible for a universal algebra to coexist peacefully with words, which could be swapped out at will: in the course of a single paragraph, he makes y mean "good" and then changes its meaning to "sheep."[140] The symbol does not replace these words; instead, it draws its meaning from them, and thus from the English language. This is different from how Leibniz and Condorcet handled meaning. Gone are the "characteristic numbers"; gone, too, is the attempt to build up complex concepts from scratch, starting with the senses. Even Boole's laws of thought themselves are supposed to be found in "the language of common discourse."[141] Language did not arise as a problem for Boole because, unlike his predecessors, he was not attempting to build a whole new way of thinking about the world from the ground up. He purchased arbitrary control over the interpretation of symbols in exchange for deference toward the common tongue.

UNINTERPRETABLE REALMS

Boole's system does not, however, leave language totally alone. To be of any use, symbols cannot bend entirely to the will of the interpreter; they must somehow be able to push back. It is not hard to find a particularly charged example. In his derivation of the laws of thought, Boole considers "the two mental operations implied by 'white' and 'men.'"[142] The *men* operation selects all things that are men; the *white* operation then selects those who are white. One thus constructs the class of "white men."[143] One can also, according to Boole, reverse the order: one starts by selecting "white objects," then one selects from those objects the ones that are men.[144] Hence he derives the law $xy = yx$. The problem here is not hard to see. He is treating the *white* in "white men" as equivalent to the *white* in his earlier example, "horned white sheep"—that is, as a literal color.[145] Boole's system treats words as fixed in their meanings, signifying the same things regardless of context. This presumed fixity enables the symbols to be tossed around like counting stones.

This tendency would not be a problem if one had a perfect analytical language of the sort Leibniz had hoped to create. The early version of Leibniz's universal characteristic—the one that was supposed to resolve questions with absolute certainty—required the discovery of a universal set of basic concepts from which all other concepts could be constructed; the meanings of the symbols would be grounded in ideas divinely inscribed on the mind and thus wholly unaffected by the niceties of words. Boole's logic system, along with all the technical forms derived from it, demands no such quixotic effort to replace language, and therein lies the danger. Does "white men and white sheep" really mean the same thing as "white objects that are men and sheep"? If not, then the symbols do not necessarily come out of Boole's symbolic process with the same meanings they had when they went in.

Boole was far from unaware of these issues. Even if his departure from language is "lawful," it is still a departure—the symbols do not follow the same rules that words do in ordinary writing. That is precisely the point; if the symbols did nothing other than reflect back what was already in the English language, they would be worthless. Boole's algebra is supposed to constitute a different sort of symbolic system from historically contingent languages such as English; as he states in the 1848 article, the logical algebra is analogous "with the forms which human speech would assume, were its rules entirely constructed upon a scientific basis."[146] Whereas the Abbé de Condillac brought algebra and language into an intimate relation, Boole opens a rift between them. The symbols may gain their interpretations from words, but they can also open new realms that words cannot describe.

Boole addresses this rift directly in one of the most notorious passages in *The Laws of Thought*. One of the prime dangers of applying algebraic methods to logic, he concedes, is that such methods might produce expressions that are "uninterpretable in that sphere of thought which they are designed to aid."[147] Such uninterpretable statements occur frequently in Boole's text. For instance, one of his examples begins with the following definition: "Responsible beings are all rational beings who are either free to act, or have voluntarily sacrificed their freedom."[148] In symbolic notation, he renders this statement as "$x = yz + yw$," with x being responsible beings, y being rational ones, z being those free to act, and w being those who have voluntarily sacrificed their freedom.[149] This equation has a clear interpretation given Boole's definitions of multiplication, addition, and equality. Yet in analyzing it, he winds up with an equation that does not have any logical meaning:

$$1 - y = \frac{z + w - x}{z + w}$$

This equation is equivalent to $x = yz + yw$ in ordinary numerical algebra.[150] But it bears no meaning that can be translated into English by Boole's method, since, unlike the other basic algebraic operators, division has no logical counterpart.[151] The only way to make sense of it is to transform it using Boole's procedure, which produces a much longer equation that can finally be translated into an English sentence, albeit a less-than-elegant one: "*Irrational persons consist of all responsible beings who are either free to act, or have voluntarily sacrificed their liberty, and are not free to act; together with an indefinite remainder of irresponsible beings who have not sacrificed their liberty, and are not free to act.*"[152]

The use of uninterpretable expressions has never been widely accepted among logicians. One of Boole's early champions, William Stanley Jevons, criticized his system for employing "dark and symbolic processes," and later practitioners of symbolic logic mostly attempted to ensure that one could always, at least in theory, understand the meanings of the symbols.[153] But uninterpretability is worth taking seriously as a side effect of the adoption of symbolic methods. As an example that, in Boole's view, proves the legitimacy of the practice, "the uninterpretable symbol $\sqrt{-1}$," although devoid of any sensible meaning, may be used "in the intermediate processes of trigonometry."[154] Although we might question whether imaginary numbers such as $\sqrt{-1}$ are really "uninterpretable" in an absolute sense, it is true that they cannot be interpreted in terms of ordinary notions of quantity or magnitude. Yet, as Gerolamo Cardano discovered in the sixteenth century, they can nonetheless be used in calculations that produce verifiably correct results about quantities. Most mathematicians from Cardano's time to the early nineteenth century were suspicious of such methods because they seemed to lack an adequate conceptual foundation. For Boole, as for Joseph-Louis Lagrange, the apparent incomprehensibility of imaginary numbers has no bearing on their legitimacy as instruments of computation; all that matters, for the purposes of mathematical validity, is that one follows the rules.

The danger of wandering into a realm of uninterpretable nonsense is the price one pays for this embrace of formal rules. Uninterpretability, Boole explains, is specific to symbolic methods, since "this apparent failure of correspondency between process and interpretation does not manifest itself in the *ordinary* applications of human reason."[155] Boole does not, however, take this as a knock against algebraization. It is valid, Boole contends, to employ symbols "in obedience to laws founded upon their interpretation, but without any sustained reference to that interpretation, the chain of demonstration conducting us through intermediate steps which are not interpretable, to a final result which is interpretable."[156] Despite Boole's protestations that one must always clearly understand the meanings of the symbols one uses, this requirement is only really relevant at the beginning and end of the deductive process, when it is necessary to translate between algebraic notation and the language of another field of study. In the midst of Boole's procedure, interpretation is both needless and, in the case of "uninterpretable" expressions involving division, futile.

This return from the abyss of the uninterpretable is only possible, however, if the procedures one is using are the right ones. It is here that the religious basis of Boole's system becomes important. As he states in a manuscript probably written in 1854, because "methods and processes are truly the consequences of laws and do not spring up arbitrarily into existence," one can presume that any complete method rests on a basis that is "not merely empirical or analogical."[157] This assumption that the "laws" of algebra are nonarbitrary enables Boole to turn deductive reasoning into a mechanical procedure without at all embracing instrumentalism. The fact that the steps of this procedure were sometimes uninterpretable could be brushed off as resulting from the finitude of human understanding. In an 1855 letter to John Penrose, Boole gives "infinite space, eternal duration, . . . perfect goodness and purity, unchanging rectitude and truth etc" as examples of terms that cannot be given clear and distinct meanings but that one can nonetheless reason about with certainty by following the laws of thought.[158] So long as we use symbols in "obedience" with these laws, we cannot produce anything that is out of sync with the order of nature, however little we mere temporal beings may be able to understand the expressions our reckoning produces.

These explanations bear the distinct markings of Kantian thought and, in particular, of the idealists' elevation of pure reason over understanding. Boole read Kant's *Critique of Pure Reason* in the original German in the 1840s, and his work contains numerous Kantian-sounding references to the idea that logic and mathematics deal only with the conditions that experience must meet rather than with the sensory content of any particular experience.[159] He also expressed such views in his poetry. Daniel J. Cohen has pointed out the Kantian influence in Boole's sonnet "To the Number Three":

When the great Maker, on Creation bent,
Thee from thy brethren chose, and framed by thee
The world to sense revealed, yet left it free
To those whose intellectual gaze intent
Behind the veil phenomenal is sent
Space diverse, systems manifold to see
Revealed by thought alone; was it that we
In whose mysterious spirits thus are blent
Finite of sense and Infinite of thought,
Should feel how vast, how little is our store;
As yon excelling arch with orbs deep-fraught
To the light wave that dies along the shore;
That from our weakness and our strength may rise
One worship unto Him the Only Wise.[160]

This sonnet, written in the late 1840s, is as much a hymn to the human intellect as to the Abrahamic God. Whereas the terrestrial wave is transient, Boole identifies "yon excelling arch with orbs deep-fraught" with the infinite and eternal. Our ability to grasp such things lies not in our finite senses but in "pure thought," through which we may gaze at "systems manifold" that exist "behind the veil phenomenal." In the final couplet, Boole asserts his confidence that our senses and our capacity for "pure thought" will harmonize both with each other and with revealed religion. Mathematics gains its peculiar power, the poem suggests, from the fact that it works "by thought alone," untainted by the influence of the empirical; as such, it can lead us infallibly to truths about quantity and the divine nature alike.

What is paradoxical in this association of symbolic methods with the "intellectual gaze"—a paradox that echoed far and wide in the nineteenth century—is that, as much as formalization elevated mathematics to the ethereal realm of pure reason, it also deepened its dependence on physical aids. Even if mathematics was supposed to exist apart from empirical knowledge, it involved the senses more than ever: the symbols consisted of marks on a sheet of paper, which, in Ohm's formulation, constituted the "objective" aspect of algebra. Mathematical rationality was moving from the brain to the page. Soon enough, the logical processes that Boole characterized as "pure thought" would be enacted by machines as well. It is one of the great ironies of intellectual history that, by delineating pure reason from sensory experience, Kant ultimately enabled logic to move out of the human mind and into the physical realm.

Boole himself showed little interest in mechanizing thought. He was aware of his near contemporary Charles Babbage, but he seemed to be more impressed by Babbage's work on the theory of functions than by his plan for a computing engine.[161] The amenability of Boole's system to mechanization does, however, become apparent in his approach to pedagogy. According to Mary Boole, he thought it important that children "should spend a great deal of time over some mechanical work which could be done without the presence of a teacher, and which they must concentrate their whole energies upon, and do with perfect accuracy."[162] Students, then, must be taught how to work a sum mechanically before the rule is explained to them; they are "to *obey* first and *understand* afterwards."[163] The contrast with the Lockean pedagogy of Sarah Porter, which leads students toward the rules of computation by explaining the reasoning behind them, is plain. After symbolic methods were freed from their dependence on concepts, mathematical and logical validity came to have less to do with how one thinks than with what one does. It is only a short way from this "directive method" to the literal mechanization of logical reasoning.

The inventor of Boolean logic would never have traveled down this road. While his students may have begun with mechanical work, they were eventually supposed to reach a moment of epiphany in which they finally grasped the leading idea behind the rules.[164] The need to appreciate the reasoning behind algorithms, which we have seen Ada Lovelace state directly, persisted. But the epiphany could just as well never come, and the machine of logic would go on cranking out symbols that are correct by the standards of the system whether one understands them or not. It was the fate of Boole's system to become such a machine. The divine sanction that was supposed to compensate for the symbols' occasional uninterpretability got lost in the shuffle. To the extent that their rules are, instead, merely arbitrary, the formal languages that dominate our modern world— most prominently, programming languages—sharpen the distinction between those who understand and those who must be content to obey.

FORMALISM AND THE PROGRESS OF CULTURE

Based on some of his remarks, one would think that Boole's logic system had no practical use at all. In the preface to *The Mathematical Analysis of Logic*, he denies that he is out "to supersede the employment of common reason, or to subject it to the rigour of technical forms."[165] His purpose, he tells us, is only to contribute to speculative knowledge. *The Laws of Thought* does include applications, but Mary Boole later claimed that her husband's intentions were entirely religious, executed in "obedience to the commands of the Pentateuch," and that he only included these examples "to show that his system was not a mere fanciful outcome of religious fervor."[166] Although they had an Anglican background, the Booles were religious eclectics; they sought spiritual insight in various traditions including Judaism, early Christianity, and Hinduism. The main value of George Boole's logical algebra, if we believe this, is as a way of contemplating the harmony of the divine plan.

But Boolean logic as it exists now is a technological matter, not a religious one. Boole's followers wasted little time in exploring the possibilities his ideas opened for mechanization. In 1869, Jevons used his simplified version of Boolean algebra to develop a "logic piano" that could automatically draw certain types of logical conclusion (figure 4.2).[167] Boolean logic also formed the basis of a different sort of machine that turned symbolic logic into a general-purpose model of signal processing. Credit typically goes to the American mathematician C. E. Shannon, later known as the founder of statistical information theory, for noticing that Boolean algebra corresponded to the structure of certain types of electrical circuits. Shannon was not the first to have this insight; C. S. Peirce had a similar idea in the 1890s, although he did not put it into practice.[168] This analogy between logic and switching circuits was an important step in the development of modern electronics and, in particular, computers, in the design and programming of which Boolean logic still plays a pervasive role.

Boole's work also inspired a revival of Leibnizian ideas about the power of symbols. Starting around 1874, the German mathematician Ernst Schröder began to extend the ideas of Boole and Ohm into a general theory of algebra.[169] Algebra, for Schröder, was the study of operations, considered apart from any conceptual interpretation. He formalized these operations as "algorithms," although his meaning is not the modern one. In Schröder's terms, *algorithms* (*Algorithmen*) are collections of equations determining the properties of invertible operations; for instance, the equations $ab = ba$, $a(bc) = (ab)c$, and $(a/b)b = a$ could be used to construct an "algorithm" of multiplication and its inverse, division.[170] This use of the word *algorithm*, following Leibniz, refers not to a step-by-step procedure but to a collection of equations that may be used to manipulate operations algebraically. Schröder later expanded his system into a "pasigraphy" that included symbols for a range of logical relations. On this basis, he planned out an "absolute algebra" that would, he wrote, constitute "a scientific universal language" fundamentally different from spoken languages.[171]

Even if it was not exactly about algorithms in the modern sense, late nineteenth-century formalism created an interchange between mathematics and logic that long would be a fruitful source for algorithmic thinking. In his 1879 pamphlet *Begriffsschrift* (*Concept Notation*), Gottlob Frege introduced a new logical notation that included several important features

Figure 4.2. William Stanley Jevons's logical piano, as depicted in the frontispiece of his book *The Principles of Science* (London: Macmillan, 1892).

that Boole's system lacks, including the quantifiers "there exists" and "for all," which made it possible to formalize the entirety of a mathematical proof in purely symbolic terms. Frege's system is not based on uninterpreted symbolism—his notation has a "conceptual" aspect—but it does share Schröder's emphasis on formal rules as a path to precision. Similarly to Schröder, Frege described his system as a step toward the realization of Leibniz's universal characteristic.[172] While Frege concedes that a logical "ideography" cannot do everything that Leibniz had intended, he nonetheless frames symbolic logic as a continuation of the long-standing struggle against language—an aid to philosophers in their quest "to break the domination of the word over the human spirit."[173]

As this statement indicates, the old Baconian hostility toward words returned in force in the late nineteenth century. This attitude differentiates Frege and other late-century formalists from Boole, who was more interested in making symbols work together with the subjective aspects of language.[174] Yet the extent of this difference should not be overstated. Boole is sometimes charged with what Frege called "psychologism," meaning conflating the rules of logic with empirical facts about the mind.[175] But as Boole makes clear in the final chapter of *The Laws of Thought*, he does not regard the laws of thought as empirical propositions like the laws of physics; they are supposed to be general strictures that people may follow more or less faithfully in different situations.[176] Boole distinguished logic from actual human thought just as much as Frege did; what changed in the late nineteenth century was the value formalists assigned to the historical processes by which thought developed.

For Boole, the progress of human thought is subject to universal, divinely sanctioned laws that direct it toward a single goal. The moral faculty, he writes in the last chapter of *The Laws of Thought*, "tends, wherever human progress is observable, wherever society is not either stationary or hastening to decay, to attach itself to certain classes of actions, consentaneously, and after a manner indicative both of permanency and of law."[177] He goes on to argue that his algebraic "laws" have occurred again and again across religions and cultures and throughout history.[178] The laws are supposed to be universal because of this general tendency, even if they have developed to varying degrees in particular historical and cultural contexts and among particular individuals.

This conception of a logical "law" cannot be disentangled from Boole's religious views. Like other algebraists in his time, Boole understood the history of mathematics in terms of divine providence. William Rowan Hamilton wrote of a "Philological" school of algebra, in which he includes Peacock, Ohm, and Gregory—thinkers who all combined symbolic methods with a deep attention to language.[179] In his article on "Arithmetic" in the *Encyclopædia Metropolitana*, Peacock presents extensive research about the development of arithmetic from prehistoric times, building on the work of philologists such as Friedrich Schlegel and Wilhelm von Humboldt.[180] While he considers the differences between a variety of cultures, Peacock clearly thinks the Hindi–Arabic system is the one goal toward which all progress naturally tends. Boole likewise reduces cultural differences to levels of progress toward a single truth. In a speech entitled "The Social Aspect of Intellectual Culture," Boole argues, drawing on Whewell's historiography, that progress is driven by "the slow but combined action of the social state," with each new advance depending on everything that has been achieved in the past in both the sciences and the arts.[181] The formation of scientific knowledge, then, depends on and is a part of culture, and yet it is guided by divine providence toward the correct theory.

This teleology is distinctive of how science and history intersected in the early Victorian period. By the time of Schröder and Frege, discussions of mathematical progress had turned toward a greater recognition of multiplicity—of the variation in practices from country to country. In some cases, this recognition was attended by racial or ethnic stereotypes. In his 1889 *History of the Study of Mathematics at Cambridge*, for instance, W. W. Rouse Ball speculates about "the influence of race in the selection of mathematical methods."[182] "The Semitic races," Ball claims, "had a special genius for arithmetic and algebra," whereas the Greeks were better suited for geometry; he thinks that the English are especially fitted for analytical methods, even though Newtonians suppressed them for more than a century.[183] The implication is that some groups of people, as classified by the racial theories of the time, are inherently attracted to certain types of mathematics over others.

It is also worth mentioning that Weierstrass's oft-quoted remark about mathematicians and poets, which appeared in a private letter to one of his students, occurred in an anti-Semitic context. Weierstrass claims that members of "the Semitic tribes" (and, specifically, the Jewish mathematician Leopold Kronecker) lack the poetic imagination needed to become true mathematicians.[184] Such racist distinctions addressed a tension that was implicit in the nineteenth-century idea of culture: how can one reconcile the apparent universality of mathematics with the fact that it develops differently among different groups of people? To the mindset of the time, an obvious answer was to rank groups of people against each other.[185] Certainly, not all mathematicians held such views; yet Weierstrass's remark shows how the concern with elevating pure mathematics above such "merely mechanical" activities as computation could reflect and reinforce the prejudices of the time.

Opinions about the nature of pure mathematics were changing in the late nineteenth century, although the result was hardly a consensus. In the 1860s, Weierstrass gave the first modern account of the *real numbers*, which he explained based on the continuity of the number line rather than notions of counting or measurement. Starting in the 1870s, Georg Cantor introduced the rudiments of set theory, which provided a new way of reasoning about the infinite.[186] In 1888 and '89, Richard Dedekind and Giuseppe Peano assembled what would become the standard axiomatization of arithmetic.[187] The work of these and other late nineteenth-century mathematicians implied that numbers were stranger and more counter-intuitive than before suspected: Cantor, for instance, proved that the real numbers, understood to have infinite decimal expansions, are uncount-able, meaning that any attempt to list them in a linear order will always leave some numbers out. The new definitions of number also raised thorny questions about the metaphysics of mathematical entities, inspiring Kro-necker (a critic of Cantor's work) to state that "God made the integers, all else is the work of man."[188]

The real number system developed by Weierstrass and his contempo-raries deepened the gap that algebra had created between mathematical and common understandings of number. Three centuries before, Simon Stevin had introduced decimal notation as a challenge to classical, geomet-ric number theories, founding number instead on the workaday computa-tional techniques of the mathematical practitioners. In the new system, by contrast, real numbers stood aloof from geometry and practical computa-tion alike. Cantor's theorem implied the existence of numbers that were utterly inaccessible by any procedure resembling the Hindi–Arabic algor-ism: strings of effectively random digits that extended to infinity without following any logic that could be described. Dedekind explained the irra-tionals by what is now called the *Dedekind cut*, a division of the rational numbers into two sets whose elements infinitely approach the position of the irrational on the number line. The theories of Cantor and Dedekind were arguably antialgorithmic in that they posited the existence of entities that could not be constructed through any finite procedure. The historian of mathematics Jeremy Gray thus contrasts Dedekind to Kronecker, who proposed an alternative view of number that placed computational meth-ods at the center.[189] Yet Cantor's work opened a philosophical question that would, in the hands of later thinkers such as Alan Turing, motivate the development of a mathematical theory of computation: if not all numbers can be computed, which ones can? Thus, as we have already seen in Boole and Lovelace, the relegation of mechanical procedures to a limited role within mathematics—the idea that such procedures cannot provide the poetic insight needed by a "complete" mathematician, as Weierstrass had put it—ultimately spurred on the development of what would eventually become the theory of algorithms.

These new conceptions of number relegated the old geometric explana-tion of calculus to the margins and, in particular, to the realm of pedagogy. In the preface to his 1872 pamphlet *Continuity and Irrational Numbers*, Dedekind admits with consternation to having "recourse to geometric evi-dences" when teaching calculus.[190] Geometric intuition, he continues, is "exceedingly useful, from the didactic standpoint, and indeed indispensable, if one does not wish to lose too much time. But that this form of introduction into the differential calculus can make no claim to being scientific, no one will deny."[191] Here, then, is the subject–object divide in plain form: making a system comprehensible to people is an entirely separate matter from making it scientific. This division remains encoded in the mathematical curriculum today: college calculus usually employs informal explanations and visual aids, whereas the truly scientific foundation is taught in Mathematical Analysis, typically a senior-level course aimed at majors.

A consequence of this division was a newly absolute alienation of mathematical and logical systems from the practicalities of communication. Although Frege framed his system as a successor to Leibniz's universal characteristic, he focused only on a portion of the total Leibnizian program—only on the use of symbols in logical reasoning, not so much on communication, with regard to which his notation is not, in any meaningful sense, actually universal. The communicational side of Leibniz's project came to perdure less in logic than in the development of international auxiliary languages (IALs) such as Esperanto. First introduced by L. L. Zamenhof in 1887, Esperanto is a spoken and written language with a simple, consistent grammar designed to be easy to learn so as to facilitate international communication.[192] Such projects had precedents in Leibniz's work just as much as symbolic logic did; one of Leibniz's projects was a simplified version of Latin. This idea would, in the twentieth century, become the basis of several attempted IALs, one of them by a mathematician: the aforementioned Giuseppe Peano. Yet Peano made it clear that he saw this constructed language as entirely distinct from mathematical notation.[193] By the turn of the twentieth century, two types of artificial language aiming at different sorts of universality—mathematical and logical notations meant for scientific purposes and IALs meant to replace spoken vernaculars—had diverged.

This divergence coincided with a sharpening of the boundary between mathematical and linguistic disciplines. In the eighteenth century, it had been a matter for debate whether languages were objects of conscious design, and the example of algebra offered evidence that they were. By the end of the nineteenth century, linguists were nearing a consensus around the opposite conclusion. As Andrea Henderson has pointed out, just as linguists came to see form as something that could be described with scientific precision, they began to see content—that is, meaning—as mysterious, incapable of being pinned down or controlled.[194] In the 1830s, the philologist Wilhelm von Humboldt had called language "an involuntary emanation of the mind" that people use "without knowing how they have fashioned it."[195] This involuntarism became a doctrine in Ferdinand de Saussure's 1916 *Course in General Linguistics*, which was written down by his students based on his lectures. This book contains a detailed argument that "the individual has no power to alter a sign in any respect once it has been established in a linguistic community."[196] Signs, that is, were governed collectively, and (contra Locke) one cannot simply choose to understand a word however one wants.

Saussure's theory of signs was not the only option available at the time, and linguistic involuntarism never went uncontested. In 1903, Lady Victoria Welby introduced a contrasting theory called *significs*, which was meant to elevate language "from the 'instinctive level' to the volitional and fully rational plane" by providing a set of techniques for developing meaning.[197] In the twentieth century, the desire to improve language would become a central concern of analytic philosophy—a predominantly Anglo-American tradition that builds upon logical formalism to develop methods of clarifying concepts. Yet Saussure's involuntarism became dominant in the field of linguistics, where it helped solidify the idea that languages were natural phenomena governed by forces that no individual could overcome. Language may follow rules that can be stated abstractly, but the purview of linguistic science is only to describe those rules, not to change them.

The symbols used in mathematics and logic did not clearly fit this emerging consensus about language. In these technical fields, meaning was governed by explicit, usually written definitions that fixed the symbols' roles within systems. Peak notation occurred with Alfred North Whitehead and Bertrand Russell's *Principia Mathematica* (1910–13), which attempts to demonstrate all of mathematics in terms of the simplest possible set of axioms and rules. Their definition of the symbol "1" gives a flavor of this immensely complex system:[198]

*52·01.	$1 = \hat{\alpha}\{(\exists x).\, \alpha = \iota'x\}$	Df
*52·1.	$\vdash: \alpha \in 1. \equiv. (\exists x).\, \alpha = \iota'x$	[*20·3.(*52·01)]
*52·11.	$\vdash:.\, \alpha \in 1. \equiv: (\exists x): y \in \alpha. \equiv_{y}.y = x$	[*52·1.*51·14]

In the first line, "1" is defined as the class of all α such that, for some x, α is equal to a class containing only x. The next two lines offer equivalent statements of the definition. In contrast to Saussure's account of signs, Russell and Whitehead treat this sort of definition as voluntary, presenting their system for consideration and leaving it to readers to choose whether or not to adopt it. To the extent that they did persuade, they were doing just what Saussure thought was impossible: exerting the power to alter signs.

The difference comes down to what a definition is supposed to accomplish. The Russell-Whitehead system is not fully formal; it employs "primitive ideas" that are left undefined, including the operations of Boolean algebra.[199] Yet its definitions deal only with relations among the symbols and take the shape (like De Morgan's proposed definition of subtraction) of equations. Unlike the linguistic prescriptivism that Saussure deplores, such formal definitions do not interfere with the semantic conventions that are established in human communities; whatever preexisting associations may accompany the symbol 1 are permitted to lurk in the background, helping students learn the system, providing intuitions to guide its use, and easing translation between the formal notation and the languages of other disciplines. This way of thinking differentiates modern formal languages from what symbolic notations were before the nineteenth century. As much as axioms masquerade as an absolute beginning, formal languages do not aim at a revolutionary break with established ways of thinking. Instead, they serve as guarantors of a disciplinary rigor that must be balanced with the cultivation of human practices—notational conventions, informal explanations, and educational methods—that make the symbols comprehensible to people.

It was the undeniable importance of these human matters that led Florian Cajori to research the history of mathematical notations in the 1920s. I have already cited Cajori's 1929 book as a secondary source, but the time has come to place it in its historical context. Cajori states that his goal is to determine what history can tell us about what makes for good notation. He sums up one of his main conclusions:

> *Individualism a failure.*—As clear as daylight is the teaching of history that mathematicians are still in the shadow of the Tower of Babel and that individual attempts toward the prompt attainment of uniformity of notations have been failures. Mathematicians have not been profiting by the teachings of history. They have failed to adopt the alternative procedure. Mathematical symbols are for the use of the mathematical community, and should therefore be adopted by that community. The success of a democracy calls for mass action.[200]

Cajori goes on to recommend an international committee on the model of the International Congress of Electricians, which, in Paris in 1881, propounded international standards for electrical units of measurement. The only way to establish uniformity, he argues, is by "breaking the infatuation of extreme individualism on a matter intrinsically communistic."[201]

In his preference for intentional planning over tradition, Cajori exhibits attitudes toward progress that we might call modernist. In language reminiscent of Ezra Pound, Cajori deplores the proliferation of conflicting notations: "O Goddess of Chaos, thou art trespassing upon one of the noblest of the sciences!"[202] He also laments the hesitancy of mathematicians to adopt new symbols, as in Descartes's notation for exponents, which was not universally adopted for more than fifty years. He attributes this hesitancy to the "force of habit"; that seventeenth-century mathematicians actively valued tradition and viewed "innovation" as a bad thing does not seem to occur to him.[203] "In such redundancy and obsoleteness," he concludes, "one sees the hand of the dead past gripping the present and guiding the future."[204] Cajori's historiography is devoid of the providential overtones we find in Peacock's early nineteenth-century history of arithmetic, instead emphasizing the practical need for mathematicians to take matters into their own hands and establish a common language. Standards institutions are supposed to liberate the present from the past by replacing haphazardly developing conventions with ones based on agreements that are both intentional and collective.

Toward standardization was, indeed, the way things would go in the twentieth century. Rhetoric much like Cajori's would recur in the work of the computer scientist Edsger Dijkstra, who called for his discipline to shed the baggage of the past. But the conversation about formal languages underwent another fundamental change just two years after the publication of Cajori's book. The advent of Kurt Gödel's incompleteness theorems in 1931, with the work of Alonzo Church and Alan Turing following close behind, dashed all hopes of a self-sufficient mathematical system of the sort Russell and Whitehead thought they had constructed. Gödel's results did not undo the distinction between formal and natural languages, but they did show that formal rules can never suffice to keep systems outside of history. This finding would ultimately have implications not just for pure mathematics but also for a technology of great social importance: the algorithm.

CHAPTER FIVE

Mass Produced
Software Components

We don't need paintings or
Doggerel written by mature poets when
The explosion is so precise, so fine.
Is there any point even in acknowledging
The existence of all that? Does it
Exist? Certainly the leisure to
Indulge stately pastimes doesn't,
Any more. Today has no margins, the event arrives
Flush with its edges, is of the same substance,
Indistinguishable. "Play" is something else;
It exists, in a society specifically
Organized as a demonstration of itself.

—JOHN ASHBERY, "Self-Portrait in a Convex Mirror"

THE PURIFICATION

Circa 1900, *algorithm* usually meant one of two things: a venerable set of calculating techniques commonly taught in elementary school or an abstract type of algebraic structure that few nonmathematicians would ever encounter. Circa 2000, algorithms were an economic engine, enacted in the servers of the recently founded Google and in a growing number of home computers. The idea of algorithm that solidified in the late twentieth century exists in a complex relation to the story I have told in this book. One thing that was new (since the plans of Charles Babbage and Ada Lovelace had gone unfulfilled) was the widespread use of programmable computers, which certainly changed the task of describing procedures. Precision takes on a new meaning when working with a machine that does exactly what it is told, no questions asked. The theory of algorithms also came to reflect the institutional pressures of the post–World War II period: the competing

163

interests of universities, businesses such as IBM, and military institutions, along with the persistent background noise of the Cold War.

But for all that changed in the twentieth century, the theory of algorithms shows the continuing influence of the subject–object divide that transformed symbolic algebra in the nineteenth. As I discussed in chapter 4, George Boole's work of the 1840s restricted logic to the formal aspects of reasoning, excluding the many other emotional and cultural factors that went into the formation of human thought. Whereas earlier mathematicians had attempted to ground their methods on clear conceptual definitions, algebraists in Boole's time reversed the terms: the rules of algebra, to paraphrase George Peacock, would determine the interpretation of the symbols and not the other way around. Much like algebra for Boole and Peacock, modern algorithms may be interpreted in multiple ways, constrained only by their formal qualities. A typical network analysis algorithm, for instance, says nothing about what the network represents; it could apply equally well to a road map or to a diagram of family relationships. As formalism detached pure mathematics from physical notions of magnitude and quantity, modern algorithmic thinking sets procedures apart from the concrete situations in which they are applied and the conceptual means by which people understand them—a division that contributed greatly to making algorithms the powerful and dangerous force they have become.

The emergence of this style of thinking is difficult to localize geographically. The first detailed study of the topic, A. A. Markov Jr.'s 1954 book *Theory of Algorithms*, came from the Soviet Union. Markov's book was translated into English, with US government sponsorship, by the Israel Program for Scientific Translations; the translation was published in 1961, although some Americans learned of Markov's work prior to this time.[1] This scientific exchange continued for decades. In 1979, Andrei Petrovich Ershov and Donald E. Knuth organized an international symposium that brought both Western and Soviet computer scientists on a "pilgrimage," as Knuth called it, to the presumed homeland of the mathematician and astronomer Muḥammad ibn Mūsā al-Khwārizmī—the Khorezm region in Uzbekistan, then part of the USSR—to discuss "Algorithms in Modern Mathematics and Its Applications."[2] Participants debated a range of questions, such as how to define *algorithm* and whether algorithmic methods were fundamentally different from algebraic ones. The thinking on the two sides of the Iron Curtain showed different affinities—Soviet computer scientists were more influenced by Markov's theory, which was probably not of decisive significance in the West—but the common ground was not insubstantial, and so the concept of algorithm cannot be said to be specific to Western countries or capitalist economies.

The theory of algorithms did, however, come to reflect tensions that existed in Cold War–era capitalist institutions. In the United States, some of the first programmable computers were developed at universities and government research facilities, although IBM had a stake from early on. As the role of private enterprise increased in the 1950s, computer researchers expressed resentment over the perceived intrusion of for-profit businesses into their field. Some major figures in early computer science, especially in Western Europe and to some extent in the United States, were academics more interested in pure mathematics than in practical applications. This devaluation of practicality fit the value hierarchies of academe. Much like algebraists in the seventeenth century, early programmers faced the accusation that their work was a mere practical skill, and they had to justify its place among the university disciplines. In industry, too, programmers faced pressure to develop their work into a form of engineering, which meant eschewing ad hoc solutions and informally shared "folklore" in favor of systematic methods based on scientific principles.

As I show in this chapter, the concept of *algorithm*, as the word came to be used in computer science, assuaged these disciplinary anxieties by creating a bridge between practical computation and the lofty theories of mathematics and formal logic. In response to the perception that their work lacked rigor, some early programmers sought to formalize computational procedures so that their correctness, efficiency, and elegance could be analyzed mathematically. This approach faced opposition from others (often working in military and commercial institutions rather than universities) who emphasized the importance of communication and teamwork—aspects of programming that formal approaches were ill suited to address. The discipline ultimately settled on a compromise. By the 1980s, programming languages had come to distinguish algorithmic logic from the "human factors," as they came to be called, by which teams could communicate and systems be made comprehensible to people. This distinction effectively bracketed the political questions that had attended symbolic methods prior to the rise of formalism, establishing, in terms similar to those employed by Boole, a disciplinary line between the putatively logical aspects of computer code and the social complexities of communication.

In assembling a theory of algorithms, computer scientists referred to a conversation about computation that had begun slightly before the computer era. In the 1930s, Alonzo Church, Emil Post, and Alan Turing had independently developed mathematical models of calculation (Church's term), problem solving (Post's), or computation (Turing's).[3] Church's theory defines mathematical functions using equations, which may then be transformed by rules that he calls an "algorithm." The models of Post and Turing are based on analyses of the mental processes undertaken by human computers. Although they responded to questions in pure mathematics, these theories contain elements that were later adopted by programming languages: Church contributed a new notation for functions, whereas Turing showed the possibility of virtualization, of a machine that could simulate another machine. Church and Turing provided, moreover, a framework for reasoning mathematically about the properties of algorithms.

The papers of Church, Post, and Turing did not, however, provide explicit definitions of *algorithm*; indeed, Turing's paper did not use the term at all.[4] One of the first modern explications of the term appeared in a 1943 paper by Church's student Stephen Kleene, who uses *algorithmic theory* to refer to methods for deciding whether propositions of certain types are true or false.[5] This sort of "theoretical conquest," as Kleene puts it, has been obtained for propositions of the form "*a is divisible by b*," since we have a reliable means of testing such statements for a given *a* and *b*.[6] Markov went into greater depth in his *Theory of Algorithms*, giving both informal and formal definitions. Informally, algorithms are procedures with three characteristics: definiteness, generality, and conclusiveness.[7] That is, they must be precisely defined, must be applicable to a range of inputs, and must reliably produce the desired output given a proper input. Markov's book presents a formal theory of algorithms as collections of substitution rules for transforming sequences of characters—he repeatedly uses the nonsensical example "papagiglema"—drawn from a given alphabet.

The definitions of Kleene and Markov indicate not just a further expansion of the concept of algorithm but also a shift in priorities. Through the late nineteenth century, "algorithms" were primarily valued either as practical methods or as ways of making new types of mathematical entity such as real and imaginary numbers accessible. Now, the central issue was *effectiveness*—that is, the property of producing the correct result through operations that are precisely specified and that can be performed in a finite amount of time. The interest in effective procedures stemmed in part from a continuation of the long-standing quest for a "directive method," as Boole had called it, for discovering truths, but it also reflected a newfound awareness of the limitations of such projects. In 1931, Kurt Gödel had shown that any axiomatic system powerful enough to represent arithmetic must either be inconsistent or leave some problems undecidable.[8] As Markov points out, there was no need for a rigorous theory of computational procedures until the *non*existence of procedures for certain problems became an issue.[9] After Gödel, what could be computed and what could not became a pressing question, and the nature of computational rules gained a newfound theoretical significance.

The interest in algorithms between the 1930s and the 1950s responded to philosophical issues about the foundations of mathematics, which initially had little direct bearing on practical computation. But the term *algorithm* also had a more workaday meaning at the time. Starting around 1900, Euclid's procedure for computing the greatest common divisor began, for the first time, to be called the *Euclidean algorithm*, and the word *algorithm* was also applied to the calculations for statistical methods such as least squares.[10] A pamphlet from 1905 presents a series of "algorithms" for use in computing the locations of railway lines (figure 5.1). This usage was atypical; such instructions were more often called *computing plans*.

ALGORITHM III.—Line Equations.

Formulas : $a = (y_n - y_k) / (x_n - x_k)$; $b = y_k - ax_k$.

		Line $P_2 P_3$	Line $P_1 V_1$
1........	x_k	1600	0
2........	y_k	600	0
3........	x_n	2600	3250
4........	y_n	1700	1610
5........	$y_n - y_k$	1100	1610
6........	$x_n - x_k$	1000	3250
7........	$\log (y_n - y_k)$	3.04139	3.20683
8........	$\log (x_n - x_k)$	3.00000	3.51188
9........	$\log a$	0.04139	9.69495
10........	a	1.1	0.4954
11........	$\log x_k$	3.20412	......
12........	$\log ax_k$	3.24551	......
13........	ax_k	1760	0
14........	b	−1160	0
15......(Eq.) $y = ax + b$		$y = 1.1 x - 1160$	$y = 0.4954 x + 0$
	A	B	C

Figure 5.1. A computational procedure for use in railroad surveying. From J. C. L. Fish, *Mathematics of the Paper Location of a Railroad* (New York: M. C. Clark, 1905), 16. This procedure computes the equation of a line based on the coordinates of two points, using base-ten logarithms for division and multiplication; one might compare it to chapter 1's figure 1.6 from more than two centuries before. The purpose of writing down the procedure step-by-step like this, as Fish explains, is to further the "economy of time and effort" by making the computations into "largely mechanical processes" (12).

Yet calling them algorithms would not have been a major conceptual leap. Even though mathematicians from G. W. Leibniz to Ernst Schröder had appropriated the word *algorithm* to refer to algebra-like systems, its original meaning was not forgotten. The medieval algorism had provided techniques for solving numerical problems, and extensions of these techniques were now of widespread importance in engineering, accounting, and especially the growing field of statistics, which by 1906 had large teams of human computers cranking away at Brunsviga adding machines.[11]

By the later twentieth century, these theoretical and practical senses of *algorithm* had collided. Algorithms, as they are studied in computer science, are at once abstractions that may be analyzed by logicomathematical means and instructions for solving problems. This conceptual hybrid had a genealogical connection to mathematics, but it was also turning into something new, and defining it became something of a cottage industry. In his multivolume work *The Art of Computer Programming*, the first installment of which was published in 1968, the computer scientist Donald E. Knuth gives a definition based on five criteria, which I will discuss in detail later: finiteness, definiteness, input, output, and effectiveness.[12] In their 1969 textbook, Alexandra I. Forsythe and her collaborators compare algorithms to recipes and dance choreographies, both of which break complex processes down into simple steps.[13] Three years later, Harold S. Stone offered another informal gloss: "any sequence of instructions that can be obeyed by a robot."[14] At the symposium in Uzbekistan, the Soviet mathematicians Vladimir Uspensky and Alexei Semenov presented a very different take on the issue, arguing that algorithms have a semantic character absent in other mathematical entities.[15] Unlike equations, that is, algorithms take the form of imperative statements, urging whoever will listen: do this! More important than definitions, these texts assembled a theory of algorithms that provided methods of reasoning about the processes that by the end of the century would become one of the dominant technologies of an age.

This chapter offers an account of how the idea of algorithm came to be enshrined in the discourse of computer programming. After considering relevant developments in the early twentieth century, I focus on the early programming language ALGOL, whose name is usually explained as *algorithmic language*. Starting around the time of ALGOL's introduction in 1958, programmers adopted the term *algorithm* to refer to something midway between folk knowledge and mathematical entity: computational procedures that could be reused by multiple people and for multiple purposes. At first, the word was treated as a synonym of *program*, but by the 1980s it had come to refer to something more ethereal, to abstract procedures considered apart from their implementations and conceptual interpretations. This abstraction depends on a disciplinary boundary: on the one side is the "hard logic" of algorithmic systems, on the other the "soft" concerns of making those systems work in social contexts. This division, I hope to show, is fundamental to the design of ALGOL and its many successors, including Python, C++ and numerous other programming languages in common use today, and it contributes to the ease with which algorithms may be transferred from one application context to another. Before I can explain this point fully, I must discuss some aspects of the logical and mathematical ideas on which early computer science drew.

PARADOX AND THE INDIVIDUAL WILL

The standard account of what happened to logical formalism goes something like this. Bertrand Russell and Alfred North Whitehead's monumental effort in *Principia* struck readers as unsatisfactory, and David Hilbert and Wilhelm Ackerman tried to do better.[16] Their project was supposed to produce a complete and consistent set of axioms along with a clearly defined procedure (one can now say with only slight anachronism, an algorithm) for deciding whether given mathematical statements were true. An early critique of Hilbert's project appeared in the work of L. E. J. Brouwer, who advocated for an antiformalist approach he called *intuitionism*, and whose work would be an influence on Turing; this school of thought rejects the law of excluded middle and thus does not accept the existence of any mathematical entity that cannot be constructed through some realizable procedure.[17] Gödel, Church, and Turing, within the span of just a few years, proved key elements of the Hilbertian program impossible. In papers published in 1936, Church and Turing proved independently that Hilbert's decision algorithm cannot exist: any formal system must leave some problems unsolvable. The advent of paradox produced, as a side effect, the first precise definitions of *computation*, which would later form a central part of theoretical computer science.

Caught up in this intellectual tumult was yet another utopian attempt to replace words with symbols. One of Gödel's mentors at the University of Vienna was Rudolf Carnap, whose 1928 book *The Logical Structure of the World*, usually referred to as the *Aufbau* (after its German title, *Der logische Aufbau der Welt*), outlines a "constructional system" in which complex concepts are reduced to simpler ones.[18] This project, which Carnap likens to Leibniz's universal characteristic, is supposed to unify the sciences by placing them all on the same conceptual basis.[19] While this project is not as significant to the technical theory of algorithms as the work of Church and Turing, it is worth considering first because it shows how the Leibnizian program was faring in the early twentieth century. Following the practice of Russell and Whitehead, Carnap defines symbols using equations:

114. *Similarity Between Qualities* (Sim)
   Construction: Sim $=_{Df}$ $\hat{a}\hat{\beta}\{a, \beta \in$ qual. $a \uparrow \beta \subset$ Ps$\}$
   Paraphrase: Two quality classes are called *similar* if each element of one of them is part similar to each element of the other.[20]

The equation is supposed to construct a concept of *similarity of qualities* by showing how statements using this concept can be transformed into ones that involve only previously defined concepts—namely, *quality* and *part similarity.* Carnap managed to create such formulae only for what he took to be very basic concepts; when he advances to physical reality and "cultural objects," he gives only a sketch.

At least in its aims, the *Aufbau* was as close as anyone came to reviving the Marquis de Condorcet's utopian project. Like the marquis, Carnap wanted to enact social leveling by expelling received dogmas in favor of scientific rationality. Carnap's early work consisted of "conceptual engineering"—an attempt to build a better system of concepts which he saw as a continuation of the Enlightenment project of the *Encyclopédie.*[21] The philosopher Richard Jeffrey later described Carnap as a voluntarist; for Carnap's purposes, language was a conceptual technology that ought to be judged by its utility for particular purposes and improved when possible.[22] Much like the marquis, Carnap regarded history as a story of progress toward both technological development and political equality, progress that would be embodied in a scientific universal language.

Yet Carnap did not reverse the narrowing of the scope of logic that occurred in the nineteenth century. Near the opening of the *Aufbau,* he notes that he will discuss some issues from the "traditional viewpoint," meaning the preexisting language in which people discussed science before his intervention.[23] He assigns no epistemological significance to this viewpoint; "merely traditional" concepts, he later states, are "objectively speaking, merely accidental (just like the historical boundaries of a state)."[24] Yet it is necessary to engage with these historical legacies to make the new language comprehensible to people. This concession to tradition differentiates Carnap from Condorcet, who wanted to avoid making any use whatsoever of prior conventions. Although Carnap was overtly hostile toward Immanuel Kant, the *Aufbau*'s negotiation with "tradition" betrays the influence of the fundamental distinction that transformed logic in the aftermath of Kant's *Critique of Pure Reason*: a distinction between the way an individual arrives at knowledge and the knowledge itself. "Even though the subjective origin of all knowledge lies in the contents of experiences and their connections," Carnap writes, one can still "advance to an intersubjective, objective world, which can be conceptually comprehended and which is identical for all observers."[25] Science, for Carnap, need not achieve total autonomy from received ways of thinking to be universal; instead, the subject–object divide provides a safe place to stow the influence of culture.

Gödel's proof forced Carnap to regroup, and his later efforts took a different tack from that of the *Aufbau*. He made a second major attempt at a universal language in his 1934 book *The Logical Syntax of Language*. The title change is significant: Carnap is no longer claiming to limn the structure of the world. Instead, he has moved further into voluntarism, maintaining "that we have in every respect complete liberty with regard to the forms of language; that both the forms of construction for sentences and the rules for transformation . . . may be chosen quite arbitrarily."[26] This "liberty" enables experimentation with forms whose meanings are not yet known: "let any postulates and rules of inference be chosen arbitrarily, then this choice, whatever it may be, will determine what meaning is to be assigned to the fundamental logical symbols."[27] Carnap presents this as a new idea, although it is not wholly dissimilar to the approach George Peacock had taken in algebra a century before. The difference is in the extent to which Carnap permitted the choice of rules to be arbitrary rather than governed by universal laws of correctness or truth. Whatever Ferdinand de Saussure may say about natural languages, formal languages work by rules that are wholly up to the individual.

In *Logical Syntax*, Carnap set the bounds of this voluntarism by means of a distinction between the *object language* and the *syntax language*.[28] The object language is the new symbolic system being constructed; the syntax language (also called the *metalanguage*) is the language used to describe it—that is, German or English, supplemented with a few special symbols. A hint of this division may be found in the Marquis de Condorcet's remark that it would sometimes be necessary to use words to explain the meanings of symbols; the marquis was also, one might further note, actually using French to describe his system. But Carnap was working within a subject–object division that was not in place for his precursor. Condorcet's universal language was supposed to be comprehensible to all because of its grounding in natural reason; how people came to understand the symbols was thus, for him, an epistemological matter just as much as the structure of the system itself. Carnap, by contrast, treated the syntax language more pragmatically: how the symbols are initially explained is a merely sub-jective matter irrelevant to the objectivity of the object language, which stems, instead, from its governance by formal rules.

Carnap's utopian plans would not be realized, although they do have an oblique relation to computer science. Along with contemporaries such as Alfred Tarski, Carnap contributed to an idea that would play a central role in the ALGOL project—the idea of a *formal language* in which, as Tarski put it, "the sense of every expression is uniquely determined by its form."[29] Carnap was, however, concerned less with computation than with pre-cisely representing empirical knowledge, and in this regard his systems are different in purpose from programming languages.[30] Turing's work, along with that of Post, Church, and Kleene, provided a bridge between the theory of formal languages and computation. Although Turing chafed at the rigidity of totalizing systems like Carnap's *Aufbau*, his model of com-putation inhabits much the same subject–object dualism by which Carnap divided the "objective" aspects of his system from historically situated hu-man thought. Just as Carnap made concessions to the "traditional view-point," Turing left room for intuition, thus enabling mechanical processes to work together with human thought.

Turing's model of computation is based on an imaginary device that later came to be known as the Turing Machine. This machine consists of a read–write head that moves left and right along a paper tape divided into discrete squares, reading, erasing, and writing symbols as it goes. At each point, the machine is in one of a finite number of states; its behavior is de-termined by a table that specifies what it will do in each state, based on which symbol it reads from its current square on the tape. The state of the machine is supposed to represent the "state of mind" of a person perform-ing a computation; Turing argues that a finite number of states is sufficient because if there were infinitely many possibilities, some of them would have to be "confused."[31] He describes the tape as "the analogue of paper," meaning that, like the notebook of a mathematician, it provides a place to record the solution along with "rough notes to 'assist the memory.'"[32] Us-ing this conceptual model, Turing proved two seemingly contrary things. One is the existence of a "universal computing machine"—a Turing Ma-chine that can simulate any other Turing Machine by reading an encod-ing of its table from the tape.[33] The other is the existence of problems that cannot be solved by any rules precise enough to be performed by such a machine.

Some popular accounts have incautiously described Turing's model as a "universal machine," suggesting that it could accomplish anything and everything, but for Turing the model functions more as a way of setting disciplinary lines—as a statement about what counts as computation and what does not. Turing's particular way of defining computation aligns it strongly with the original algorism. To begin with, the form of "writing" employed by Turing Machines involves sequences of discrete symbols, not any sort of pictorial representation—not (say) lines whose relative lengths represent numerical relations. In this regard, Turing's theory is aligned with the Hindi–Arabic numeral system and against the classical geometric tradition founded in the work of Euclid. Moreover, a Turing Machine is limited to a fixed, finite vocabulary of symbols—there is no possibility for the machine to coin, as it were, a new symbol in the course of its operation. Turing justifies this limitation on the grounds that the basic symbols may be combined into infinitely many "compound symbols," a practice that he links to numerals and alphabetical writing: "an Arabic numeral such as 17 or 999999999999999 is normally treated as a single symbol. Similarly in any European language words are treated as single symbols (Chinese, however, attempts to have an enumerable infinity of symbols)."[34] The more distinct symbols we introduce, as Turing has it, the greater the chance that two of them will look alike and thus become confused; expressing data through a small number of basic symbols thus maximizes clarity.

Turing's view of writing has a further idiosyncrasy. When the machine "reads" a square of the tape, all it has to do is identify the type of symbol written there and follow the corresponding rule: if the symbol is "A," do this, if the symbol is "D," do that. The shapes of the symbols only matter in their distinguishability from each other; their specific visual qualities—the fact that 9 resembles an inverted 6—are of no relevance, nor are any cultural traditions that may exist regarding their use, except insofar as those traditions are in some way reflected either in the data or in the design of the rules. As a result, one can easily set up a Turing Machine that violates established semantic practices—one can create an adding machine, say, that swaps the digits 1 and 2, so that 19 + 1 = 32. Such a system might be so counterintuitive as to be effectively useless, but such considerations have no place in Turing's theory of computation. In their use of symbols, Turing Machines are, to coin a word, *aconventional*: nothing, not even social agreement, places any restriction on which symbols may play which roles in the system.

This aconventionality must, however, have limits. Clearly, some aspects of computational practices (be they human or mechanical) are influenced by the conventional meanings of symbols: people do not usually choose to construct systems that swap 1 and 2, even if they theoretically could. It is here that the influence of the subject–object divide becomes apparent. With respect to the programming of his imaginary machine, Turing was, like the Carnap of *Logical Syntax*, a voluntarist: he viewed computational rules as objects of individual choice, to be selected based on how well they suit a particular purpose. Turing states this view most explicitly in his doctoral thesis, published in 1938. In the thesis, he considers a potential response to Gödel's incompleteness theorem. Given a logic system L, one constructs another logic system L_1 that adds some additional element; one then applies the same procedure to L_1 to produce L_2, and so forth. By exerting judgment about how far up this ladder to ascend, one can solve at least some problems that cannot be addressed within the original system L.

This argument responds to incompleteness, in essence, by backing away from formalism. As Turing explains, his purpose is "choosing a particular constructive system of logic for practical use"; as a result, formalizing his method "would be putting the cart before the horse."[35] This argument resonates with the work of Turing's mentor Ludwig Wittgenstein, who would later argue that rules can never be self-sufficient—one always would need more rules for how to follow the rules.[36] Turing even considers some rules that would be impossible to follow, such as a hypothetical Turing Machine with access to an "oracle" that can instantly answer certain (possibly uncomputable) questions.[37] Which computational systems are usable and useful cannot be settled by applying more formal rules, which would only increase the level of abstraction without resolving anything. (It is easy to prove by contradiction that if the decidability of a problem is itself undecidable, we can never find that out.) Like late-period Wittgenstein, Turing resolves this regress by abandoning the search for absolute foundations and instead situating systems within their social contexts. Ultimately, for Turing, computational rules answer to the human purposes of the people who use them, which they either serve or do not.

While some later commentators have presumed that Turing saw the human mind as a sort of biological Turing Machine, the thesis suggests that, at least in 1938, he saw it more as a somewhat unreliable oracle machine—as a Turing Machine supplemented by an additional element that cannot be reduced to mechanical rules.[38] Determining the truth or falsity of mathematical propositions, he writes, involves "a combination of two faculties, which we may call *intuition* and *ingenuity*."[39] Intuition involves "making spontaneous judgments which are not the result of conscious trains of reasoning," whereas ingenuity involves "aiding the intuition through suitable arrangements of propositions, and perhaps geometrical figures or drawings."[40] The purpose of formal logic, according to Turing, is to remove the "arbitrariness" introduced by the fact that intuition varies from person to person.[41] In "pre-Gödel times," he continues, it was believed that formalism could be taken so far as to eliminate intuition altogether, but this is now seen to be impossible.[42] His alternative tactic is to embrace the role of intuition and develop practices in which "not all the steps in a proof are mechanical, some being intuitive."[43]

This aspect of Turing's work would have an underrated influence on the disciplinary practice of computer science, which, as we will see, grants an important role to something called "intuition" or "common sense" in justifying algorithms. Turing's model of computation was, however, only one of several to appear around the same time, and there was more to the emerging theory of computation than his generalization of the Hindi–Arabic algorism. The contributions of Church and his student Kleene were based, instead, on the idea of a recursive function—that is, a function defined in terms of itself in a way that is ultimately resolvable into a single value. A simple example would be the following definition of a factorial:

$$factorial(x) = x \cdot factorial(x - 1) \qquad x > 1$$
$$factorial(x) = 1 \qquad\qquad\qquad\quad x = 1$$

Church's 1936 paper introduces a notation called the λ-calculus (or lambda calculus), which provides a way of constructing functions on the spot within a formula: instead of defining f with the separate equation $f(x) = M$, one could use $\lambda x[M]$ to refer to this function without giving it a name. Variants of this notation are now used to construct "anonymous" functions in a number of programming languages, including Python. Turing found out about Church's work while his 1936 article was under review and added an appendix proving that the two models are equivalent.[44] In spite of this formal equivalence, the models are based on different types of symbolic method. If Turing follows the original algorism by describing step-by-step instructions for arranging digits, Church's calculus has more kinship with algebra, involving the transformation of symbolic equations that include letters representing unspecified values.

It is notable that Church, not Turing, is the one who used the word *algorithm* in his 1936 paper. Church's use of this term is not quite in line with the modern sense, and so it is worth examining. The λ notation is often taken as a means of formalizing algorithms by representing them as recursive functions, but Church does not present his calculus this way. "It is clear," he writes, "that for any recursive function of positive integers there exists an algorithm using which any required particular value of the function can be effectively calculated"; specifically, the algorithm consists of performing a reduction procedure until the desired result is found.[45] If we think of recursive function definitions as computer programs, then this "algorithm" is not the program itself—it is a method for running programs. In this regard, Church's work exhibits the continuing association of the word *algorithm* with the manipulation of formulae. While Church was undoubtedly aware of the original sense of the word, he was also deeply engaged with the work of Schröder, for whom "algorithms" were systems of algebraic relations.[46] Church does not use the word in Schröder's idiosyncratic sense, but his usage can still be placed in the algebraic lineage going back to Leibniz.

That one must think of computing machines in terms of either of these models is not self-evident. Babbage had already imagined a programmable computer a century before Church and Turing, and the designers of some early computers, such as Konrad Zuse's Z1 (1936–38) and the IBM Mark I (1939–43), were initially unaware of their work.[47] John Von Neumann, often held up as the designer of the standard computer architecture, knew Turing personally and was deeply familiar with his paper on the decision problem, but it is not clear that Turing's imaginary machines had any strong influence on his plan.[48] Yet Church and Turing did eventually become common reference points for the discipline of computer science. The most important thing they provided was less a paradigm for the design of actual machines than a theoretical framework for reasoning mathematically about what came to be called "algorithms."

This new discipline would not be firmly established until well after Turing's death in 1954. Historians of computation write of a "software crisis" brought on, starting in the 1960s, by a perception that programs were growing too complex to be trusted. One triggering event was the failure of the Mariner 1 spacecraft in 1962, apparently due to a coding mistake.[49] In 1968 and '69, NATO held conferences on "software engineering" at which participants proposed ways of preventing such errors and making large systems more manageable.[50] It was amid this space-age ferment that the word *algorithm* became a major disciplinary term. The idea of algorithm, which in the time of Church and Turing had been associated more with pure mathematics than with computing machines, promised to transform programming from a practical skill into a science by subjecting procedures to rigorous mathematical analysis. This transformation was closely tied to the development of one of the most influential systems ever to claim the mantle of formalism: the programming language.

AN ALGORITHMIC LANGUAGE

The first programmable computers were controlled through punch cards, paper tape, or direct manipulation of electrical connections. In modern computers, most programming is done in abstract languages such as C++, JavaScript, and Python that are designed not just for directing machines but also for presenting computational processes in a human-readable form. It would be a mistake to frame programming languages too uncritically as a successor of the symbolic methods I have discussed in this book; computer code, as Mark Priestley has emphasized, serves a different purpose from the notation used in mathematical proofs.[51] One difference is the assignment operators that, in some programming languages, instruct the computer to change the value of a variable—a practice that is absent in modern algebra.[52] Warren Sack has argued for an alternative genealogy in which programming languages descend from the "work languages" by which Enlightenment thinkers codified knowledge of the mechanical arts.[53] But algebraic symbols did play a major role in the design of programming languages, and computer code raises some of the same semiotic questions that arose in universal algebra schemes. Programming languages must combine technical rigor with communicational clarity, and the particular way they came to divvy up these functions draws on the subject–object dualism that first emerged in the time of Boole.

While programming languages always combined computation and communication, the boundary between the two was initially malleable. Very early programmers represented computational processes through notations such as flowcharts, which originated in the work of Herman H. Goldstine and John Von Neumann in the late 1940s.[54] Yet these notations could not be fed into a computer without first being translated into numerical codes. In her 1952 article "The Education of a Computer," Grace Hopper argues that the need for this translation has become a tedious distraction; rather than focusing on the problem at hand, the mathematician "has become a programmer."[55] Hopper thus proposes the creation of what we now call a *compiler*—a program that automatically translates a human-readable sequence of instructions into a machine-executable form. With the introduction of this translator, she writes, "the programmer may return to being a mathematician."[56] As Priestley, Nofre, and Alberts have shown, the terminology was different in this early phase of electronic computation.[57] Hopper only uses the word *program* when referring to machine code; her term for the human-readable form is "computer information." The "programmer," in her language, is the agent—typically, at this point, a woman—whose task is to translate mathematics into machine code. Once this process is automated, the machine, not the human, is the programmer.[58]

Hopper's article marked an early step toward enabling people to control machines through something resembling language. She concludes that the computer she had been working on, the UNIVAC, currently has "a well grounded mathematical education fully equivalent to that of a college sophomore" and hopes that it will soon advance to the graduate level.[59] The capacities of computers would indeed expand, although not always in the direction she predicted. Starting in 1955, Hopper directed the development of two programming languages, MATH-MATIC and FLOW-MATIC; the latter evolved into COBOL, which stood for "*CO*mmon *Bus*iness *O*riented *L*anguage."[60] In spite of her focus on mathematics in "The Education of a Computer," FLOW-MATIC and COBOL code contains little mathematical notation, instead taking the form of English imperative sentences such as "`ADD a TO b`." As the name COBOL indicates, the language was primarily intended for business uses, and its design reflected the communicational conventions of bureaucrats at institutions such as the Pentagon.

As the example of COBOL shows, programming languages have no necessary connection to mathematical symbols. Code could just as well draw on the English language. Mathematical notation did, however, become a major source for programming languages. In 1954, an IBM group led by John Backus drafted the first specification of the programming language FORTRAN, which stood for "The IBM Mathematical FORmula TRANslating system"; the language became commercially available in 1957.[61] FORTRAN arose in part in response to the question: "Can a machine translate a sufficiently rich mathematical language into a sufficiently economical program at a sufficiently low cost to make the whole affair feasible?"[62] The concern with economy is crucial. Some computer operators were skeptical of "automatic programming" systems because, at least with very early compilers, automatically generated machine code was often slower both to prepare and to execute than handwritten code. Despite these concerns, compiled languages such as COBOL and FORTRAN had major advantages in regard to human intelligibility that encouraged their widespread adoption. What is more, a standardized programming language would (at least in theory) enable the same instructions to be used with different models of computer, making code easier to share and reuse.

This standardization became a primary focus of another, somewhat different programming language: ALGOL. Like LISP (*LIS*t *P*rocessor), which appeared around the same time, ALGOL was developed largely by academics rather than employees of business or military institutions, and it reflected a desire to apply the rigor of mathematics and formal logic to computer programming. The first version, ALGOL 58, was designed by a committee organized at the 1957 joint meeting of the US-based Association for Computing Machinery (ACM) and the German Society of Applied Mathematics and Mechanics (GAMM) in Zürich.[63] ALGOL went through two later revisions known as ALGOL 60 and ALGOL 68. The name ALGOL is usually explained as *algorithmic language*, although the original name was International Algebraic Language; a 1958 proposal described it as a "universal language for the description of computing processes" that followed the conventions of "mathematical formula notation."[64] ALGOL was supposed to serve not just as a compilable language like FORTRAN but also as a notation for publishing computational procedures in the journal *Communications of the ACM*, which launched in January 1958. While never as widely adopted as COBOL and FORTRAN, ALGOL influenced numerous programming languages that remain popular, including C, C++, and their many derivatives.

The adoption of the phrase *algorithmic language* (in German, *algorithmische Sprache*) was probably due to influence of the ALGOL committee member Heinz Rutishauser. Rutishauser had previously worked on the ERMETH, a computer developed by the Swiss Federal Institute of Technology starting in the late 1940s. In a 1955 journal article, he described a form of "algorithmic writing" by which users of the ERMETH could specify the computations they wanted done.[65] Drawing on the work of Konrad Zuse, who had developed a sophisticated notation for computing plans called *Plankalkül*, Rutishauser's system contained a number of ideas that would become elements of ALGOL, such as **for** loops and **if–then** statements.[66] That such languages should be called "algorithmic" was controversial; Backus preferred the term "algebraic."[67] Yet the phrase *algorithmic language* eventually came to refer to a certain style of computer language— one based on branching, looping sequences of instructions, written (in contrast to early versions of COBOL) in something resembling mathematical notation.

ALGOL's official adoption by *Communications of the ACM* was, I would venture, a major factor in securing the widespread adoption of *algorithm* as a general term for computational procedures.[68] Early ACM publications included numerous procedures, but they are referred to by various terms, including *procedure*, *method*, or *subroutine*, and often described in a combination of prose and symbolic notation. In the February 1960 issue of *CACM*, the journal added a new section called "Algorithms," initially edited by Joseph Henry Wegstein of the US National Bureau of Standards. The section solicited readers to share programs and procedures as well as "certifications" of their effectiveness and critical feedback; to be eligible for inclusion, procedures had to be written in ALGOL and accompanied by an informal comment indicating what they do and how to use them. In a precursor of open-source licenses, the section included a copyright notice permitting free reuse. Most of the procedures addressed numerical problems that would have interested mathematicians, although not all of them specifically dealt with numbers. It was in this venue that the famous Quicksort algorithm, which would later become a textbook mainstay, was first published.[69]

The goal of this section, as an editorial note explained, was to create a "library" of standard procedures that programmers could draw on.[70] This accumulation of standard procedures—an underappreciated aspect of the concept of algorithm—would become a central principle of software engineering. At the 1968 NATO conference, Malcolm Douglas McIlroy called for "mass produced software components," which would make development more systematic by providing a shared collection of modules that could be reused in various contexts.[71] Although the *CACM* collection did not fully live up to McIlroy's standards, it fit this general way of thinking by encouraging programmers to design procedures that could be repurposed beyond their initial applications. It is notable that, while they were published in ALGOL, the "algorithms" were not only used with ALGOL-based systems—readers reported testing the procedures in other languages, such as FORTRAN.[72] Creating this open forum encouraged people to view algorithms as a common stock of general solutions that could be reused in various programming languages, with various hardware, and for various purposes. It also encouraged reflection about what made for a good algorithm, since some editorial standards were needed; the section thus spurred on the development of a theory of algorithms.

The rise of ALGOL coincided with the increasing intellectual status of programming. Early in the computer era, hardware was viewed as a subject fit for engineers, whereas programming was denigrated as tedious and often coded as a feminine, secretarial practice.[73] Now, academics were turning serious attention to it, and they had to justify its inclusion among the university disciplines. In a later article, Niklaus Wirth recounts the atmosphere of the University of California, Berkeley in 1963, where he worked on ALGOL. Wirth's research group, he writes, faced resistance because "word passed that some of these people did neither know Ohm's law nor Maxwell's equations"—that is, they did not understand the electrical principles on which the machines were designed.[74] Using a rhetoric that Wendy Hui Kyong Chun has discussed in detail, Wirth describes this situation as involving a kind of wizardry.[75] "Looking at it from the distance, that compiler was a horror; the excitement came precisely from having insight into a machinery that nobody understood fully," Wirth recalled. "There was the distinct air of sorcery."[76] This situation is the converse of the one algebraists faced in the mid-nineteenth century, when critics worried that, with the adoption of symbolic methods, the science was becoming a mechanical practice rather than an intellectual one. Now, the physical basis of computation threatened to vanish in a cloud of abstraction.

In spite of this initial pushback, computer scientists eventually did overcome their reservations about abstracting out machinery; "the essence of programming," Wirth would later declare, "is abstraction."[77] The trend among ALGOL programmers at Berkeley, Wirth recalls, "was to discover the fundamental concepts of algorithms, to extract them from their various incarnations in different language features, and to present them in a pure, distilled form, free from arbitrary and restrictive rules of applicability."[78] ALGOL was sometimes described as a problem-oriented language as opposed to a computer-oriented language.[79] Its purpose was to enable computer operators to turn their attention from circuitry to algorithms—that is, general procedures that do not depend on the nature of any particular machine. An example of this trend, according to Wirth, was Adriaan van Wijngaarden's description of the hypothetical Generalized ALGOL, which aimed (very much in the spirit of Charles Babbage) at "simplicity in generality."[80]

Generalized ALGOL was never implemented, but the three main versions of ALGOL enact a form of generalization that would make Babbage proud. Its basic elements will be familiar to readers who know common modern programming languages, although there are some subtle differences.[81] I will focus here on ALGOL 60 because it was the most popular version. ALGOL 60 programs usually employ variables, which are used to store data that (unlike algebraic variables) can change over time. The variables have types that must be declared, as in "**integer** A." The values can be set using the assignment operator $:=$, as in "$A := 5$." One can employ algebraic expressions on the right side of an assignment, although not on the left. The language also includes, among other things, **if–then** statements that perform operations only on certain conditions and **for** loops, which iterate through sequences of numbers. For instance, "**for** $i := 2$ **step** 1 **until** A **do**" will sequentially set i to each value from 2 to A (which we previously set to 5) by increments of 1, and perform the operation following **do** for each. We might assemble the above into the following complete program:

```
begin
    integer A, B, i;
    A := 5;
    B := 1;
    for i := 2 step 1 until A do
        B := B × i;
    print B;
end
```

This will compute the factorial of 5—that is, the product of all the integers from 1 to 5—and then print its value to the screen.[82]

While this program can only compute the factorial of one specific number (e.g., 5), ALGOL code is usually organized into "procedures" that are intended to be applicable as widely as possible. Here, for instance, is a procedure from the "Revised Report on the Algorithmic Language Algol 60":[83]

> **procedure** *Absmax* (*a*) size: (*n*,*m*) Result: (*y*) Subscripts: (*i*,*k*) ;
> **comment** *The absolute greatest element of the matrix a, of size n by m is*
>   *transferred to y, and the subscripts of this element to i and k* ;
> **array** *a* ; **integer** *n*, *m*, *i*, *k* ; **real** *y*;
> **begin integer** *p*, *q* ;
> *y* := 0 ;
> **for** *p* := 1 **step** 1 **until** *n* **do for** *q* := 1 **step** 1 **until** *m* **do**
> **if abs**(*a*[*p*, *q*]) > *y* **then begin** *y* := *abs*(*a*[*p*, *q*]) ; *i* := *p*; *k* := *q* **end end**
>   *Absmax*

This procedure is a variant of the "high water mark" algorithm, the procedure of finding the largest number in a list by going through the numbers one by one and keeping a record of the largest one yet. What makes the procedure applicable to a range of inputs and thus (by Markov's informal definition) an algorithm is a certain kind of abstraction: the matrix to be processed is not specified but rather referred to only as *a*. This practice is the culmination of the analytical method François Viète introduced in 1591—the use of arbitrary letters to refer to values that are left unspecified so as to present a result in its full generality.

In explaining this system, the creators of ALGOL drew on the idea of formal language that had solidified in the early twentieth century. One of ALGOL's most influential innovations was the formal notation used to describe its syntax, which later came to be known as the *Backus–Naur form*. For instance, the **if–then** statement is defined as follows:[84]

> ⟨if clause⟩ ::= **if** ⟨Boolean expression⟩ **then**
> ⟨unconditional statement⟩ ::= ⟨basic statement⟩ | ⟨compound statement⟩
> | ⟨block⟩
> ⟨if statement⟩ ::= ⟨if clause⟩ ⟨unconditional statement⟩

The first line indicates how an "if clause" may be constructed by placing a Boolean expression between an **if** and a **then**; the second provides a list of three options (separated by "|") for constructing an "unconditional statement"; the third combines these two into a definition of "if statement." Indicating an emphasis on precision, the section on syntax in the *Revised Report* takes its epigraph from Wittgenstein's *Tractatus Logio-Philosophicus*: "What can be said at all can be said clearly, and whereof one cannot speak thereof one must be silent."[85] Drawing on the work of logicians such as Tarski, Backus also sketched out a formal semantics (never fully completed) that would specify the computational "meanings" of ALGOL programs— that is, what they tell the machine to do.[86]

This approach to formalization has a peculiarity worth emphasizing. Like FORTRAN and COBOL, ALGOL allows for the use of words in identifiers—the names of variables and other program elements, such as the procedure name *Absmax* in the foregoing example. Yet the meanings of these words play no role in the computational semantics. The authors of the *Revised Report* list five example identifiers that seem calculated to emphasize their total arbitrariness: "q," "Soup," "V17a," "a34kTMNs," and "MARILYN."[87] "Identifiers," they explain, "have no inherent meaning, but serve for the identification of simple variables, arrays, labels, switches, and procedures. They may be chosen freely," provided that they do not interfere with built-in features.[88] The sample code mostly follows the conventions of algebraic notation by using single letters for variable names, but it also annotates those letters with English words such as "size" and "Result," for obvious reasons.[89] The code, then, contains words, but its computational "meaning" is wholly separate from whatever those words might convey to people.

On account of its standardizing ambitions, ALGOL was sometimes described as an "Esperanto" of computing, a universal language for describing algorithms.[90] A closer comparison, however, would be John Wilkins's real character—the system of written symbols that were supposed to express universal ideas that were shared by everyone and that could therefore be read aloud in any spoken language. ALGOL 58 and 60 existed in three distinct "representations": a standard "reference" form, a plaintext form designed for entry into machines, and a publication form that allows for niceties like Greek letters. The foregoing example is in the reference form. As the *Revised Report* explains, the three forms differ "only in the choice of symbols," whereas "structure and content" are the same.[91] Identifiers, likewise, could be swapped out at pleasure without affecting this structure. Like the real character, then, ALGOL seeks unity in multiplicity, representing ideas that are common to all regardless of the particular words used to express them.

The limitations of this approach to universality became more apparent in later discussions of ALGOL. The final version of the language,

ALGOL 68, was intended to exist in localized variants based on different languages—one could write **if** in the English version and **если** (*jésli*) in the Russian version.[92] In some cases, English ALGOL ended up being used in contexts where other languages were spoken; one thus finds in Van Wijngaarden's ALGOL 68 code a mixture of English and Dutch. Consider these lines from a puzzle-solving program, which construct an array for keeping track of positions that are currently being used (*gebruikt*), then records the current time (*tijd*) using the system clock:[93]

```
[1 : 12] 'BOOL' GEBRUIKT;

'FOR' K 'TO' 12 'DO' GEBRUIKT[K]:= 'FALSE' 'OD';

'REAL' TIJD:= CLOCK;
```

This mottled history of internationalization shows the extent to which ALGOL's status as a universal language depends on the idea that algorithms exist independently of the particular words or symbols used to express them. Identifiers like GEBRUIKT may make no difference to the semantics of the program, but the words are still there, and it is not hard to see why Van Wijngaarden preferred typing them in his native tongue.

The idea of creating one "algorithmic language" to rule them all would not come to fruition. Programming languages, dialects, and variants proliferated. Later languages mostly continued the use of words (often in English), although there were some exceptions. In 1962, Kenneth E. Iverson published APL ("A Programming Language"), which eschews words in favor of symbols such as ← and ⋀ ; his purpose, as he later explained, was to combine the cognitive advantages that thinkers such as Boole and Florian Cajori had found in mathematical notation with the "universality" of programming languages.[94] But ALGOL's elevation of formal semantics over the details of presentation had a lasting influence. Some might favor the look of ALGOL's a + b over LISP's (SUM, A, B), but the difference is merely what the ALGOL programmer Peter Landin called "syntactic sugar"— linguistic sweetener to help the logical medicine go down.[95] The result was a disconnect between an epistemologically privileged formal view in which words and symbols were wholly arbitrary and a practical view in which they played undeniably important cognitive and communicational functions. In working out how to manage this disconnect, the emerging discipline of computer science was forced to grapple, once again, with the questions about symbols and language that had bothered mathematicians for centuries. This time, the idea of algorithm became pivotal.

CULTURE AND THE SWITCHING CIRCUIT

What, exactly, is the algorithmic "content" that is supposed to be common to all the variants of ALGOL? The *Revised Report* states that ALGOL 60 is "suitable for expressing a large class of numerical processes in a form sufficiently concise for direct automatic translation into the language of programmed automatic computers."[96] Yet the designers of programming languages had, from the very beginning, ambitions beyond mere number crunching. The founder of Stanford's computer science program, George Forsythe, stressed in a lecture on "Educational Implications of the Computer Revolution" that computers could not be contained within the field of mathematics: "Machine-held strings of binary digits can simulate a great many *kinds* of things, of which numbers are just one kind."[97] The ones and zeroes may, he continues, "simulate automobiles on a freeway, chess pieces, electrons in a box, musical notes, Russian words, patterns on a paper, human cells, colors, electrical circuits, and so on."[98] But a field cannot do everything; there had to be something distinguishing it from others. As Chad Wellmon has pointed out, modern disciplines are defined less by their objects of study than by the communities in which scholars participate.[99] The disciplinary community of computer science (CS) was formed around a set of practices that, while they could apparently be applied to any object of study, were distinct from those of other sciences.

A central element of this new discipline (although how central depended on whom one asked) was a theory of algorithms. In 1963, Juris Hartmanis and Richard E. Stearns—then General Electric employees, although both later became academics—laid out the rudiments of computational complexity theory.[100] They present this theory as an extension of Turing's work that considers not just whether functions are computable but also how hard it is to compute them. They measure this difficulty by how many operations must be performed by a Turing Machine for input of a given size—an abstract conception of time that does not depend on the specifics of computing hardware. Suppose, for instance, that we are sorting a list with n items. The most obvious sorting algorithms require, on average, a number of operations proportional to n^2, and they can thus become very slow for long lists. With some ingenuity, we can devise algorithms that follow the slower-growing curve $n \log n$, and that can thus handle much larger amounts of data in a reasonable amount of time (figure 5.2). The same thinking can also be applied to memory usage. Complexity theory developed into a classification system that is used to judge both the efficiency of algorithms and the difficulty of problems. It also

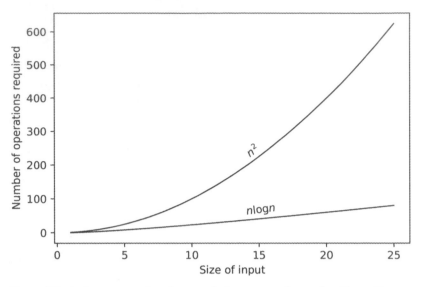

Figure 5.2. A chart comparing the complexity curves of two algorithms. The performance is similar when the input is very small, but the time requirement for the n^2 algorithm grows rapidly as the input size increases. The $n \log n$ algorithm scales much better.

opened a range of difficult questions about the relationships between complexity classes, thus setting a long-standing research agenda for theoretical computer science.

The work of Church and Turing provided a basis for thinking about such issues and a rationale for treating algorithms abstractly. Turing's proof that his imaginary machines were equivalent to Church's λ-calculus was only the first of a number of equivalence proofs showing the interchangeability of various forms of computation. Kleene later articulated what came to be known as the Church–Turing Thesis, the unprovable but widely supported proposition that all effective methods of computation are equivalent to these two models (and thus to each other).[101] This thesis (although attended with philosophical difficulties) provided a rationale for viewing algorithms as capable of preserving their fundamental identity regardless of what formal system is used to express them and what physical process is used to enact them.

This mutual convertibility applies to programming languages, at least provided that they are powerful enough to simulate a Turing Machine. In a well-known 1976 Massachusetts Institute of Technology (MIT) memo titled "Lambda: The Ultimate Imperative," Guy Lewis Steele Jr. and Gerald Jay Sussman show how a number of common ALGOL techniques can be translated into Scheme, a LISP-like programming language inspired by Church's model of computation. These translations, they tell us, "capture the essence of the semantics of the construct."[102] A premise of this project is that algorithms can retain their fundamental "essence" even when they are rewritten in disparate styles (although the authors clearly have an opinion about which style is best). The memo has an epigraph attributed (probably apocryphally) to Abraham Lincoln, suggesting that the choice between these styles is a matter of personal preference: "People who like this sort of thing will find this is the sort of thing they like."[103]

But Turing equivalence only extends to the range of mathematical functions a language can theoretically compute. In other ways, programming languages are not all the same. ALGOL is associated with the paradigm of structured programming, whereas LISP encourages a functional style. Other languages such as Simula and Prolog tread entirely different paths. These paradigms were integrated into institutional culture: ALGOL was used at many European universities, whereas MIT was LISP territory and FORTRAN and COBOL ruled the private sector. The proliferation of languages and coding standards sparked what are sometimes described as "holy wars"—bitter exchanges between partisans of particular approaches, sometimes over issues as trivial as how many spaces should make up a tab.[104] Although these disputes were widely mocked, the depth of the rancor shows that matters affiliated with subjectivity—aesthetics, clarity, intuitiveness of organization—remained crucial to the practice of programming, even as they were excluded from formal semantics. Disciplines cannot eliminate the human altogether. What they can do is place certain questions outside of bounds, leaving them to individual choice, institutional culture, or pedagogical expediency rather than subjecting them to rigorous epistemic norms.

In setting up this boundary, computer scientists underwent yet another iteration of the ages-old debate about how much continuity should exist between symbols and ordinary language. In 1965, Jean E. Sammet (who later became the first woman president of the ACM) gave a talk entitled "The Use of English as a Programming Language," in which she called for a natural-language interface that would make computing more efficient and accessible by enabling people to "communicate directly with the computer without having to learn some specialized intermediate language."[105] The idea was that programming languages would eventually incorporate the whole of English grammar, thus completing the process of automation that Hopper had begun with her compiler—letting the machine take over the tedious parts of programming so that people can focus on the problems at hand. By advocating for natural language coding, Sammet continued Hopper's centering of human purposes and institutions, which encouraged a view of programming languages primarily focused on communication.

This view sat uneasily with the formal approach Backus had taken in the ALGOL project, and it faced pushback from those who embraced the aconventionality of Turing-style computation. The transcript of Sammet's presentation records the following, somewhat tongue-in-cheek, response by the Finnish computer scientist Reino Kurki-Suonio:

> The main reason I find it convenient to use English words and phrases for strictly defined programming purposes is that English is not my native language (laughter). English provides a huge store of symbols to which any meaning can be assigned (more laughter) without referring to my former knowledge.[106]

This comment exemplifies a dismissive attitude toward "human factors" that would become common among academic computer scientists. At stake was the potential to establish voluntary control over the semantics of computer code. Viewed through a formalist lens, programming languages promised the liberty to assign values to symbols at will in the manner of the algebraic *let*, thus establishing independence from the linguistic practices of other people.

This individualistic attitude toward meaning was consanguineous with a commitment to logicomathematical methods for analyzing systems. In a 1962 article, John McCarthy had proposed a "mathematical science of computation" that would make it possible "to prove that given procedures solve given problems."[107] The idea was to provide a more rigorous method of software development than the trial-and-error processes often used in practice. Testing programs by running them on particular inputs and checking the output was unsatisfactory since, quite commonly, procedures will work in some cases and fail in others. Program proofs would enable code to be evaluated not by empirical tests but through a mathematical analysis of the algorithm itself, thus ensuring reliability in all cases. This goal required algorithms to be expressed in a formal notation, and so it conflicted with the natural-language approach advocated by Sammet.

Although mechanically parsing English grammar proved harder than Sammet initially thought, formalist approaches to programming raised difficulties of their own. People, after all, must be able to understand code, and for this purpose it was hard to resist the siren call of language. In 1988, the Dutch computer scientist Edsger W. Dijkstra, then in an endowed chair at the University of Texas at Austin, expressed his views on the matter in the provocatively titled "On the Cruelty of Really Teaching Computing

Science."[108] A common approach to understanding new things, he writes, is that "by means of metaphors and analogies we try to link the new to the old, the novel to the familiar."[109] Computers, he argues, present a situation so radically novel that this approach breaks down; "though we may glorify it with the name 'common sense,' our past experience is no longer relevant, the analogies become too shallow, and the metaphors become more misleading than illuminating."[110] Dijkstra continues a long tradition by linking these received ways of thinking to language. "Coming to grips with a radical novelty," he continues, "amounts to creating and learning a new foreign language that can not be translated into one's mother tongue."[111]

In effect, creating this new language would mean eliminating metaphors, examples, and other traces of the real world from pedagogy. As an example, Dijkstra discusses arithmetic, in regard to which he sounds much like the presumptuous older child in Sarah Porter's 1835 *Conversations on Arithmetic*. "Instead of teaching 2 + 3 = 5," Dijkstra writes, "the hideous arithmetic operator 'plus' is carefully disguised by calling it 'and', and the little kids are given lots of familiar examples first, with clearly visible objects such as apples and pears."[112] He calls this a "silly tradition"; it is better, according to him, to teach arithmetic using the "hideous" language of symbols.[113] He suggests a similar austerity in teaching programming. "In order to train the novice programmer in the manipulation of uninterpreted formulae," he suggests, professors should teach logic "more as boolean algebra, familiarizing the student with all algebraic properties of the logical connectives. To further sever the links to intuition, we rename the values {true, false} of the boolean domain as {black, white}."[114] The point of switching from logical terms to colors is to remove any hint that the symbols are anything other than wholly arbitrary, and thus to protect students from being tempted into thinking in terms of meaning.

Dijkstra's hopes for this new symbolic language were (as such hopes always have been) lofty. Computing, he predicts, will one day "transcend its parent disciplines, mathematics and logic, by effectively realizing a significant part of Leibniz's Dream of providing symbolic calculation as an alternative to human reasoning."[115] This transcendence, according to Dijkstra, has a number of enemies, among them business and military communities too enamored with the idea that computers can make their lives easier, as well as "all soft sciences for which computing now acts as some sort of interdisciplinary haven."[116] He also differentiates his proposal from the field of artificial intelligence (AI), which in 1988 was at something of a nadir: "The effort of using machines to mimic the human mind has always struck me as rather silly: I'd rather use them to mimic something better."[117] To realize Leibniz's Dream (which Dijkstra imperiously capitalizes), students must learn to think of programs less as descriptions of things machines do than as symbolic formulae that must be reasoned about abstractly. This rigorous practice would require a painful exorcism of the meanings, meta-phors, and analogies that fooled people into thinking code had anything to do with received notions.

By 1988, when Dijkstra was writing, this austere view of code had al-ready lost a great deal of ground to what was, by then, coming to be called "user friendliness."[118] Taking much the opposite approach to language compared to Dijkstra, Knuth developed the idea of *literate programming*, in which code contains a prose explanation of itself.[119] Grace Hopper con-tinued to advocate for the importance of considering the social contexts in which programming languages were deployed.[120] Sandy Payette has noted an undercurrent of sexism in Dijkstra's description of Hopper's approach as "soft"; a concern with people was treated as feminine and thus extrin-sic to science.[121] While such "soft" matters continued to be devalued in the academy, they were harder to ignore in the private sector. Especially after the founding of Apple Computer in the late 1970s, the computer indus-try made room for creativity and aesthetics, which were valued as a way of pleasing customers and opening new markets by encouraging people to incorporate computers into more aspects of their lives. In commercial contexts, the "cruelty" Dijkstra advocated would not do—making software that sold meant constructing systems that worked for people, for which purpose a radical break with existing languages was not on the table.

One programming language that explicitly privileged "soft" matters was Ada. Starting in 1977, the US Department of Defense ran a contest to de-velop a standard programming language for its operations. The winning entry was developed by a team led by Jean Ichbiah, who worked for the French company CII Honeywell Bull. Like many modern programming languages, Ada is a descendant of ALGOL, but it placed a heightened em-phasis on "concern for the human programmer."[122] The design of Ada, as a rationale document states, rests on the belief "that our understanding of programs can be greatly simplified if our intuition is able to rely on tex-tual forms that convey the logical structure of the program"; as a result, the Ada language gives "major consideration to readability and teachability."[123] Influenced by Barbara Liskov's CLU programming language, Ada divided code into modules called "packages" that included both procedures and structures for storing data.[124] This approach (a precursor to what is now called *object-oriented programming*) was tailored less toward mathemati-cal proof than toward facilitating teamwork by making it easier to divvy up tasks and communicate the purposes of specific parts of a complex system.

As the choice of the name "Ada" indicates, programmers were by this point assembling a canon of historical precedents for their profession. Ada Lovelace, Lord Byron's daughter and the writer of notes on Charles Babbage's computing engines, came to stand in part for the role women played in the early days of the computer age and in part for the need to supplement mathematical methods with something else. An important figure in the canonization of Lovelace was Betty Alexandra Toole, who in 1992 published a heavily annotated edition of some of Lovelace's letters that Toole edited in collaboration with some of the developers of the Ada programming language. According to Toole, it is Lovelace's combination of mathematical acumen with poetic imagination that makes her an appropriate icon for the digital age. Modern mathematics, Toole writes, requires both "digital skills such as objectivity, observation and experimentation, and analogue skills such as imagination, visualization and the use of metaphor."[125] As Toole reports, a poll of Ada programmers found that its most important attribute was "connectivity, the ability and ease of teams of people working in the language to communicate."[126] The design of Ada continued the focus of nineteenth-century mathematicians like Lovelace on balancing the advantages of mechanical methods with an organically cultivated intuition and an ability to explain ideas clearly and elegantly.

In spite of the general recognition of the importance of communication, "soft" approaches faced pushback for a perceived lack of rigor. A prominent criticism of Ada appeared in the 1980 Turing Award lecture of the Oxford computer scientist C. A. R. Hoare, best known as the inventor of the Quicksort algorithm. Inverting the story of the Emperor's New Clothes, Hoare implies that Ada is in danger of becoming a pile of clothes without an emperor. While his immediate point is that the language is too complicated to rely on, the comparison also suggests superficiality: "Gadgets and glitter prevail over fundamental concerns of safety and economy."[127] Ada, he argues, contains "dangerous" features that lead to errors because they create confusing interdependencies between distant parts of programs.[128] Dijkstra had made a similar criticism of the **go to** statements that were common in early programming languages.[129] Like Dijkstra and McCarthy, Hoare sought to ensure the correctness of code with demonstrative certainty—to make it "logically impossible for any source language program to cause the computer to run wild."[130] Some techniques along these lines have become a part of mainstream software engineering, including strong typing and memory protection. But in practice, code did sometimes behave unpredictably, and correctness could not always be judged with mathematical certainty.

The idea of proving programs correct culminated in the Haskell programming language, which was conceived at the 1987 International Conference on Functional Programming and first released in 1990.[131] Haskell was named in honor of Haskell B. Curry, a professor of mathematics at Pennsylvania State University who had died a few years before. Along with William Alvin Howard, Curry had discovered a method of converting between computer programs and mathematical proofs that came to be known as the Curry–Howard correspondence.[132] This correspondence provides a way to prove that a program will have certain behaviors. This approach only works, however, if the language is purely functional, meaning that a unit of code cannot affect anything beyond its formally defined output. This limitation requires Haskell to encapsulate interactions with other systems (such as user interfaces) within a layer of abstraction—called, in what is apparently not a reference to Leibniz's metaphysics, a *monad*—to prevent the outside world from compromising the logical purity of programs.

Haskell is still actively developed and has an ardent community of users, but it is far less widespread than such languages as C++, Python, and Go, which all contain elements that are not amenable to strict proofs.[133] Even in Haskell, formal rules cannot (as Church and Turing proved) wholly determine the right way of proceeding. The typical experience of programming is less like Dijkstra's analysis of uninterpreted formulae than like Toole's balance between "digital" and "analogue." A large part of software development involves communication: writing code that other team members can read, following house styles, portioning out work in sensible ways, dealing with "legacy" systems inherited from past employees, and documenting new systems for future maintainers and users.

Such matters were incorporated into the academic discipline of computer science. In their influential 1985 textbook *Structure and Interpretation of Computer Programs*, the MIT-affiliated computer scientists Harold Abelson, Gerald Jay Sussman, and Julie Sussman take the position that "programs must be written for people to read, and only incidentally for machines to execute"; they de-emphasize "the mathematical analysis of algorithms and the foundations of computing" in favor of "the techniques used to control the intellectual complexity of large software systems," including an appreciation for style and aesthetics.[134] It is telling that the book includes an epigraph from John Locke's *Essay Concerning Human Understanding*, indicating a focus on the cognitive act of abstraction.[135] Far from the purely formal "algorithmic language" envisioned at the ALGOL conferences, modern programming languages are complex beasts that serve a range of purposes and draw on a range of influences, some logico-mathematical and some not.

Algorithms thus became one topic among many within the discipline.

In the early 1960s, computer scientists treated *algorithm* as a synonym for *program*; the framing of the "Algorithms" section in *CACM* reinforced this.[136] By the 1980s, algorithms and programs had diverged. In his forward to *Structure and Interpretation of Computer Programs*, Alan J. Perlis holds up the algorithm as an ideal to which programs should aspire:

> Among the programs we write, some (but never enough) perform a precise mathematical function such as sorting or finding the maximum of a sequence of numbers, determining primality, or finding the square root. We call such programs algorithms, and a great deal is known of their optimal behavior, particularly with respect to the two important parameters of execution time and data storage requirements.[137]

Even if only an elite few programs could achieve the status of true algorithms, one could still apply the theory of algorithms to parts of programs—to procedures that perform clearly defined tasks, considered in isolation from the broader software systems in which they are used. Abstracting algorithms from programs enabled computer scientists to analyze them with at least some of the rigor ALGOL's creators envisioned, even when that rigor could not extend to actual production code. This rigor was and is genuine—it contributed greatly to the reliability and efficiency of software—but it came at a cost. The theory of algorithms, as it came to be taught in computer science, threw up a barrier between computation and the world of human life whose full implications would not become apparent for decades.

PROBLEMS AND SOLUTIONS

In critical studies of technology, there has been some discussion about the difficulty of locating algorithms.[138] Are they in the code, in the executable files, or in the running circuitry? How can one tell which parts of a system are algorithms and which are not? One answer would be that algorithms are not in computers at all—they are on the pages of books and academic journals. Computer science degree programs typically require a course on Data Structures and Algorithms, in which students learn a canon of well-known computational procedures along with techniques for reasoning about them. The ALGOL project's goal of providing a standard notation for these procedures did not obtain: one of the earliest CS textbooks, Alexandra I. Forsythe and colleagues' *Computer Science: A First Course* (1969), uses flowcharts, while others use either pseudocode (a semiformal notation that varies from author to author) or real programming languages.[139] Programmers are expected to know some algorithms by heart. A common job interview question, for instance, involves recounting an algorithm for determining whether a network contains a loop. Studying algorithms is important not just because of their individual utility but also because the theory of algorithms is crucial to developing software that can operate efficiently at scale.

Students hoping to learn about machine learning and predictive modeling are sometimes surprised at what they encounter in algorithms courses. These courses largely cover what we might call *classical algorithms*— human-designed procedures for solving precisely defined problems. The paradigmatic problem is sorting, which is ubiquitous in textbooks because it illustrates a number of issues that arise in the theory of algorithms. Suppose that one has to sort a shelf of books. A simple approach would be to swap all pairs of adjacent books that are in the wrong order, then repeat until the books are all sorted. The 1987 edition of the textbook *Data Structures and Algorithms* by Aho, Hopcroft, and Ullman presents this method, which is called *bubble sort*, in a notation based on the PASCAL programming language:[140]

```
(1)    for i := 1 to n−1 do
(2)        for j := n downto i+1 do
(3)            if A[j].key < A[j−1].key then
(4)                swap(A[j], A[j−1])
```

The procedure uses two **for** loops, which repeat the blocks of code indented beneath them. Loop (2) goes through the unsorted portion of the array in reverse order and, in lines (3) and (4), swaps adjacent items that are misordered. Loop (1) repeats that process the necessary number of times. Note that this algorithm can be used to sort any type of data whatsoever, provided that the desired order (represented by *key*, which assigns each item a numerical value) is specified. This method, however, is extremely slow when applied to large inputs, on account of which it is taught as an example of a bad algorithm. The desire for more efficient sorting led to the development of a number of less obvious sorting procedures. The most famous is Quicksort, which involves moving the items to one side or the other based on their ranking relative to an arbitrarily chosen item, then performing the same procedure recursively for each side; this much-studied procedure became a paradigm of good (meaning efficient) algorithm design.

While computer scientists never reached a consensus definition of *algorithm*, the procedures in algorithms textbooks do share some common features. One is that the procedures are, in Knuth's terms, *finite, definite,* and *effective*—that is, they consist of an eventually terminating sequence of clearly defined operations that can each be completed in a finite amount of time.[141] As Aho, Hopcroft, and Ullman point out, whether a procedure meets these criteria can be debatable, since "what is clear to one person may not be clear to another, and it is often difficult to prove rigorously that an instruction can be carried out in a finite amount of time."[142] Although it may be hard to prove that a procedure is an algorithm, it is easier to show that some procedures are not. For instance, suppose we wanted to verify Michael Stifel's result, from over four centuries before the computer era, that $\sqrt{8} + \sqrt{18} = \sqrt{50}$. We might come up with something like this (in pseudocode):

$$a = \sqrt{8} + \sqrt{18}$$
$$b = \sqrt{50}$$
if $a = b$ **then**
    return "Stifel was right!"
else
    return "Something is wrong."

But this is not an effective procedure, at least if one interprets it as a solution to the problem we have set out. The issue is that, as Stifel observed with such consternation, the values of those roots can only be approximated.[143] We can compute as many digits of those approximations as we like ($\sqrt{50} = 7.0710678\ldots$) but it would take infinitely many steps to compare all the digits of a and b. This point is subtle, but it illustrates an important difference between algorithms and other forms of mathematical reasoning, such as the geometric method Stifel used to arrive at that equation. Classical algorithms inhabit the world of Turing Machines, in which the fundamental objects of operation are digits, not numbers.[144] They are thus haunted by the specter of infinity, by the possibility that a procedure will go on forever, the specified stopping criterion never coming true.[145]

A related point is that algorithms are only able to interact with other systems in certain limited ways—namely, through the input and output. As Knuth puts it, algorithms must have zero or more inputs "taken from specified sets of objects" and one or more outputs that "have a specified relation to the inputs."[146] Specifying this relation does not mean knowing the output ahead of time—in many cases, that would make the computer redundant—but we must still define the output in some precise way, as an equation defines its roots whether we know them or not. In some cases, the algorithm includes a random factor or supplies only an approximate solution, but the desired output must still be specified. Algorithms in this sense cover only a small slice of the things computers do; in particular, systems that require continuous feedback from an environment, such as robot control systems and user interface loops, are not technically algorithmic. While computer scientists have long recognized the limits of this conception and sought to expand the concept of algorithm to include processes that are more dynamic, some narrowness of definition is needed to make the theory of algorithms tick.[147] Complexity theory, in particular, only works when that input is clearly circumscribed; otherwise, the outcome could depend on any number of intractable factors ranging from the air temperature to the whims of the person sitting at the keyboard. It is easier to get away with vagueness as to the output, but a precisely specified goal is crucial if we hope to prove an algorithm correct.

Just how to determine this goal is a complex matter. "Half the battle," Aho, Hopcroft, and Ullman declare, "is knowing what problem to solve."[148] Their textbook suggests a two-phase method: first define the problem precisely, then figure out an algorithm to solve it. Others have preferred a more flexible approach. At the 1968 NATO conference, Dijkstra argued that incomplete problem specifications were better than complete ones because they made it easier to develop general solutions that could be reused in other cases.[149] This statement points to another characteristic of algorithms that Knuth does not include in his definition but that Markov does: generality. Typically, algorithms are not meant to solve only a single instance of a problem but to be reused with as wide a range of inputs as possible. Another early textbook puts it this way: "Although one may easily learn how to solve problems of many types, it is more difficult to recognize the common structure of problems which are seemingly quite different and to develop general methods for their solution. A precise description of how to solve a class of problems is called an *algorithm*."[150] To the extent that they claim generality, classical algorithms carry on the imperial program Viète enacted in the 1590s: like his equation-solving theorems, they are supposed to conquer problems by the hundreds rather than attacking them one by one.

Classical algorithms such as Quicksort should not be conflated with the predictive models that later came to be associated with the word *algorithm*. The procedures compiled in algorithms texts are mostly unconnected to statistics, nor are they necessarily intended to predict anything. A classical algorithm is also distinct from the function it computes, meaning the specification of what the output should be for a given input. This distinction is sometimes overlooked in critical scholarship, but it is crucial for understanding the disciplinary practices that attend classical algorithms.[151] Often there are multiple algorithms that compute the same function by different means and that are thus, as far as the output goes, interchangeable. For instance, as long as there are no ties in the ranking, standard sorting algorithms such as bubble sort, merge sort, and Quicksort will all produce exactly the same output, given the same input; they differ primarily in their speed and memory use.[152] Conversely, standard sorting algorithms make no assumptions about what order items should be in, instead producing whatever ordering the user specifies.[153] Classical algorithms deal primarily with how to compute things, not with what to compute, and, at least ideally, one can apply them without having to worry about their details, just as countless programmers have used Python's list.sort() method without needing to ask which sorting algorithm it employs. (It uses, as it so happens, an algorithm developed by Tim Peters called *Timsort.*)

This interchangeability only works, however, to the extent that the problem is precisely defined. Sorting problems are typically precise enough, but other problems can be much messier. A good illustration is the procedure that came to be known as "Dijkstra's algorithm." This procedure, which Dijkstra first published in a 1959 article, finds an optimal path between two points. His original application was to find routes between cities in the Netherlands; much later, this algorithm came to be celebrated, fairly or not, as the foundation of GPS navigation.[154] Dijkstra's algorithm draws on the methods of graph theory, a mathematical theory of networks that was developed in the early twentieth century (figure 5.3). In this sense, a *graph* consists of a set of nodes, or *vertices*, along with *edges* that connect pairs of nodes. In Dijkstra's formulation, the edges also have defined lengths. Dijkstra's method makes use of the fact that if a given path through a graph is the shortest one possible, any subpath of that path must also be the shortest possible from its starting point to its end. (Dijkstra does not refer to his procedure as an *algorithm* in the article; he calls it a "process" and a "solution."[155]) Most other common pathfinding algorithms are variants of Dijkstra's procedure; for instance, the A* algorithm, which is sometimes used to guide the motion of video game characters, adds a heuristic that biases the search toward moves in the general direction of the goal.[156]

From a mathematical perspective, it is hard to deny that Dijkstra's algorithm is correct: it does indeed find an optimal path, as one can prove by mathematical induction. But applying such an algorithm requires rep-

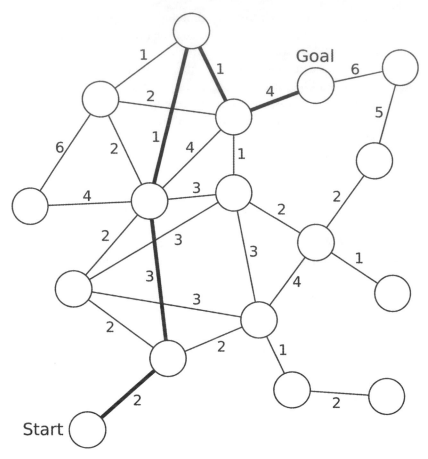

Figure 5.3. An illustration of the pathfinding task. The network, or *graph*, consists of a set of nodes, or *vertices* (represented by circles), connected by means of *edges* (represented by lines). Each edge has a *weight* that represents its cost. The goal is to find a path from the start to the goal that minimizes the total cost. The optimal path in this instance is shown with edges in boldface.

resenting the problem as a network, and the choices involved can make real-world differences. In 2017, residents of Leonia, New Jersey, noticed a sudden increase in traffic through their residential town, to the point of total gridlock.[157] As it turned out, construction had led to backups around the George Washington Bridge, and navigation apps such as Google Maps and Waze were sending cars through side streets as an alternate route. While outsiders cannot know exactly which pathfinding algorithms these programs were using, the algorithms themselves are less important to this issue than how the problem was defined. To employ a network-based pathfinding algorithm, one must, first of all, choose which locations constitute the nodes. For Dijkstra, these were cities, whereas on a street map they are typically intersections. Second, one must decide what values to assign to the edges. Dijkstra called these values "lengths," which suggests that they represent physical distances, but there are numerous other ways one could mark the relative desirability of edges. Navigation apps typically account for some factors such as speed limits and current traffic, but in the New Jersey case, they did not consider the cost of routing cars through a particular area for the local community.

This incident illustrates the dangers and the opportunities created by the classical paradigm.[158] By containing the messy details of the application within the formally defined input and output, the paradigm opens computational logic to rigorous study within a rarefied mathematical realm even as its broader implications remain (as they must be) unfathomably complex and incompletely understood. An IBM promotional video from 2018 states the logic of formalization with striking bluntness: "Researchers are helping AI systems understand human values by defining them in engineering terms."[159] Whether human values can even in principle be defined in engineering terms is questionable. But algorithmic thinking demands the construction of formal definitions that exist alongside the conventional meanings of words. From a technical point of view, whether these definitions really do capture the values they are meant to embody is secondary to the correctness and efficiency of the algorithms, which is what ultimately determines how well they will work when implemented.

The extent to which this disconnect is a problem hinges on how, exactly, things like ideas and values come to be translated into those formal definitions.[160] While such questions fall outside the official scope of the theory of algorithms, they cannot be ignored. As Warren Sack has pointed out, computational systems require rhetoric: their designers must persuade people to accept the particular way they have chosen to frame the problem.[161] In academic computer science, this rhetoric borrows heavily from the dualistic view of knowledge that began to attend symbolic methods with the rise of logicomathematical formalism in the nineteenth century. An illustrative case is the much-discussed algorithm that formed a key part of the original Google search engine: PageRank.[162]

The purpose of PageRank is to assign web pages numerical scores that may be used to prioritize search results. (To be clear, PageRank is a ranking algorithm, not a sorting algorithm; a separate procedure such as Quicksort must be used actually to arrange the items in order.) In 1998, when the Google founder Larry Page was a graduate student at Stanford, he and his collaborators explained their method in a white paper titled "The Page-

Rank Citation Ranking: Bringing Order to the Web" as well as in a journal article that more broadly describes the prototype of the Google search engine and in two patent applications.[163] The white paper spends less time on explaining the algorithm itself than on expounding a rationale for its use as a ranking of web pages. In doing so, the paper grapples, in a way characteristic of modern algorithmic thinking, with the conceptual problems that have arisen time and again over the centuries when mathematicians have traded words for symbols.

Building on the tradition of citation analysis, PageRank scores web pages by counting *backlinks*—that is, counting how many other pages link to a given page.[164] Compared with simple citation counting methods, PageRank gives more significance to the network structure of the web. Like Dijkstra's algorithm, it formalizes the problem in terms of graph theory, representing web pages as nodes and hyperlinks as edges. Going back to the 1950s, it had been common to measure the relative importance of nodes in a graph using various "centrality" measures based on their connections; these methods were initially used in sociometric studies of communication networks.[165] PageRank combines this sort of network analysis with citation counting. Rather than treating all connections equally, it gives more weight to backlinks that come from pages that themselves have higher PageRanks. Similar ideas were proposed decades before by both Leo Katz and Phillip Bonacich and around the same time by Jon M. Kleinberg.[166] Weighting the links by PageRank makes the definition circular: just as Leibniz's monads each represent the universe, the rank of each page depends on the ranks of all the other pages on the web. This circular definition reduces to a system of equations that may be solved using techniques from linear algebra, thus producing the ranking.

Page and his white paper coauthors state their purpose in developing this metric at the beginning of their abstract: "The importance of a Web page is an inherently subjective matter, which depends on the readers [*sic*] interests, knowledge and attitudes. But there is still much that can be said objectively about the relative importance of Web pages."[167] By ranking the importance of pages "objectively and mechanically," PageRank enables the search engine to prioritize results that are shown to the user.[168] This phrase invokes what Lorraine Daston and Peter Galison have called "mechanical objectivity"—mechanization as a way of taking all human judgment out of a practice.[169] But merely using a machine cannot be a sufficient condition for objectivity for a simple reason: computers can be programmed to say anything. (The true antagonist of algorithmic objectivity is not subjectivity but arbitrariness.) The authors are thus tasked with convincing the reader that their calculation bears a strong enough relevance to the existing idea of "importance" to make it worth adopting as a measurement.

This argument takes the form of a gradual progression from words to symbols. First, the authors review previous work on citation analysis and make a case for how this method could be applied to web links as well as to academic citations. Second, they argue that simple citation counting does not always correspond to "our common sense notion of importance," since links from major websites are more important than links from obscure ones.[170] In the next section, which consists of only two sentences, they state an "intuitive description of PageRank": "a page has high rank if the sum of the ranks of its backlinks is high."[171] After this verbal definition, they present a symbolic equation that "formalizes the intuition."[172] Let B_u be the set of pages that link to the page u, N_u be the number of links on page u, and c be a constant. The ranking R of a page u may be defined as follows:

$$R(u) = c \sum_{v \in B_u} \frac{R(v)}{N_v}$$

Because of its grounding in an intuitive rationale, this equation is supposed to constitute a computable definition of importance.

There then follows some backtracking. While the authors maintain that it corresponds to "common sense," this initial equation has a drawback: it is possible for groups of pages to mutually cite each other, which would form "rank sinks" that skew the scores.[173] As a result, they add an additional corrective factor E, resulting in the following modified equation, which (unlike the foregoing one) is crowned with the formal heading "**Definition 1**":

$$R(u) = c \sum_{v \in B_u} \frac{R(v)}{N_v} + cE(u)$$

This equation, it should be made clear, is not itself an algorithm, but rather a definition of the function to be computed. This definition is not recursive in Church's sense because it is circular: in expanding the formula for $R'(u)$, one will often come back around to that same value. The equation can, however, be solved through an iterative method that constitutes the Page-Rank algorithm, which they describe, in a brief section titled "Computing PageRank," by means of a terse pseudocode consisting almost entirely of symbols.

This approach to justifying an algorithm bears some resemblance to the verbal definitions that came to attend symbolic methods in Leibniz's wake. Recalling Jean Le Rond d'Alembert's detailed discussion of what *negative* means, Page and colleagues are careful to provide a conceptual basis for their symbolic equation; they even echo Francis Maseres in their invocation of "common sense." Yet, in a way that would have been unacceptable in the eighteenth century, this conceptual basis falls away once symbols arrive. Not only do Page and his coauthors revise the intuitively justified equation based on mathematical concerns; they also reinterpret it. Whereas they initially explain PageRank in terms of citation counting, they later present a different rationale based on a simulation of people "surfing" (browsing) the web. A "surfer" moves from page to page by randomly clicking links; occasionally the surfer "gets bored" and chooses a page at random.[174] PageRank is equivalent to a measure of how much time this surfer is likely to spend at each page over a large number of trials.

This reinterpretability contributes greatly to the power of classical algorithms. A few years after the Page group's white paper was published, Alon Altman and Moshe Tennenholtz reinterpreted PageRank once more, this time as a simulation of a voting process.[175] The algorithm found another application further afield in biochemistry, where it can be used to analyze protein interaction networks.[176] The idea that algorithms can retain a stable logical core even as they are transferred to entirely different disciplinary contexts continues the privileging of formal rules over conceptual interpretations that transformed symbolic algebra in the time of William Roman Hamilton and George Boole, and that is implicit in ALGOL's distinction between formal semantics and "syntactic sugar." The rationale is figured as mere window dressing upon the logic, crucial to the practical application of algorithms but capable of being swapped out whenever it is convenient to do so.

A difficulty of this approach is that it creates an epistemological divide between the technical definitions presented (often in the form of equations) in algorithms textbooks and research papers and the nontechnical perspectives of users. The PageRank paper may include an intuitive rationale for the method, but user interfaces typically skip this step. When faced with search results sorted by "relevance," for instance, the user has to trust that whatever calculation the system is using corresponds somehow to the meaning of *relevance*.[177] Numerous commentators have expressed a concern that algorithms are often kept hidden, cranking away in the server farms of companies such as Facebook and Google, concealed from public scrutiny. There is nothing in the classical paradigm that says algorithms must be hidden; going back to the "Algorithms" section in *Communications of the ACM*, it has been a common practice to publish algorithms, and explanations of Quicksort, Dijkstra's algorithm, and PageRank are up on Wikipedia for all to see. But the practice of keeping some algorithms as trade secrets is only one expression of a deeper divide. By subordinating conceptual explanations to mechanical operations, the classical paradigm enables programmers to enact their own technical definitions without the input of the broader community that, in ordinary writing and speech, has some say as to what a word means.

The classical algorithm has proven workable as a set of disciplinary practices, even if its limitations are substantial. The barrier that early computer scientists created between "human factors" and algorithmic logic—a barrier between the choice of what to compute and the technical details of how to compute it efficiently—enabled methods that have borne practical fruits, among them the highly complex procedures used to parse the syntax of computer languages like such as Python and HTML.[178] But computation began to depart from the classical paradigm in the early twenty-first century. In the first two decades of the century, the word *algorithm* came increasingly to refer not to human-designed procedures such as Quicksort but to machine learning (ML) systems. Machine learning always involves some sort of predefined goal or at least measure of success, but the means of defining this goal is entirely different: in ML, the goal is laid out through data rather than through a direct specification of the mathematical function the computer is to compute or approximate. Machine learning has decidedly not replaced classical algorithms, but it did gain a newly prominent role in computational systems in the 2010s. In the coda to this book, I consider how the conversation about algorithms changed in the first two decades of the twenty-first century, as ML gained prominence both in technical fields and in the public imagination of what an "algorithm" is.

The Age of Arbitrariness

"But what if x should turn out, after all,
not to be the unknown quantity?"

—ATTRIBUTED TO JAMES CLERK MAXWELL

BLACK BOXES AND ADJUSTABLE KNOBS

If classical algorithms are divided from the human understanding, they are also divided from data.[1] As far as the effectiveness of Quicksort goes, it makes no difference *what* one hopes to sort—provided the order is well defined, the procedure would work just as well with videos to be sorted by running time, names to be sorted alphabetically, or images to be sorted by brightness. Because complexity theory works solely through the analysis of computational instructions, it is possible to study a classical algorithm rigorously without ever even turning on a computer. Machine learning (ML) changes this. The "algorithms" are no longer designed by engineers but instead tuned by machines based on large amounts of data. While these two types of "algorithm" are often discussed together, ML is not a straightforward continuation of the disciplinary practices established by early computer scientists such as Edsger Dijkstra and Donald E. Knuth. The ML sense of *algorithm* emerged less from computer science than from the newer discipline of data science, and the use of the same word in both contexts has led to a great deal of confusion. ML departs from the definitions of algorithm put forth in the twentieth century, and the difference has wide-ranging implications for the epistemology of computation.

The term *machine learning* can refer to a wide range of techniques that vary greatly in sophistication.[2] The idea goes back to the 1940s, when researchers attempted to formalize the structures of human cognition.[3] In the 1950s, this effort came to be associated with the field of cybernetics, which dealt with dynamic relationships between systems, and with artificial intelligence (AI).[4] Although there was some overlap, these research programs were largely separate from the subfield of algorithms; rather than seeking solutions to precisely defined mathematical problems, they considered interactions of machines with their environments and sought to automate tasks that were more open-ended, such as translation and pattern recognition. Much of the technical apparatus of modern ML was already in place by the 1980s, but the practice gained greatly in prominence early in the first decade of the twenty-first century, owing in large part to an increase in computing power and in the easy availability of very large data sets.

As an example that illustrates just one of these techniques, consider a common application: image classification. To train an ML system to distinguish images of horses from images of dogs, one would first need to assemble a collection of image examples, all annotated with correct labels; these labels are sometimes called, suggestively enough, the *gold standard*. One would also need to construct a model, meaning a mathematical function whose parameters are to be determined in the training process. Image recognition models typically consist of artificial neural networks (ANNs), which are inspired by the structure of the human brain, although they ultimately consist of little more than sequences of matrix multiplications and other simple mathematical functions. Next, one must specify a *loss function* that measures the quality of the output, lower numbers being better. Since the correct labels are known in this case, the loss function would measure how badly the model misclassifies the training data.

Once these elements are assembled, one can begin the training. Through an optimization procedure such as *gradient descent*, which uses differential calculus to determine how local changes in each parameter would affect the output, the machine adjusts the parameters little by little so as to decrease the loss. (Optimization was, incidentally, just the problem G. W. Leibniz was addressing when he called the method of differentials an "algorithm" in 1684, although the approach is not quite the same.) The result is a trained model that can be used to make predictions. The excitement over this method stems from the fact that it can apparently handle tasks that had formerly been outside the scope of what computers could do. A human being would have a hard time devising explicit criteria for whether an image depicts a horse, but having a machine "learn" the criteria based on a large number of image examples works remarkably well (although it certainly does not work perfectly).

ML "algorithms" have some similarities to the procedures cataloged in algorithms textbooks. Trained ANNs can, with no particular problem, be expressed as clearly defined steps—namely, sequences of matrix multiplications and other mathematical operations. These procedures also share classical algorithms' independence from specific embodiments; one can convert them between file formats such as ONNX and TorchScript while preserving their fundamental logic (although, as anyone who has actually tried to do this can attest, the conversion is not always lossless in practice). More broadly, ML shares the classical paradigm's tendency to judge processes primarily by whether they produce the desired output. An ML system's performance can, in theory, be evaluated automatically without regard for whether the operations it employs make any sense to people. In this way, ML takes the subject–object divide that I have identified as characteristic of algorithmic thinking—a divide between standards of correctness and the formation of human thought—to an extreme.

The content of these models, however, is rather different from the problem-solving procedures compiled by Knuth, and the gap ML can create between computational logic and human understanding has increasingly appeared as a problem. Classical algorithms usually employ all-or-nothing rules: if the value of x is this, do that. Some ML models, such as decision trees, contain such conditional rules, but ANNs employ calculus-based optimization methods to which strict either/or logic is inimical. For gradient descent to work, every operation must (with some restricted exceptions) be differentiable, roughly meaning that the transitions between states must follow smooth curves.[5] As a result, ANNs can only represent categorical distinctions fuzzily: an image might be classified as 90 percent dog and 10 percent horse. It is true that the models must be reduced to ones and zeros to run on computers, but in theory they work with continuous functions over real numbers, not with the discrete symbols of Turing Machines.[6]

Apart from these mathematical differences, ML raises distinct epistemological questions from the ones that led early computer scientists to develop a theory of algorithms. In realistic cases, the training data stand in for a broader range of phenomena: a collection of photographs may be very large, but it cannot encompass every form a photograph of a horse may take. As a result, even if the model gets perfect results on the training data, it can still go wrong when tested on other data that have not yet been seen. The technical term for this problem is *generalization*.[7] To reason about generalization mathematically, ML employs idealizing assumptions such as i.i.d. (independent and identically distributed), which asserts that the data all come from a single probability distribution and do not affect each other.[8] But this approach is only as good as the data used in computing evaluation metrics. Proving that an image classifier can generalize not just to the test data but to any data whatsoever would require nothing short of Leibniz's universal characteristic: a complete and correct theory of everything. Absent such a theory, a model will remain open to dispute, and whether it can ever attain the level of trustworthiness C. A. R. Hoare and John McCarthy sought—an issue especially important in life-and-death applications such as autonomous cars—will be debatable.

The rise of machine learning coincided with yet another shift in the usage of the word *algorithm*. Both in the popular discourse and in technical fields, *algorithm* is now sometimes used to refer to ML systems rather than to step-by-step procedures such as Quicksort and A*. The shift stems in part from the word's adoption in a field with disparate priorities from those of software engineering: statistics. In a 2001 article titled "Statistical Modeling: The Two Cultures," Leo Breiman contrasts two schools of thought that he calls "data modeling" and "algorithmic modeling."[9] The data modeling approach, which he claims is dominant in academia, attempts to explain the data by producing a model that corresponds to a comprehensible theory. Algorithmic modeling, which he claims is more common in business, aims only to make predictions, treating the model as a "black box" to be judged solely by the accuracy of its output.[10] Breiman argues that academics have overlooked the potential of black box models, sacrificing predictive accuracy for the sake of what he calls "irrelevant theory."[11] In Breiman's wake, the word *algorithm* came to be associated less with the abstract procedures compiled in computer science textbooks and more with a results-oriented flavor of statistical data crunching.

Breiman's article uses the word *algorithm* in two distinct ways, neither of which exactly corresponds to how it is used by Knuth and other early computer scientists. At times, he uses it to refer to what Bernhard Rieder has called "algorithmic techniques": general methods for designing computational systems.[12] Breiman refers, for instance, to "neural nets and decision trees" as "powerful new algorithms for fitting data."[13] Breiman's claim to fame is inventing one such technique: *random forests*.[14] In this sense, an *algorithm* is a set of techniques used to construct and train a model, not the model itself. Yet Breiman's "Statistical Modeling" article contains some usages that suggest otherwise. The goal of algorithmic modeling, he writes, is "to find an algorithm $f(\mathbf{x})$ such that for future $\mathbf{x}$ in a test set, $f(\mathbf{x})$ will be a good predictor of $\mathbf{y}$."[15] The *algorithm*, in this case, would seem to be the predictive function found through the training process. It is in this second sense that one talks about algorithms recognizing horses or predicting loan defaults: the algorithm is not a general technique but the particular result of a computationally intensive process of data crunching.

The coexistence of these two senses of *algorithm* led to a misunderstanding when Breiman moved to defend his position. The "Statistical Modeling" article was published along with a collection of responses and rebuttals. One of the respondents, the Stanford statistician Bradley Efron, notes that the article looks at first glance "like an argument against parsimony and scientific insight" and states: "The new algorithms often appear in the form of black boxes with enormous numbers of adjustable parameters ('knobs to twiddle')."[16] Breiman replies:

> This is a perplexing statement and perhaps I don't understand what Brad means. Random forests has only one adjustable parameter that needs to be set for a run, is insensitive to the value of this parameter over a wide range, and has a quick and simple way for determining a good value. Support vector machines depend on the settings of 1–2 parameters. Other algorithmic models are similarly sparse in the number of knobs that have to be twiddled.[17]

The disagreement, it seems, comes down to whether "algorithm" refers to the function whose parameters are set through the training process (Efron) or to the method used to produce this function (Breiman, at least in this instance). Efron argues that, because of the trade-off between bias and variance, models with too many parameters will lead to biased estimates. In this regard, it makes little difference whether a human or a machine is making the choices. Breiman gives much more significance to the machine/human (or at least algorithmic/nonalgorithmic) distinction. An ML model may have an enormous number of learned parameters (GPT-3 reportedly has 175 billion), but these parameters are determined within the black box and so Breiman does not count them as "adjustable."

The dispute between Efron and Breiman is not merely a matter of terminology. As I argued in the discussion of PageRank in chapter 5, applying computational methods to loosely defined problems such as ranking web pages can inspire an anxiety about arbitrariness—a sense that people can coax out whatever results they want by tweaking settings. The PageRank authors address this anxiety with a conceptual rationale, arguing that their definition corresponds to "common sense." Similarly, Efron writes from the traditional position that statistical models should have theoretical bases. Breiman's article is overtly an argument for practicality over such theorizing, but it also provides a way of assuaging the anxiety about arbitrariness by placing the exclusive focus on predictive accuracy as measured through empirical trials. This focus has led (on websites like Kaggle) to a gamified form of research in which, rather than justifying their methods rhetorically in the manner of the PageRank paper, data scientists vie for top positions on metric-based leaderboards.[18]

While the usage of the term *algorithm* in the ML literature varies, the equivocation as to which part of an ML system constitutes the "algorithm" has become widespread. The 2016 textbook *Deep Learning* by Ian Good-fellow, Yoshua Bengio, and Aaron Courville, for instance, gives the following informal definition: "A machine learning algorithm is an algorithm that is able to learn from data."[19] Later in the text, they give a more specific "recipe" for the construction of deep learning algorithms: "combine a specification of a dataset, a cost function, an optimization procedure and a model."[20] This recipe suggests that a "machine learning algorithm" is a reusable set of techniques used to construct and train models, which is mainly how the authors use the term: they refer, for instance, to linear regression as "a simple machine learning algorithm."[21]

The statement that algorithms are "able to learn," however, sits uneasily with this conception—it suggests, at least equivocally, that the algorithm changes over time, expanding its capacities as it combs through data. This proposition only makes sense if the "algorithm" encompasses not just the techniques used in constructing an ML system but also the parameter values as they are adjusted by the machine during training. The authors reinforce this assignment of agency by their use of the phrase *learning algorithm* rather than *training algorithm*. The ambiguity is far more prevalent in informal discussions, where "algorithms" are commonly said to possess a range of acquired abilities, from predicting fraud to identifying faces to translating text. The blurring of the lines around algorithms has likely been reinforced by new development tools such as Jupyter Notebook, which enables programmers to edit a program while its variables are loaded in memory; training procedures and learned parameters can thus seem like parts of a single dynamically changing system. This type of "algorithm," if the concept really is a coherent one, is a very different beast from Quicksort, and the definitions put forth by Markov and Knuth are of limited value in explaining it.[22]

There is some reason to lament this semantic shift. Associating *algorithm* specifically with machine learning leaves behind no good term for the precisely defined procedures covered in algorithms courses, which, while less flashy than ML, are and will remain important to the discipline of programming. Indeed, classical algorithms play a role in ML itself: the *Deep Learning* textbook includes numerous human-designed procedures for use in training and inference, set off with such headings as "Algorithm 6.3."[23] The book presents these procedures in a pseudocode form that basically follows the classical paradigm, although it is notable that the authors usually do not specify the desired output. The ambiguity as to whether "machine learning algorithm" refers to these human-designed training procedures or to trained predictive models has led to a great deal of cross talk and confusion. Amid the widespread hype about these methods, the word *algorithm* has taken on a range of connotations more aligned with marketing than with serious analyses of technology, and commentators would do well to meet imprecise uses of the term with skepticism.[24]

Yet the broader shift in thinking about computation could be a positive development. The classical paradigm abstracts out social issues by containing them in formal problem definitions, thus encouraging a narrow focus on technicalities to the exclusion of the broader contexts in which systems will be employed. Machine learning research can take a similarly blinkered view to the extent that it sets its sights on fixed benchmarks; yet the practical adoption of ML methods has made social "externalities" harder to dismiss. Someone, after all, must set the goals, and, in the absence of strong rhetorical justifications, the numerous decisions that go into the design of models, performance metrics, and data sets create opportunities for arbitrariness to slip back in. It is not surprising, as a result, that ML has reacted so explosively with identity politics: the anxiety about arbitrariness naturally leads one to question *who* is arbitrating, *who* is turning the knobs that determine what ML models do. For better or worse, ML is breaking down disciplinary boundaries, and it opens an opportunity to rethink, once again, fundamental assumptions about how computation relates to human thought and language. A case in point is GPT-3.

MACHINE LEARNING AND THE LIVING LANGUAGE

Unlike the algorithms I discussed in chapter 5, GPT-3 is black-boxed in a strong sense. The trained model, as of this writing, is inaccessible to the public; it runs on servers with which one can interface only by permission. It is, however, possible to download GPT-2, and there is much to be said about the general methods behind these technologies, as described in research papers. As "multitask learners," the GPT models are supposed to be applicable in a range of domains, but not through the sort of conceptual reinterpretation we saw with PageRank. Instead, one is supposed to induce the program to perform tasks by feeding in inputs written in the English language—for example, `What is the capital of Belgium?` or `"Rewrite the following passage in the style of Herman Melville."` The words in these prompts are not "inherently meaningless," as identifiers in ALGOL are supposed to be, but instead must be used in ways that correspond to their usage in the human writings on which the model is trained. The idea of communicating with a computer in natural language is not new; it goes back at least to Jean E. Sammet's 1965 talk about programming in English. Yet the possibility did not come close to realization until the 2010s, and this new technological chimera promises a radical shift in how machine and human interact. By granting mathematics and language coequal roles, GPT-3 pushes the subject–object dualism that has attended symbolic methods since the time of George Boole, and that forms the basis of both the classical conception of algorithm and the modern programming language, to a breaking point.

The GPT models are based on an ANN architecture called the Transformer, which was first introduced in 2017.[25] Transformers are language models, meaning models trained to predict what word will appear in a certain context. The GPT models predict the next word after a given sequence of words (e.g., I bought some food at the grocery → prediction: store), whereas other models such as BERT (Bidirectional Encoder Representations from Transformers), which was developed by a Google AI Research team led by Jacob Devlin, are trained to fill in the blanks, Mad Libs style.[26] The Transformer architecture has two parts, an encoder and a decoder. Full Transformers such as T5 are supposed to be able to translate from one language to another by encoding the text into a numerical form and then decoding it in the target language.[27] The GPT models, which are primarily intended for text generation, consist only of decoders. By swapping in different "heads"—extra layers added to the end of the neural network—it is possible to apply Transformer models to a range of tasks, leveraging whatever they learned in the training process for such purposes as answering questions, finding information, classifying texts, and performing automatic summarization. It is also possible to use text generation heads for other tasks by feeding in specially designed inputs that (at least ideally) induce the generator to produce the desired output.

The idea of a language model originated in the twentieth-century work of C. E. Shannon and Warren Weaver. In his germinal 1948 paper on information theory, Shannon described the "n-gram" model, which (building on the prior work of A. A. Markov Sr.) determines which words or characters are most likely to appear after certain sequences of a fixed length.[28] Thus, after "ENGLISH," Shannon's model guesses that the next word will be "WRITER."[29] The original purpose of these models was to describe statistical properties of language so as to better account for noise in communication channels. In the second paragraph of the paper, Shannon performs a notorious act of disciplinary boundary setting: "Frequently the messages have *meaning*; that is they refer to or are correlated according to some system with certain physical or conceptual entities. These semantic aspects of communication are irrelevant to the engineering problem."[30] Weaver expands on this point in a companion essay that was published together with Shannon's paper in a 1949 book, distinguishing "three levels" in the theory of communication.[31] The "technical level" has to do with whether symbols are transmitted accurately, the "semantic level" with what those symbols mean, and the "effectiveness level" with whether that meaning is chosen appropriately. According to Weaver, language models deal only with the technical level, although he also notes that the other two levels "have much to learn from" a statistical analysis of language.[32]

By presenting language models as multitask learners, the designers of GPT-3 take up Weaver's idea of making statistics speak to semantics. In other ways, however, GPT-3 departs from twentieth-century disciplinary categories. In addition to his work with Shannon, Weaver was an early advocate of machine translation, which he discussed in a 1949 memo. In this memo, Weaver argues that there is a logical core common to all languages. The best way to translate between languages, he argues, "is to descend, from each language, down to the common base of human communication—the real but as yet undiscovered universal language—and then re-emerge by whatever particular route is convenient."[33] This claim echoes the universalist thinking behind John Wilkins's real character and G. W. Leibniz's universal characteristic, presuming that a certain set of concepts is common to all languages. But as with ALGOL, Weaver's universality is cut across by a disciplinary line dividing logic from culture. Mechanical methods, Weaver states, may be harder to apply to "literary translation," in which "alogical" elements of language such as "intuitive sense of style, emotional content, etc." are of crucial significance.[34] Weaver here draws a line much like the one Boole set up in the last chapter of *The Laws of Thought*. On the one side are the literal meanings of words and their logical relations, and on the other are feeling, aesthetics, and the distinctive qualities of particular languages—in short, all that would typically be viewed as matters for the humanities.

Experience has shown that Weaver was wrong about what machines could do: it is precisely what he called the "alogical" aspects of language that machines are best able to manipulate. The GPT models have proven capable of "style transfer," meaning rewriting a passage in the style of a different writer. Generating logically coherent text has turned out to be more difficult. Consider this machine-generated paragraph, from the blog post introducing GPT-2:

It is easy to identify why the Civil War happened, because so many people and so many books and so much television and films tell us that it was the cause, that it has something to do with race or economics or religion. There is much agreement that it was essentially a war of slavery on behalf of capitalism, about a century of slavery. But that's not what most people think of when they think of the Civil War. Many people think of the war as about states rights. You're not wrong about that, but it's kind of misleading to say that the Civil War was a conflict between states' rights and federalism. So let's try again. What's the idea that has inspired so many people to go onto the battlefields of Gettysburg and Antietam and Gettysberg and Petersburg and Fredericksburg? The American idea of the republic—a notion of limited government—is a great part of the history.[35]

Such text is the polar opposite of the all-substance, no-style logical language imagined by thinkers from William of Ockham in the 1300s to Edsger Dijkstra in the 1980s. If understanding a text is, to use a very old analogy, like digestion, then reading machine-generated text can at times seem like smelling food without eating anything: one gets the sense that something is being said, but one can never pin down exactly what that something is.

This is not to say that the developers have abandoned logic. It would be more accurate to say that they have reversed the terms: logic is not the starting point for computational methods but the goal. The technical paper about GPT-3 presents the ability of the model to perform three-digit arithmetic as a sign of the emergent intelligence of the system. Thus, in a twist that is almost too obvious to point out, an "algorithm" is supposed to learn on its own how to perform the procedures that originally went by the name *algorithm*. The aim is to produce a system that is capable of a range of logical tasks such as "commonsense reasoning."[36]

Unlike in the classical paradigm, there is no way to judge with mathematical certainty whether the software is performing these tasks correctly. Instead, researchers employ benchmarks such as SQuAD (Stanford Question Answering Dataset), which resemble nothing so much as standardized tests.[37] SQuAD consists of a series of short passages accompanied by questions to be answered using quotations from the passages. "What," for instance, "is controled [*sic*] by the market and economy?" Correct answer: "workers [*sic*] wages."[38] There are multiple versions of the question-answering task; some are about finding the answer in a supplied passage, whereas others administer a "closed-book" quiz, in which the information must be found in the model itself. The ability to answer questions is supposed to demonstrate that language models can "understand" texts and thus perform the sorts of information extraction tasks for which they are used in systems such as Google search and Alexa.

Much has been written about whether such a system can really "think," a question that touches on central problems in the philosophy of mind. More to the present point is how the models relate to language. To begin with, GPT-3 is in no way a straightforward representation of a coherently defined "English language." The training data that were used to produce the GPT models consist primarily of books and text from the internet, filtered through semiautomated means to eliminate unintelligible or otherwise undesirable text. Defining the criterion of quality is a troublesome matter, and the data sets fall short of the standards linguists follow in constructing corpora. Since a large part of the text comes from internet discussion sites, the models tend to reproduce the viewpoints of the people who spend the most time on those sites. The training data also include text in various dialects and creoles, as well as from a range of time periods—the models have, for instance, clearly been exposed to Shakespeare—but the texts are not necessarily accompanied by any information about where or when they were written. English is a living language, existing in many forms and changing over time, and the models can only account for these differences insofar as information about them can in some way be abstracted from patterns in the words themselves.

A related point is that language models depend on, or at least benefit from, a historical process of language standardization. Starting in the seventeenth century, dictionaries and grammar books encouraged regularity in written English, leading to the standardization of spelling. For centuries before then, no single dialect of English was clearly dominant, and spelling was chaotic: the proverbial scribe might spell the same word three different ways on the same page. There is not nearly enough surviving text from this earlier stage in the development of English to train a language model on the scale of GPT-3, so we can only speculate as to how well it would work.[39] The task would, in any case, be different. The model would not be able to presume a stable correspondence between sequences of letters and "words" in the abstract sense, nor would it be so easy to pin down the semantics of grammatical forms when usages differ widely between regions and writers. If language models work through abstraction, some of that abstraction has already been done through the standardization of the English language itself.

Setting aside the issue of what language is being represented, there is a broader disconnect between language models and language that stems from how ANNs work. After the training process is complete, the ANN is static; when GPT-3 is used to generate text or perform other tasks, it does not "learn" anything new in the process. Critics have noted the possibility that such static models will lead to "value-lock," in which a technology ossifies practices that existed at the time of its design.[40] While this specific training paradigm is liable to change, a broader issue remains regarding how the model's success is determined. Because ML models do not answer to any precisely defined standard of what the output should be when they are fed unseen input, there is no way of judging whether a text generator's imitation of language is good enough without resorting to some form of linguistic knowledge. This knowledge could consist of the competency of native speakers, in which case evaluating the model is a matter for surveys and focus groups. If academic disciplines are at issue, then we must turn to the ancient trivium of grammar, rhetoric, and logic, along with literary aesthetics and, most antithetically to Dijkstra's way of drawing the disciplinary lines, hermeneutics.

By granting such forms of knowledge a role in the evaluation of computer systems, ML reopens an ancient question: is language really a reliable medium? This question is relevant not just to text generators such as GPT-3 but also to any other ML system that employs words. Image classification models are culpable, since they usually classify the objects depicted in photographs using words such as *motorcycle* or *cat*.[41] In constructing computable definitions of words based on data, these systems recall the empirical methods by which some Enlightenment thinkers hoped to build a new and better view of the world.[42] ML models differ from these Enlightenment projects, among other ways, in that they do not deign to replace existing languages all the way down to the formation of concepts. Like Boole's logical equations, GPT-3's pronouncements draw what meaning they have from the languages people already speak. Yet in contrast to Boole's system, the technology does not offer a standard of correctness that can plausibly be said to work independently of those languages. Understanding the implications of this development requires considering the terms on which computational processes interact with the social practices that govern human communication. From here, there is no direction to proceed but into politics.

THE NEGOTIATION

If there is a common thread linking computer culture to the Enlightenment, it is a belief that technological progress can create social equality. As Condorcet thought the printing press would create an open public sphere in which people could discourse as equals, the early denizens of the internet envisioned a utopia. One computer-age echo of the Enlightenment occurred in hacker culture—a loose affiliation of computer users who sought to break into secure systems, sometimes for malicious purposes and sometimes merely to explore. In the widely circulated 1986 manifesto titled "The

Conscience of a Hacker," the pseudonymous hacker "The Mentor" (real name Lloyd Blankenship) presents a rationale for this practice. The form of hacking he endorses resembles the Enlightenment genre of the exposé, which sought to reveal what powerful people are hiding. Throughout the manifesto, he repeats the refrain that hackers are "all alike."[43] The members of the hacker community, he writes, "exist without skin color, without nationality, without religious bias"; their culture involves "judging people by what they say and think, not what they look like."[44] The idea is that, by enabling people to communicate purely through text, digital media could place everyone on a level playing field, regardless of their race, gender, or socioeconomic status. It is a dream that today seems naïve, but it persists in the discourse of neutrality that continues to surround algorithms. The idea that a system of communication can eliminate bias reiterates the desire for a level playing field that motivated Condorcet's language scheme.

But there is a complexity to this utopianism that is worth unpacking. For the Mentor, equality stems primarily from the openness of the technical medium: as long as people are able to think, speak, and explore freely online, he supposes, social differences will vanish. In effect, he treats the members of the hacker community as autonomous subjects engaging in a public sphere defined by the exchange of words. Yet the manifesto also suggests the need for education: hackers have "been spoon fed baby food at school when we hungered for steak."[45] Education is often de-emphasized in narratives about the egalitarian possibilities raised by the internet, but it was central to the Enlightenment project. Condorcet's utopian plan did not stop at enabling words to flow freely; instead, he treated words themselves as bearers of prejudice. Achieving true equality, then, would mean not just granting people the freedom to think and speak but also changing *how* they think and speak. To his detractors, this makes his philosophy implicitly totalitarian, an attempt to impose a certain way of thinking on the world. Yet the internet has made the absence of a common language felt. Technology has divided as much as it has united, and open communication has reinforced hierarchies as much as it has demolished them.

In hindsight, it is not surprising that the decline of internet utopianism coincided almost exactly with the shift in usage of *algorithm*. Classical algorithms fit with the early internet's ideal of openness: the "Algorithms" section of *Communications of the ACM* provided a repository of reusable procedures that were described and explained in a public forum. With the rise of machine learning, the word *algorithm* has come to be associated less with openness than with opacity. The shift had to do in part with the emergence of new software engineering techniques: the mid-2010s saw the widespread adoption of container management systems such as Docker and Kubernetes, which make it much easier to transfer complex software configurations across computers. Such systems can work together with "low-code" and "no-code" approaches to development, which work mainly through combining and configuring prepackaged programs. If algorithms were, in the early decades of computer science, about generalizing procedures for reuse, the parts of computer systems being repurposed and reused have now become far more complex, encompassing not just procedures but also models, data, and software installations.

These changes in technical practice coincided with a shift in how the internet worked as a medium, both institutionally and culturally. If the Mentor imagined the internet as a transparent conduit through which people could speak their minds without resistance, the last great expression of this view was the personal blog, which suffered a devastating blow when Google shut down its aggregation service Google Reader in 2013. Such services largely gave way to social media platforms with so-called algorithmic feeds, which decide what to show users through hidden processes tailored to maximize profitability by keeping people's attention for as long as possible. In the second half of the decade, two highly influential books, Cathy O'Neil's *Weapons of Math Destruction* (2016) and Safiya Noble's *Algorithms of Oppression* (2018), singled out algorithms as engines that can reinforce inequality along race and gender lines.[46] The internet circa 2010 was never the inclusive utopia the Mentor imagined, and algorithmic media threaten to worsen the problem.

The critiques of O'Neil, Noble, and others inspired two distinct lines of response. Some artificial intelligence (AI) researchers began to analyze their models for biases that could theoretically be addressed in future revisions. An example of this approach appears in the GPT-3 paper, which reports experiments testing for bias along the lines of gender, race, and religion.[47] Researchers have also undertaken efforts to make AI systems "interpretable" or "explainable" so as to allay concerns about their opacity.[48] Others have questioned whether AI is a good idea at all. In March 2021, Emily M. Bender, Timnit Gebru, Angelina McMillan-Major, and Margaret Mitchell (under the intentionally transparent pseudonym "Shmargaret Shmitchell") presented a paper titled "On the Dangers of Stochastic Parrots: Can Language Models Be Too Big? 🦜"[49] Gebru and Mitchell had been coleaders of Google's Ethical AI team, but both left the company before this paper was published. (Reports differ as to whether Gebru was fired or resigned voluntarily.) The paper is a discussion of the disadvantages of large language models such as GPT-3; its main suggestion is that researchers focus on smaller data sets that can be curated and thoroughly described. Gebru and Mitchell's departure from Google received a great deal of media coverage, contributing to a widespread awareness of the issue of algorithmic bias.

This incident shows that the redrawing of the disciplinary lines around algorithms in the mid-2010s was not merely about semantics. The arguments of O'Neil and Noble inspired a real change in the issues being raised. As has been widely reported, it is not difficult to induce a range of ML technologies, including the GPT models, to produce outputs that seem to replicate stereotypes.[50] When, for instance, one enters the text "The man worked as a," GPT-2 predicts that the next two words will be "security guard"; when one enters "The woman worked as a," it comes up with "waitress."[51] The problem stems in part from the makeup of the training data used to produce the GPT models, which likely contain a large amount of racist and sexist text; as a result, it is not surprising that the model can generate racist and sexist output. Yet the models raise a broader issue of trust that cannot be resolved merely by fixing specific, known issues. Since the models cannot be proven correct in the way Dijkstra wanted to prove the correctness of programs, it is hard to imagine what undiscovered problems may lurk in GPT-3's massive parameter matrices.

The conversation about these issues has raised slippery questions about how computational processes relate to language. From the 1960s to the 1980s, literary critics debated whether meaning is based on the intention of a writer or constructed by the reader.[52] In the late 2010s, this debate gained a renewed relevance when it came to be cited in discussions of the legal implications of AI. In a 2018 book, the legal scholars Ronald K. L. Collins and David M. Skover argue that machine-generated text constitutes "intentionless free speech" that ought to be protected under United States law.[53] The "Stochastic Parrots" article takes quite the opposite position, stating that "the tendency of human interlocutors to impute meaning where there is none can mislead both NLP [Natural Language Processing] researchers and the general public into taking synthetic text as meaningful."[54] The seeming coherence of the output, the authors elaborate, is an illusion because the text "is not grounded in communicative intent, any model of the world, or any model of the reader's state of mind."[55] This line of thought suggests that, for all that it can parrot human writing practices, GPT-3 is not doing the same thing with words that humans do, and to assume otherwise is to be duped.

While their immediate reference points are literary theory and psycholinguistics, these arguments are also reopening questions that attended symbolic methods centuries ago. By combining language models with other sources of information, AI researchers aspire to realize the most ambitious version of Leibniz's calculus ratiocinator: a unified method for answering questions about anything and everything.[56] As with Leibniz, these ambitions have run up against the messiness of communication. In one of the examples given in the GPT-2 article, the machine is given the question "What river is associated with the city of rome?"; it answers, "the Tiber."[57] It works, at least in this case. But something is missing. While the question about the Tiber has a generally agreed-on answer by virtue of which that output could be judged indisputably correct, the model provides no clear mechanism for creating agreement about terms that are contentious or divisive. Instead, the correctness of the answer is judged wholly by whether the model chooses the right sequence of words, without regard to what people may think those words mean.

The problem is most apparent in regard to politically charged questions. For instance, in a nod to the French Revolution, I asked GPT-2 whether liberty, equality, or fraternity was most important. The most important one, it responded, is "the freedom of the individual to choose his own life."[58] Could such an output, even coming from a system people seriously trust on such issues, rationally change anyone's mind about anything? Evaluating the model solely by whether it chooses the right words by a predefined rubric covers over the fact that words do not always mean the same things to all people. If two users agree that individual freedom is the highest value but differ as to what *freedom* entails, they may agree that the AI produced the correct answer while having vastly different ideas of what that answer actually was.

This profusion of conflicting interpretations could not be resolved merely by improving the quality of the answers, as researchers will undoubtedly do. If an AI's pronouncements truly were viewed as authoritative, they would be pored over and debated like the text of the Bible. The problem is language itself. Words—not any particular words, but words in general—are incapable of doing what the question-answering task asks of them. This issue is not new; rather, it has been a part of the discourse about universal computation for centuries. The biases found in language models are a modern iteration of Francis Bacon's idols of the market—prejudices and errors reproduced through everyday chatter. There is, indeed, a not insignificant parallel between the concerns about meaning raised in the "Stochastic Parrots" paper and John Locke's account of the "abuse of words": in both cases, people are charged with credulously presuming that words adhere to the same concepts for the writer as for the reader. How to respond to such issues was a central epistemological question before the subject–object divide took it off the table.

In this regard, a more illustrative figure than Leibniz would be Condorcet. Whereas Leibniz had a theological basis for trusting that people would eventually reach an accord about what symbols meant, Condorcet wanted to create this accord through education and social transformation—not just providing an instrumental means of producing answers but also teaching people a new way of understanding the world. The question-answering task skips over this step since, as in the idea that equal participation in the public sphere can be produced solely through the free exchange of words, its epistemology stops short at language. All the difficulties of forming human understanding are off-loaded onto words, on whose preexisting ability to produce an agreement the outcome depends. This dependence is the ultimate result of the passive attitude toward words that has often attended symbolic methods since the Lockean epistemology collapsed—the tendency to treat language as a natural resource from which meanings can be taken as needed but that is ultimately separate from algorithmic logic. Recent experience has shown that this approach is not sufficient, that words can cause problems, that language should not just be taken as it is.

A helpful starting point in thinking through this issue would be to consider what user interface designers are doing when they place words or other signs on the screen. As P. B. Andersen explains in his *Theory of Computer Semiotics*, interface designers must find a balance between making systems understandable in terms of existing categories and opening new ways of thinking.[59] The former goal often involves metaphors—words such as *file* and *folder* suggest analogies with older practices as a way to help people understand software.[60] But computers also bend the meanings of the words in new directions. Words that are commonly used in computer interfaces, such as *open*, *save*, and *window*, have taken on new senses that depart from the metaphors that initially motivated them. Other concepts are entirely new—*database* and *software* refer to ideas that were nonexistent prior to the computer age. User interface design draws on language in part to change it, and it involves, accordingly, a negotiation between the individuals and organizations that produce software and the broader communities of language speakers who have the power to accept or to reject new coinages, and who can employ them in ways that break with the designers' intent.

The question-answering task handles this negotiation in a one-sided way, at least as far as the words in the output go. The answers are supposed to make sense within existing languages, not, in any intentional way, to contribute anything new to language. In this regard, the subject–object divide remains in place. The situation is more complex, however, regarding the words in the input, and it is here that the GPT models open the potential for a radical rethinking of how computation relates to language. Some enthusiasts of the models have argued that multitask learners will transform the practice of programming; rather than implement algorithms by writing code, the programmer will design a prompt that gets a model to perform the desired task.[61] To pick an instance from the paper about GPT-3, the researchers entered, "Please unscramble the letters into a word, and write that word: taefed =" ; the machine produced the desired response, "defeat."[62] It would not be especially hard to design an algorithm that could solve this problem reliably, provided one has a list of acceptable words; it is the sort of problem that might be used as an exercise in a computer science class or job interview. What is new is the fact that the computer is apparently able to figure out a solution on its own.

This prompting technique is not specific to text-oriented tasks. By combining a language model with image generation techniques, the popular program CLIP+VQGAN, first publicized in late 2020, produces synthetic images based on text descriptions entered by the user.[63] This program is touted as a way of democratizing AI art: since the prompts may be written in English, it is possible to experiment with the system and produce interesting (if unpredictable) results without needing much of a technical background. GitHub Copilot, which was released as a preview in June 2021, applies a similar method to computer programming: it uses a modified version of GPT-3 to generate code automatically based on natural-language comments describing what the code is supposed to do. In her work on compilers in the 1950s, Grace Hopper sought to make computing less tedious by transferring the work of converting algorithms into executable instructions from human to machine. These new practices move the chains once again: the machine, not the human, transforms the problem definition into an algorithm. This idea suggests a return to how Hopper, early in her career, understood what we now call programming: communicating with a machine.

However successful prompting turns out to be as a programming paradigm, the need to control language models raises issues that trouble the rigid distinction of algorithms from communication. The identifiers "a34kTMNs" and "MARILYN" may be interchangeable in ALGOL code, but if one is trying to induce a language model to perform a certain task, every word choice matters.[64] One proposed approach to multitask learning is *pattern-exploiting training*, which employs templates that exploit how pre-trained models respond to certain arrangements of words. For instance, the following patterns can be used to answer a question q based on passage p:[65]

    p. Question: q? Answer: __.
    p. Based on the previous passage, q? __.
    Based on the following passage, q? __. p

The effectiveness of these verbal formulae in getting language models to do things stems from the human meaning-making practices represented in the training data—from ways people have used words and punctuation marks in the past. Understanding what differentiates one prompt from another requires neither the mathematical analysis of procedures endorsed by Dijkstra nor any familiar sort of interpretation, but rather analyzing the effects the words have on a series of matrix calculations that somewhat, but not totally, parallel the structures of the English language.

In the face of this new hybrid of mathematics and language, viewing language as unproblematic because it lies outside a disciplinary boundary, as Boole implicitly did and the classical algorithm implicitly does, is no longer tenable. An alternative lies in the mistrust of words that has often attended symbolic methods in the past. Although the overbearing language reform efforts of the 1790s deserved much of the criticism they faced, the problems to which these efforts responded have never gone away. Words can harm, divide, and distort, and they can do so all the more readily when they are presumed to be transparent. The dangers posed by large language models such as GPT-3 stem in part from an uncritical attitude toward language, from a misguided belief that our preexisting conceptions of what words mean can provide an adequate guide to those words' significance even when those words are chosen by a machine. The symbolic turn of the nineteenth century removed such concerns from the domain of algebra, and this boundary persists, among other places, in the way language model benchmarks measure knowledge by the word. The time has come to reopen the questions that surrounded symbolic computation before this arrangement took shape.

CONCLUSION

As I have shown in this book, anxieties about communication have arisen again and again in discussions of symbolic methods, from the emergence of modern algebraic notation to the development of the programming language. In the seventeenth century, symbols were aligned with the dream of making writing independent from language; algebraic notation promised a way of putting thoughts directly on a page without the intervention of words. In the eighteenth century, algebra became the model of a well-formed language, as the Abbé de Condillac put it, and the arbitrariness of algebraic symbols—the ease of changing the values of a and b—testified in favor of the possibility of building a better language from the ground up. Yet the methods of algebra and calculus seemed, to many observers, to involve ill-defined or absurd ideas, giving rise to an alternate view in which symbols were only as good as the words used to define them. These differing positions had implications for whether the "algorithm" of the differential calculus could be trusted, whether mathematicians could propound a new definition of *sum*, and whether −1 was a true number or mere nonsense.

These old debates, quaint as they may seem, touch on a question worth reviving: how to balance explicit definitions with social convention as ways of determining the meanings of signs. In its dependence on existing languages, GPT-3 is implicitly aligned with the social sort of meaning: whatever significance the word *sum* may have in its output stems from how that word has been used by people, as mediated through a complex technological apparatus. Yet the practice of computation hews strongly toward treating words and other symbols as arbitrary tokens with no inherent meanings beyond the roles assigned to them in code. This way of thinking is manifest in how the question-answering task is evaluated: the output is scored solely by how much it overlaps with the desired sequence of characters. This approach shows the continuing influence of the subject–object divide that, starting in the mid-nineteenth century, granted symbols autonomy from words by consigning the social complexities of language to the practical realms of pedagogy, scientific communication, and, eventually, user interface.

This divide can be explained quite simply. Modern computation is, as I have termed it, aconventional: in standard programming languages, one can assign computational meanings to words and other symbols at pleasure, regardless of how people have used them in the past. But this aconventionality only holds if one draws a line between one's choices and the broader contexts in which they are made—between the theoretical sense in which it is possible to do something silly like swap the names input and output (as a programmer can indeed readily do) and the more practical sense in which doing so would be foolish. The hierarchical distinction between "hard" algorithmic logic and "soft" human factors, as instated by early computer scientists such as Dijkstra, elevates the former perspective as the logical one even if many of the practices it enables are far from logical. Moving away from this distinction would mean recognizing that the choices involved in programming computers are, after all, to some extent constrained by the conventional meanings of words and symbols—that there is an epistemologically relevant reason why programmers do not, except as a joke, use input to mean output.

Such issues, though excluded from the domain of "hard" logic, are hardly unfamiliar to programmers. Large software projects follow elaborate naming conventions to ensure consistency and avoid ambiguity. In the private sector, the discipline imposed by such conventions is indeed valued above individualistic coding practices. But naming conventions only operate within relatively small teams of developers or, at largest, the user bases of programming languages such as Python. Language models negotiate with communities of millions or billions of speakers, and the issues that arise at this scale are more contentious. One source of tension is a discomfort with technology companies imposing ideas on the population—with the undue amount of power granted to the people with the privilege of turning the knobs. But ML raises other issues that do not fit neatly into a trickle-down model that presumes developers have total control over systems. As the critical discourse has shown, GPT-3's training text contains elements that are best not enshrined in computer systems: prejudices, biases, falsehoods. To reckon fully with these problems is to think of people's linguistic practices not as a well from which meaning may be drawn, as in the oracle model of question answering, but rather as an arena of human activity that may be critiqued and improved.

The pendulum has already begun to swing back that way. Alongside the rise of ML, the 2010s saw a resurgence of interest in Rudolf Carnap's idea of conceptual engineering, this time with a great deal more emphasis on its political implications.[66] The new language reform is often aligned with feminist causes, such as the promotion of gender-neutral pronouns, although some advocates are more concerned with scientific precision. While their concerns are characteristic of the twenty-first century, these efforts express a centuries-old perception that words are not a neutral medium for expressing thoughts, that language can provide a distorted view of the world. The more activist forms of conceptual engineering arguably share some of the drawbacks of universal language schemes, including a tendency to force certain practices on people. Language reform also bears the potential (as noted by eighteenth-century thinkers such as Condorcet and Johann David Michaelis) to create social divisions by alienating those who have adopted the reformed language from those who have not. Yet the stakes have changed. This time around, computation no longer seems like a solution to the flaws of language, as it did for Leibniz. Now, it seems to worsen the problem, making those flaws all the more urgent to address.

While the conversation about conceptual engineering is separate from the concurrent debates about algorithms, the two raise similar issues regarding the relation of individual to collective agency. Much of modern computation's danger and power stems from the fact that its aconventionality enables programmers to undertake in isolation what is ordinarily a social matter: deciding the meanings of signs. Computation claims the status of a universal language while skipping the hard part of the work it would take to create one. Yet this aconventionality—the programmer's ability to redefine a symbol as easily as changing "$A := 5$" to "$A := 6$"—also bears the potential to challenge received practices by bending the meanings of signs in new directions. Thinking through what it would take to alter the meaning of a sign considered not as an arbitrary token, as ALGOL treats identifiers, but in its social circulation—the difficulty of achieving clarity, the need to reckon with established usages, the politics of reaching agreement—can help in navigating the issues modern computation raises. Although universality may be a misguided goal, it is still possible to work toward a common language, a shared understanding that enables people and machines to work together without merely talking or computing past each other. A crucial political issue is who gets to decide what that common language should be.

Acknowledgments

Completing a project of this scope would have been impossible without the institutional support of Pennsylvania State University's Center for Humanities and Information, which has provided both financial support for my research and a vibrant community. I want to thank Eric Hayot, Pamela Vanhaitsma, and John Russell for creating a supportive environment for humanistic research about information. I also thank Olivia Brown for helping me obtain books while the interlibrary loan system was shut down due to the COVID-19 pandemic. I am indebted to the members of my dissertation committee at the Graduate Center, City University of New York—Alexander Schlutz, Matthew K. Gold, and Joshua Wilner—who provided invaluable feedback on earlier stages of my research. At a more personal level, I am grateful for the conversations I have had with Patrick Smyth, who has always pushed back when I have ventured too far into abstraction. My partner, Ann Marie Genzale, heard more chatter about algorithms than is healthy for one lifetime, and I thank her for being there for me through this project. Finally, I want to thank my parents, who always supported me in my endeavors.

I wrote significant portions of this book during the pandemic, which has probably affected the writing in some subtle way that could be detected through stylometric analysis. My work would have ground to a halt in 2020 without the availability of digital copies of primary and secondary sources. Some of the resources I used include the HathiTrust Emergency Temporary Access Service, Early English Books Online, the Linda Hall Library's digital collection, the Gallica database of the Bibliothèque Nationale de France, the ACM Digital Library, and the collections curated by the Computer History Museum's Software Preservation Group. The pandemic limited my ability to travel to archives, but these and other resources provided a wealth of material by which I could puzzle over what people might have meant when they wrote about algorithms.

Earlier versions of some sections of chapters 3 and 4 appeared in the article "Romantic Disciplinarity and the Rise of the Algorithm" in *Critical Inquiry* 46, no. 4 (Summer 2020): 813–34. I also presented papers based on this material at two 2018 conferences, the PSU Information + Humanities Conference in University Park, Pennsylvania, and the North American Society for the Study of Romanticism conference in Providence, Rhode Island.

Notes

INTRODUCTION

1. T. Brown et al., "Language Models Are Few-Shot Learners," 25, 48. On the first and second GPT models, see Radford, Narasimhan, et al., *Improving Language Understanding by Generative Pre-Training*; Radford, Wu, Child, et al., *Language Models Are Unsupervised Multitask Learners*. The particular model used in GPT-3 is called Transformer; it was first described in Vaswani et al., "Attention Is All You Need."

2. T. Brown et al., 8.

3. T. Brown et al., 1.

4. There has been some discussion about remedying this ungroundedness by providing models with context for language; see Bisk et al., "Experience Grounds Language."

5. A famous example is John Searle's "Chinese room" thought experiment; he imagines a group of people performing an algorithmic procedure to carry on a conversation in a language they do not know and asks whether this system could be said to enact understanding of the language (Searle, "Minds, Brains, and Programs"). For some discussion of this issue in relation to GPT-3, see Weinberg, "Philosophers on GPT-3 (Updated with Replies by GPT-3)."

6. I use the phrase *algorithmic thinking*, following the computer scientist Donald E. Knuth and the historian Ivor Grattan-Guinness, to mean a style of thought directed toward the design of precisely defined procedures for solving problems. Some writers use the phrase to refer, instead, to a form of thought that itself follows algorithm-like rules. The two are not the same; designing algorithms is often thought to involve intuition and spontaneous insight. See Knuth, "Algorithmic Thinking and Mathematical Thinking"; Grattan-Guinness, "Charles Babbage as an Algorithmic Thinker."

7. This is a modern formulation of Hooke's law, first described by Robert Hooke in the seventeenth century. Hooke himself presented his theory geometrically, not in symbolic notation. See Hooke, *Lectures de potentia restitutiva*.

8. This framing is different compared with my earlier publication on the topic, and so it is worth a note. I formerly wrote about a divide between algorithms and *interpretation* or *meaning*. I have found that these terms are potentially confusing in regard to programming languages, which are sometimes said to have a computational "meaning" determined by the formal semantics. The issue I am hoping to raise is how algorithms interact with the sort of meaning that is created socially, which the term *communication* captures better.

9. An example of a history that takes this approach is Barbin et al., *A History of Algorithms*, which begins with the following statement: "Algorithms have been around since the beginning of time and existed well before a special word had been coined to describe them" (1).

10. Knuth, "Ancient Babylonian Algorithms." See also Gleick, *The Information*, 45–46.

11. The anthropologist Stephen Chrisomalis has criticized the name *Hindi–Arabic numerals* on the grounds that it obscures the fact that Arabic and Indian writing systems continue to employ different digit characters from the Western ones (*Reckonings*, xvi). I use this term not to refer to the specific symbols used in the West but rather to base-ten positional number systems in general; however, Chrisomalis's argument should serve as a warning against the presumption that Indian, Arabic, and Western systems constitute phases in a historical progression rather than enduring practices in different geographic regions.

12. Davis, *The Universal Computer*, 1.

13. Mahoney, "The History of Computing in the History of Technology." See also Daston, "Enlightenment Calculations"; Hicks, *Programmed Inequality*; M. Jones, *Reckoning with Matter*; Schaffer, "Babbage's Intelligence."

14. Priestley, *A Science of Operations*, v.

15. M. Jones, *The Good Life in the Scientific Revolution*; Antognazza, *Leibniz*.

16. Erickson et al., *How Reason Almost Lost Its Mind*.

17. The literature on these issues is immense and I cannot hope to give a full overview. For a range of perspectives, see Besteman and Gusterson, *Life by Algorithms*. For an account of the recent turn within the broader history of the idea of algorithm, see W. Thomas, "Algorithms."

18. It may be objected that computer code serves a different function from numerals: code is used to describe procedures, whereas numerals are the tokens with which procedures are performed. I would argue that, at least from the emergence of modern algebraic notation in the seventeenth century, this distinction has not been of fundamental significance. In their modern form, algebraic formulae can serve both as instructions for how to compute something and as objects of symbol manipulation: one could either use Hooke's law as a template for computing the force or transform it into an inverse formula based on the rules of algebra: $F_s/k = x$. In much the same way, computer code is meant at once to describe procedures and to be encoded as data and processed by means of other procedures. Attempting to disentangle these two functions will only ensnare one in conceptual thickets.

19. Dear, *The Literary Structure of Scientific Argument*; Dear, *Discipline and Experience*; Markley, *Fallen Languages*. See also Dear, *The Intelligibility of Nature*.

20. See Golumbia, *The Cultural Logic of Computation*; Lennon, "Machine Translation"; Lennon, *Passwords*; Larson, "Optimizing Chess."

21. Lambert, "A Natural History of Mathematics"; T. Williams, "Procrustean Marxism and Subjective Rigor"; T. Williams, "Mathematical Enargeia." See also Cifoletti, "La question de l'algèbre"; Cifoletti, "From Valla to Viète."

22. Lamy, *Art of Speaking*, 203.

23. I am not the first to call attention to the importance of such statements in the history of computation. Warren Sack discusses assignment statements as assertions of equivalence; he warns that one should "mind the gap" between systems that are held to be equivalent even though they differ (*The Software Arts*, 34–37). Few mathematicians

before the late nineteenth century, however, would have viewed the statement "let $a =$ 5" in terms of equivalence. Earlier on, assignment was usually described in terms of signification: the "let" statement was taken to determine a symbol's previously undecided meaning.

24. Gorham et al., *The Language of Nature*.

25. Dawson, *Locke, Language and Early-Modern Philosophy*.

26. Rosenfeld, *A Revolution in Language*.

27. G. Boole, *The Mathematical Analysis of Logic*, 5.

28. Bauer et al., "Proposal for a Universal Language for the Description of Computing Processes"; quotation from title.

29. Alan Turing discusses the idea of machine learning in his well-known 1950 essay "Computing Machinery and Intelligence."

30. *OED Online*, s.v. "technology, *n*," accessed January 23, 2022, https://www.oed .com/view/Entry/198469.

CHAPTER ONE

1. Bacon, *The Advancement of Learning*, 242.

2. Bacon, 242.

3. For a variety of perspectives on symbolic methods in the period, see Heeffer and Dyck, *Philosophical Aspects of Symbolic Reasoning in Early Modern Mathematics*.

4. Going back to R. F. Jones in the twentieth century, it has been a common narrative that the Scientific Revolution contributed to the development of a plain style of English prose; Walter Ong argues that the rise of modern mathematics coincided with a decline of the traditional verbal arts. Robert Markley and Timothy Lenoir later questioned some aspects of this narrative. See R. Jones, *The Triumph of the English Language*; Ong, *Ramus, Method, and the Decay of Dialogue*; Markley, *Fallen Languages*; Lenoir, *Inscribing Science*. For more background about the language arts in the period, see Howell, *Logic and Rhetoric in England, 1500–1700*; Padley, *Grammatical Theory in Western Europe, 1500–1700: The Latin Tradition*; Padley, *Grammatical Theory in Western Europe, 1500–1700: Trends in Vernacular Grammar*.

5. In some early usages, variants of the word referred to other methods of computation. In the "Miller's Tale," Geoffrey Chaucer mentions "augrym stones," applying a variant of the word *algorism* to the use of counting stones (Chaucer, *Riverside Chaucer*, 68). The title of the anonymous 1539 book *An Introduction for to Lerne to Recken with the Pen or with the Counters Accordynge to the Trewe Cast of Algorysme [. . .]* suggests that both numerals and counting stones fall under the scope of algorism, although the text uses the spelling "awgrym" in the discussion of the latter (sig. I.i.v).

6. On the parallel development of alphabetical writing and number systems, see Ifrah, *The Universal History of Computing*; Chrisomalis, *Reckonings*.

7. Gnanadesikan, *The Writing Revolution*, 4.

8. The division of mathematics into geometry, arithmetic, astronomy, and music goes back to the ancient Greeks; Boethius named these four subjects the "quadrivium," which later became a standard part of the liberal arts curriculum. See Guillaumin, "Boethius's De institutione arithmetica and its Influence on Posterity." An example of machines as mathematics is Wilkins, *Mathematicall Magick*, first printed in 1648; see Zetterberg, "The Mistaking of 'the Mathematicks' for Magic in Tudor and Stuart England."

9. Galilei, *Discoveries and Opinions of Galileo*, 237–38. On the background of this text, see Wootton, *Galileo: Watcher of the Skies*, 9, 157–70. On Galileo's focus on geometry, see Jesseph, "Ratios, Quotients, and the Language of Nature."

10. On the history of algebra, see V. Katz and Parshall, *Taming the Unknown*; Pycior, *Symbols, Impossible Numbers, and Geometric Entanglements*; Stedall, *From Cardano's Great Art to Lagrange's Reflections*; Waerden, *A History of Algebra*.

11. Nesselmann, *Versuch einer kritischen Geschichte der Algebra*, 301–3. See also Rodet, *Sur les notations numériques et algébriques antérieurement au XVIe siècle*, 69–70.

12. Recorde, *The Whetstone of Witte whiche is the Seconde Parte of Arithmetike*, ff1v. On Recorde, see F. Smith and Roberts, *Robert Recorde*.

13. Heeffer, "On the Nature and Origin of Algebraic Symbolism."

14. Hope, *Shakespeare and Language*, 5.

15. Bright, *Charac[terie.] An Ar[te] of Shorte, Swift[e], and Secrete Writing by Character*; quotation from title.

16. Bright, n.p.

17. Fleming, *The Mirror of Information in Early Modern England*, 87.

18. See Geneva, *Astrology and the Seventeenth Century Mind*, 17–23. On cryptography in relation to mathematical symbols, see Pesic, "Secrets, Symbols, and Systems."

19. There is a famous passage in Augustine's *Confessions* in which he describes someone reading in silence and offers several explanations for why he may have chosen to do so (114). This passage suggests that, even if (as it appears from other passages) Augustine was already familiar with silent reading, reading aloud was more typical. On the history of silent reading, see Saenger, *Space between Words*; Jajdelska, *Silent Reading and the Birth of the Narrator*.

20. Aristotle states that names "that are in vocal sound are signs of passions in the soul, and those that are written are signs of those in vocal sound." In his commentary on the text, Thomas Aquinas tells us that, from the three elements of "*writing, vocal sounds*, and *passions of the soul*," a fourth element "is understood": the "things" to which those passions refer. Aristotle and Aquinas, *On Interpretation / Commentary by St. Thomas and Cajetan (Peri hermeneias)*, 23–24. On the Scholastic controversies over this interpretation, see E. Ashworth, "'Do Words Signify Ideas or Things?'"

21. Isermann, "Letters, Sounds and Things."

22. Bacon, *Advancement of Learning*, 247–59. In the early twentieth century, it was common to describe the Chinese characters as *ideographs*, which suggested that the symbols referred to ideas (Mason, *A History of the Art of Writing*, 154; Creel, "On the Nature of Chinese Ideography"). This term has its defenders, but it is no longer standard. In their linguistic anthropology textbook, Ottenheimer and Pine state that the term *ideograph* "is controversial as well as confusing," since it is difficult to prove that signs can call ideas to mind without also calling to mind the associated words (*The Anthropology of Language*, 208). The consensus is now that the Chinese characters should be understood as logographs—symbols representing words or parts of words—and that they are not as independent of spoken language as Bacon thought. For further critiques of the idea of ideography, see O'Neill, *Ideography and Chinese Language Theory*; Unger, *Ideogram*. See also Eco, *The Search for the Perfect Language*, 158.

23. Bacon, *Advancement of Learning*, 248–49.

24. Bacon, 249.

25. The real character idea is sometimes conflated with the idea of a "natural language," meaning a language in which signifiers are nonarbitrary. This idea appears, for

instance, in the work of John Amos Comenius (Sadler, *J. A. Comenius and the Concept of Universal Education*, 140). Another related idea is the "primitive language," meaning the language that existed before the fall of the Tower of Babel, which was (in some accounts) given to human beings by God. While Markley has argued that mathematical symbols were associated with such thinking (67), it should be stressed that seventeenth-century writers gave a great deal of significance to the fact that algebraic symbols could be chosen at will, which differentiated them from the idea of "natural" signification.

26. Wallis, "A Letter of Dr. John Wallis to Robert Boyle Esq," 1091. Wallis also draws a comparison between Chinese characters and numerals. Wallis is elsewhere critical of some of the more extreme claims about real characters, but in the course of his criticism he concedes that algebra, "as to so much, is a kind of Real Character" (*A Defence of the Royal Society*, 16–17). For another example, see John Wilkins's 1641 book *Mercury*, which mentions roman and arabic numerals as examples of real characters (107). John Webster makes a similar claim about numerals in his 1653 *Academiarum examen* (25), as does Robert Boyle in a letter to Samuel Hartlib (quoted in Markley, 66).

27. Hooke, "Some Observations, and Conjectures Concerning the Chinese Characters," 63.

28. Maat, *Philosophical Languages in the Seventeenth Century*, 23.

29. In his book on the classification of writing systems, Barry B. Powell argues that numerals and algebraic symbols are, in Ignace Gelb's term, *semasiographs*—that is, a system that "does not make direct use of the resources of speech" (Powell, *Writing*, 36). Gelb, unlike Powell, associates semasiography with an early stage in the development of writing (*A Study of Writing*, 11). Calling symbols semasiographs differs from calling them real characters in that it does not imply that they function differently from languages. "It is easy to see," Powell writes, "that mathematical semasiography is itself a language (but not speech), capable of achieving blinding feats of logic and abstract thought and even describing the governing forces behind appearances in the material world" (34). Early modern thinkers, by contrast, generally treated real characters as distinct from languages.

30. Dawson, *Locke, Language and Early-Modern Philosophy*, 62; Tillery, "Engendering the Language of the New Science."

31. Plofker, *Mathematics in India*, 140–57.

32. Al-Khwārizmī, *The Algebra of Mohammed ben Musa*. The computer scientist Heinz Zemanek gives an extensive, although somewhat hagiographic and now dated, overview of al-Khwārizmī's life and work in Ershov and Knuth, *Algorithms in Modern Mathematics and Computer Science*, 1–81.

33. On the House of Wisdom, see Aksoy, "Al-Khwārizmī and the Hermeneutic Circle."

34. See al-Khwārizmī, *Robert of Chester's Latin Translation of Al-Khwārizmī's Al-Jabr*.

35. See Ifrah, *The Universal History of Numbers*, 531.

36. In 1857, Baldassarre Boncompagni published an edition of "Dixit Algorizmi" text as al-Khwārizmī, *Algoritmi de numero Indorum*; Kurt Vogel produced another edition in 1963 (*Mohammed Ibn Musa Alchwarizmi's Algorismus*). For an English translation, see Crossley and Henry, "Thus Spake Al-Khwārizmī."

37. Crossley and Henry, 114.

38. Crossley and Henry, 114.

39. See the title of Johann Widman's *Algorithmus linealis*, published in Leipzig c. 1488 (D. Smith, *Rara arithmetica*, 36). Crossley and Henry (105) state, citing an older version of the *Oxford English Dictionary*, that the spelling with a *th* did not appear in English until the twentieth century. This is incorrect; the 2012 edition of the *OED* cites instances from 1658 and 1699. See *OED Online*, s.v. "algorithm, *n.*," accessed January 27, 2022, https://www.oed.com/view/Entry/198469. As I discuss in chapter 2, the English word *algorithm* was used in eighteenth-century texts to refer to the methods of the Newtonian calculus.

40. Boethius, *Boethian Number Theory*. On Boethius's influence on algorism, see Evans, "From Abacus to Algorism"; Taylor, *The Mathematical Practitioners of Tudor and Stuart England*, 320.

41. Dee, preface to Euclid, *The Elements of Geometrie of the Most Auncient Philosopher Euclide of Megara*, *.j.v. Although Dee's text is a preface to a translation of Euclid, it has reference to arithmetic as well as geometry; in particular, the influence of Boethius's theory of numbers is apparent.

42. Taylor, *Mathematical Practitioners*. See also Cormack, "The Role of Mathematical Practitioners and Mathematical Practice in Developing Mathematics as the Language of Nature"; Cormack, Walton, and Schuster, *Mathematical Practitioners and the Transformation of Natural Knowledge in Early Modern Europe*; Long, *Artisan/Practitioners and the Rise of the New Sciences, 1400–1600*; Harkness, *The Jewel House*, 97–141; Hill, "'Juglers or Schollers?'"

43. Napier, *Mirifici logarithmorum canonis descriptio ejusque usus, in utraque trigonometria*.

44. Oughtred, *The Circles of Proportion and the Horizontall Instrument*. Oughtred was involved in a bitter priority dispute over this invention with the mathematical practitioner Richard Delamain.

45. As was often the case in early modern mathematics, the terminology was highly inconsistent. In classical arithmetic, *numeration* meant counting, which did not necessarily have anything to do with particular symbolic representations; this usage persists in some seventeenth-century texts. In his 1650 arithmetic textbook, Jonah Moore uses *numeration* to refer to the operations of addition, subtraction, multiplication, and division; he calls the representational use of numerals *notation* (*Moores Arithmetick Discovering the Secrets of that Art in Numbers and Species*, 2–3).

46. Lantz, *Institutionum arithmeticarum libri quatuor*, 1–2. See Leibniz, *Dissertation on Combinatorial Art*, 91n.

47. Otis, "'Set Them to the Cyphering Schoole'"; see also Wilde, "English Numeracy and the Writing of New Worlds, 1543–1622."

48. On Alsted's encyclopedia, see Blair, "Revisiting Renaissance Encyclopaedism," 392–97; Hotson, *Commonplace Learning*. On Leibniz and Alsted, see Loemker, "Leibniz and the Herborn Encyclopedists"; Antognazza, *Leibniz*, 40–45, 54–59. On Alsted's views on language, see McMahon, "The Semantics of Johann Alsted"; McMahon, "The Semantics of Post-Medieval Lullism"; Padley, *Grammatical Theory: Trends in Vernacular Grammar*, 1:248–50.

49. Quoted in Loemker, "Leibniz and the Herborn Encyclopedists," 333.

50. Alsted, *Encyclopaedia septem tomis distincta*, 803.

51. Alsted, 807. He then states that numeration can be done with either numerals or stones (*calculis*).

52. Alsted, 807–8. This figure for the size of the army comes from Herodotus.

53. Alsted, 808.

54. Ong argues that the new science involved "a movement away from a concept of knowledge as it had been enveloped in disputation and teaching [. . .] toward a concept of knowledge which associated it with a silent object world, conceived in visualist, diagrammatic terms" (150–51). Markley writes, citing Foucault, that "the ideal of 'arithmetical Characters' underwrites a kind of semiotic absolutism that represses dialogical communication" (69). I agree with Markley that the association of mathematics with real characters presumes a view of meaning as invariant, but I would question whether this view served to shut down dialogue.

55. Plato, *Plato's Phaedrus*, 158.

56. Ong, 151. This preference for oral communication over text characterizes modern mathematics as well. The disciplinary conventions of mathematics departments are strongly against reading presentations off paper, as is common in the humanities; presenters are expected to remember proofs well enough to be able to perform them on a chalkboard and explain them on the fly.

57. V. Katz and Parshall, *Taming the Unknown*, 12–57, 81–104.

58. V. Katz and Parshall, 58–80.

59. Al-Khwārizmī, *Algebra*, 5.

60. On al-Khwārizmī's use of geometry, see V. Katz and Parshall, *Taming the Unknown*, 138–44.

61. Al-Khwārizmī, *Algebra*, 86–174.

62. Al-Khwārizmī, 10.

63. Al-Khwārizmī, 10.

64. Al-Khwārizmī, 10.

65. V. Katz and Parshall, *Taming the Unknown*, 7.

66. V. Katz and Parshall, 134.

67. See Fibonacci, *Fibonacci's Liber abaci*.

68. Ulivi, "Masters, Questions and Challenges in the Abacus Schools"; Parshall, "A Plurality of Algebras, 1200–1600."

69. On mathematical teaching in early modern universities, see Feingold, *The Mathematicians' Apprenticeship*.

70. See Pycior, *Symbols*, 149–65.

71. See Bos, "Differentials, Higher-Order Differentials and the Derivative in the Leibnizian Calculus"; Bos, *Redefining Geometrical Exactness*.

72. As a result, some algebraists, including Viète, required that coefficients have a dimension to make up the difference. In his *Geometry*, René Descartes employs a trick to get around this requirement. Suppose one wants to multiply the lengths of the lines BD and BC. Choose a point A along BD such that AB has the length taken to be unity. Draw a line from A to C. By drawing a parallel line originating at D and finding its intersection E with the line BC, one can construct a line segment BE with a length equal to the product. One can use this method to square a number. This construction was sometimes viewed as inelegant because it required choosing an arbitrary unit of measurement, which was not usually required when constructing squares in Euclidean geometry. See Descartes, *The Geometry of René Descartes*, 4–5.

73. Euclid, *Elements, Books I–XIII*, 697–954.

74. The Greek *logos* famously means both "idea" and "word," but it also had a mathematical meaning. In book 5 of the *Elements*, Euclid defines *logos* as "a sort of relation in respect of size between two magnitudes of the same kind"—that is, a ratio (372). *Alogos*,

then, suggested both a lack of a ratio and a lack of speech. Arabic translators rendered this as *asamm* (أَصَمّ), meaning "deaf," which was in turn translated into the Latin *surdus*, which means either "deaf" or "silent." See *OED Online*, s.v. "surd, *n*," accessed February 10, 2022, https://www.oed.com/view/Entry/194860. See also Pesic, "Hearing the Irrational," 504.

75. See Aydin and Hammoudi, "Root Extraction by Al-Kashi and Stevin."

76. Al-Khwārizmī, *Algebra*, 199–200n.

77. See Pantin, "La représentation des mathématiques chez Jacques Peletier du Mans."

78. On Stifel and irrationals, see Pesic, "Hearing the Irrational"; Pesic, *Music and the Making of Modern Science*, 55–72.

79. Stifel, *Arithmetica integra*, 103r.

80. Stifel, 103r.

81. Stifel, 103r.

82. On the intersection of religion and mathematics for Stifel, see Koetsier and Reich, "Michael Stifel and his Numerology."

83. Stifel, 114r, 122v. Stifel's notation differs from the modern in that it includes cossic characters (explained in the next section) indicating the degree of the root and lacks the solidus or bar over the expression whose root is taken. I have omitted the cossic characters here for the sake of simplicity.

84. Stifel, 116r.

85. Stifel, 122v.

86. The pamphlet was published in Flemish as well as in a French version appended to his book *L'arithmetique*, which he published the same year; an English translation is available in Stevin, "On Decimal Fractions."

87. Stevin, *L'arithmetique*, 140; Stevin, "On Decimal Fractions," 24. This notation was adapted from the way Rafael Bombelli had notated powers of unknown quantities, which Stevin himself uses elsewhere; see Struik, "Simon Stevin and the Decimal Fractions," 476.

88. Stevin, *L'arithmetique*, 37; translation from Klein, *Greek Mathematical Thought and the Origin of Algebra*, 196.

89. Stevin, *Les oeuvres mathematiques de Simon Stevin de Bruges*, 88–89. He frames the method in terms of proportions, but it can be applied to solving equations. He first published this method in a pamphlet titled *Appendice algébraique contenant règle générale de toutes equations*, the only known copy of which was destroyed when the Germans bombed the University of Louvain Library during World War I. The text, however, was added to the second edition of *L'arithmetique*, so a version of it survives. On this transmission history, see Sarton, "Simon Stevin of Bruges (1548–1620)," 253. Fowler links this procedure to the later work of Cauchy ("Dedekind's Theorem: $\sqrt{2} \times \sqrt{3} = \sqrt{6}$," 733); for a critical response, see K. Katz and M. Katz, "Stevin Numbers and Reality."

90. Chaitin, *Information, Randomness and Incompleteness*, 132.

91. This is the fifth of al-Khwārizmī's six cases; see al-Khwārizmī, *Algebra*, 39–40.

92. V. Katz and Parshall, *Taming the Unknown*, 167–72.

93. Cardano, *The Rules of Algebra*, 98–99; translation modified. Witmer translates "dimidium numeri quod iam in se duxeras" as "one-half the number you have already squared"; however, this translation attaches the subordinate clause starting with *quod*

to *numeri*, which does not agree with it in gender. The difference is significant because Witmer's translation suggests that one is supposed to halve the actual value that was squared (e.g., to compute one-half of $b/2$), which is mathematically incorrect. Witmer also uses the modern notation x to express algebraic variables, which I have replaced with a more literal translation of Cardano's notation.

94. Pycior, *Symbols*, 22.

95. Cardano, 20.

96. Pycior, *Symbols*, 23.

97. Cardano, 106. The quotation is from Stifel, *Arithmetica integra*, 122v.

98. Bombelli quoted in Waerden, *History of Algebra*, 61.

99. Moore, 14.

100. Letters are, to be sure, not the only way of achieving the indeterminacy of modern algebraic variables; Jocelyn Rodal has argued that some forms of literary language achieve a similar effect with words ("Patterned Ambiguities").

101. Al-Khwārizmī, *Robert of Chester's Latin translation of al-Khwārizmī's al-Jabr*, 67.

102. Frege, "Sense and Reference."

103. V. Katz and Parshall, *Taming the Unknown*, 200–204.

104. Rudolff, *Behend unnd hubsch Rechnung durch die kunstreichen regeln Algebre, so gemeincklich die Coss genennt werden.*

105. Heeffer, "On the Nature," 22. See also Neal, *From Discrete to Continuous*, 49–62.

106. Travis D. Williams argues that Recorde's presentation of the rule of false position anticipated the idea of naming an unspecified quantity (T. Williams, "Mathematical Enargeia," 199–200). This is a fair reading of the passage, but it does not entail that Recorde saw his actual notation as indeterminate. In the section on cossic numbers, he makes it quite plain that the cossic characters are supposed to be interpreted as units. The symbol indicating the "name" of the unknown thus takes the form of a denomination, not of a variable as we would understand it now.

107. Recorde, sig. S.i.r.

108. Recorde, sig. S.iii.r.

109. Recorde distinguishes "nombers *Abstracte*" from "nombers *Contracte*, or *Denominate*" (sig. S.i.r). The latter he divides into "*nombers denominate vulgarely*"—that is, numbers with concrete units, such as "10. shillinges," "10. men," or "1000. yeres"—and "nombers *denominate Coßikely*" (sig. S.i.r). He explains the latter as numbers that "bee contracte unto a denomination of some *Coßike* sign as 1. nomber, 1. roote. 1. square 1. Cube" (sig. S.i.v). Thus, the cossic characters stand in for units to which another number is "contract." After introducing the cossic symbols, he states that when one assembles two cossic numbers with the plus or minus sign, the result is "a compounde nomber" (sig. S.ii.v). Jeffrey Oaks makes a similar point to mine in relation to the algebra of the period more broadly; Oaks, "François Viète's Revolution in Algebra," 269.

110. Viète, *In artem analyticem isagoge*, Biiv. An English translation of this text is included as an appendix in Klein, *Greek Mathematical Thought*; I quote this translation in what follows. Another translation by T. Richard Witmer is available in Viète, *The Analytic Art*, along with translations of related works. Witmer's is the more readable translation, but it modernizes the notation, thus losing some of the specific character of Viète's method.

111. To explain further, *planum* and *quadratum* both indicate two-dimensional

magnitudes. The difference is that *quadratum* refers to the square of a one-dimensional value, whereas *planum* refers to a magnitude already considered to be two-dimensional. See Oaks, 270.

112. Klein, 340.

113. In favor of there not being a connection, I would point out the structural similarity between cossic characters and Diophantus's notation, from which Viète specifically wanted to differentiate his system.

114. Wallis, *A Treatise of Algebra, Both Historical and Practical*, 66; Wallis argues, in particular, that Viète's use of the word *species* derives from a legal usage. On the intersection of mathematics and law, see Cifoletti, "La question de l'algèbre."

115. Klein, 175. Michel Serfati argues that Viète started a "dialectic of indeterminacy" by employing symbols for undetermined given values; see Serfati, *La révolution symbolique*.

116. Oaks, 285. Another account comes from Peter Pesic, who links the symbols to Viète's experience in cryptanalysis ("Secrets, Symbols, and Systems"). For a response, see Stedall, *From Cardano's Great Art to Lagrange's Reflections*, 20.

117. See V. Katz and Parshall, *Taming the Unknown*, 156–58, 227–46.

118. V. Katz and Parshall, 58–80; Bashmakova, Smirnova, and Shenitzer, "The Birth of Literal Algebra."

119. Klein argues that Diophantus's sign $\mathring{M}$, which was appended to numbers not dependent on the unknown, reflected the fact that the Greeks thought of "numbered assemblages" rather than of bare numbers as we would understand the term now (177–78). But the cossic notation developed in the early sixteenth century worked much the same way; this indicates that the way of thinking Klein takes to be specifically Greek also existed in sixteenth-century Europe prior to Diophantus's revival. On the relation of Diophantus to medieval algebra, see Christianidis and Oaks, "Practicing Algebra in Late Antiquity."

120. V. Katz and Parshall, *Taming the Unknown*, 66.

121. Klein, 318.

122. Klein, 153–54.

123. Viète makes this claim about Diophantus in Klein, 345.

124. Klein, 154–55.

125. Klein, 321. Klein argues that the exegetic phase is geometric and the rhetic numerical; Oaks argues (278) that Viète does not clearly distinguish the two.

126. Viète, *Opera mathematica*, 54. An English translation of this passage appears Viète, *Analytic Art*, 108–9. The translations I have given are my own based on the Latin text.

127. Viète, *Opera mathematica*, 54.

128. Viète, 54.

129. Viète leaves it to the reader to perform these last steps.

130. Klein, 353.

131. On Harriot and Viète, see Stedall, "Notes Made by Thomas Harriot on the Treatises of François Viète."

132. On the background of this text, see Stedall, *The Greate Invention of Algebra*.

133. Harriot, *Artis Analyticae Praxis*, 21–22.

134. Malcolm and Stedall, *John Pell (1611-1685) and His Correspondence with Sir Charles Cavendish*, 56–57.

135. Malcolm and Stedall, 37–39.

136. Malcolm and Stedall, 54–56.

137. Malcolm and Stedall, 55.

138. On Viète and Fermat, see Mahoney, *The Mathematical Career of Pierre de Fermat*, 26–71; on Ghetaldi and Vaulezard, see Oaks, 279.

139. See Sasaki, *Descartes's Mathematical Thought*.

140. On Dary and Hooke, see Taylor, *Mathematical Practitioners*, 217.

141. Bacon, *The New Organon*, 116.

142. Something similar could be said of mathematical instruments, which Deborah E. Harkness compares to the "black boxes" of modern technology (130).

143. V. Katz and Parshall, *Taming the Unknown*, 173.

144. Only six of the original thirteen books of Diophantus's *Arithmetic* are known to have survived in the original Greek; however, four others survive in an Arabic translation probably produced in the ninth century. An English translation of these texts is available in Diophantus, *Books IV to VII of Diophantus' Arithmetica: in the Arabic Translation Attributed to Qustā ibn Lūqā*.

145. Hooke, *The Posthumous Works of Robert Hooke*, 64.

146. Hooke, 64. On this proposal, see Hesse, "Hooke's Philosophical Algebra."

147. Oughtred, *The Key of the Mathematicks New Forged and Filed*, B5v.

148. Ward, *Idea trigonometriae demonstratae*, A2v.

149. John Guillory makes a similar point in relation to prose writing, observing a tension between competing tendencies toward clarity and technicity; Guillory, "The Memo and Modernity," 129–32.

150. Descartes, *Geometry*. Douglas Jesseph argues that seventeenth-century mathematicians may be divided into geometric and algebraic foundationalists; see "The 'Merely Mechanical' vs. the 'Scab of Symbols.'" Descartes, in Jesseph's view, rejected the idea that algebra could be founded on geometry on the grounds that algebra ought to be kept pure of all traces of physical reality. Descartes's algebraic foundationalism was not, however, the same as the formalism of the late nineteenth century on in that he did not treat symbol-manipulation rules as themselves foundational.

151. Descartes, *A Discourse on Method*, 15.

152. M. Jones, *Good Life*, 2–3.

153. Descartes, *The Philosophical Works of Descartes*, x.

154. Descartes, 4.

155. Descartes, 10–12.

156. Descartes, 11.

157. Like Viète, Descartes distinguished analysis from synthesis and viewed the latter as superior; put simply, one was supposed to use symbolic analysis as a means of discovery, then construct a geometric proof of the result that would demonstrate it synthetically, independently of any symbolic expressions. See Sasaki, *Descartes's Mathematical Thought*; Schmitter, "Mind and Sign"; Guicciardini, *Isaac Newton on Mathematical Certainty and Method*, 31–58.

158. Descartes, *Philosophical Works*, 33.

159. Descartes, 40.

160. Descartes, 67.

161. Descartes, 67.

162. Descartes, 67.

163. Klein argues that Descartes's idea of number is based on a "*symbol-generating abstraction*" (202); as a result, Klein later writes, modern numbers "can be immediately

grasped in the notation" (224). This account overlooks the fact that, from a Cartesian perspective, the "symbols" must be understood as mental constructs, not merely physical marks.

164. Descartes, *Philosophical Works*, 1.

165. Descartes, 63, 73–76.

166. Descartes, 63.

167. Cajori, *A History of Mathematical Notations*, 200. On the relation between Barrow and Hérigone, see Pycior, *Symbols*, 152–53.

168. Hérigone, *Cvrsvs mathematicvs, nova, brevi et clara methodo demonstratvs, per notas reales & vniuersales, citra vsum, cuiuscunque idiomatis intellectu faciles*, quotation from full title. On Hérigone's use of symbols, see Massa Esteve, "Symbolic Language in Early Modern Mathematics."

169. As Travis D. Williams has argued ("Mathematical Enargeia," 169), making sense of early modern mathematical notation required further writing: the reader was expected to work out the problems on a sheet of paper so as to fully understand the ideas. Such practices fit with Descartes's description of notation as an aid to the intellect, but they do not explain Hérigone's use of symbols to link equations together.

170. Hérigone, quotation from full title (see note 168), translated from Latin.

171. Hérigone, vol. 1, sig. bvr.

172. Hérigone, vol. 1, sig. biiiir.

173. Bliss introduced *Blissymbolics*, a language based on visual symbols. A survivor of the Holocaust, Bliss wanted, like countless others over the centuries, to mend the rifts between groups of people who spoke different languages; Bliss, *Semantography (Blissymbolics)*.

174. Hérigone, vol. 1, sig. biiiiv.

175. The definitive account of the Hobbes–Wallis controversy is still Jesseph, *Squaring the Circle*. Jesseph has more recently revised some of his positions ("Geometry, Religion and Politics"). See also Alexander, *Infinitesimal*; Pycior, *Symbols*, 146.

176. Hobbes, *Six Lessons to the Professors of the Mathematiques*, 23.

177. Hobbes, *Leviathan*, 12–13.

178. Hobbes, *Six Lessons*, 28. Hobbes may have meant this as a riposte to Seth Ward's statement that algebra was a "design against language," which I discuss in chapter 2.

179. Hobbes, *Elements of Philosophy the First Section, Concerning Body*, 2.

180. Hobbes, *Elements of Philosophy*, 1.

181. Dreyfus, *What Computers Still Can't Do*, 69; Haugeland, *Artificial Intelligence*, 23–28.

182. Ward, *Idea trigonometriae demonstratae*, 1.

CHAPTER TWO

1. Wiener, *Cybernetics*, 12.

2. Davis, *Universal Computer*, 1; W. Thomas, "Algorithms," 40.

3. Leibniz, *Selections*, 12–17, 50–58. See Rabouin, "'Analytica Generalissima Humanorum Cognitionum.'"

4. Leibniz, *Selections*, 51.

5. See M. Jones, *Good Life*, 248.

6. Leibniz discusses these ideas in a number of places, including "New System of

the Nature of Substances and their Communication, and of the Union which Exists between the Soul and Body" (1695) and "Monadology" (1714). Leibniz, *Philosophical Texts*, 143–52, 267–81.

7. Antognazza, *Leibniz*.

8. French took a turn toward institutionalized standardization with the founding of L'Académie Française in 1635. The first English dictionaries appeared in the early 1600s, although large dictionaries did not exist until later in the century and the first comprehensive ones arguably appeared in the eighteenth. The construction of a standard dialect of British English is generally regarded as having taken place in the eighteenth century, but I would still maintain that ideas about language shifted in the late seventeenth century toward the idea that the language was governed by rules codified in dictionaries and grammar books.

9. Wilson, *Heart of Europe*, 258–64.

10. Mattheier, "German," 235.

11. Leibniz, "Nova Methodus pro Maximis et Minimis."

12. On Leibniz and public debate, see Dascal, *The Practice of Reason*.

13. Guicciardini, *Isaac Newton on Mathematical Certainty and Method*, 102.

14. Translation from Slaughter, *Universal Languages and Scientific Taxonomy in the Seventeenth Century*, 127.

15. See Slaughter; Eco, *Search for the Perfect Language*; Eco, "The Language of the Austral Land"; Knowlson, *Universal Language Schemes in England and France, 1600–1800*; Salmon, *The Study of Language in 17th-Century England*; Cram and Maat, "Universal Language Schemes in the 17th Century"; Maat, *Philosophical Languages*; Lewis, *Language, Mind and Nature*; Fleming, *Mirror of Information*. On Leibniz's attempts at a universal language, see Pombo, *Leibniz and the Problem of a Universal Language*. For further background on Leibniz's thinking about language, see Dascal, *Leibniz, Language, Signs, and Thought*.

16. Ockham developed the Scholastic position of nominalism into the idea of a mental language that is not based on any conventional language. See Panaccio, *Mental Language*. Antognazza connects this idea to Leibniz's universal characteristic (93).

17. See Leibniz, *Dissertation on Combinatorial Art*, 185–87.

18. Wilkins, *Mercury*, 108.

19. Wilkins, 109.

20. Wilkins, 110.

21. Ward, *Vindiciae academiarum*, 19. This statement was a response to the suggestion one of Ward's opponents, John Webster, that grammarians should study other modes of communication besides words, including the symbols used in "*Algebraick Arithmetick*" (Webster, *Academiarum examen*, 24); Ward takes this as an insult to algebraists. It is worth noting that Webster cites Hérigone's book, suggesting that he took a similarly pragmatic view of symbols' role in communication.

22. Ward, *Vindiciae academiarum*, 19.

23. Vivian Salmon, for instance, states that Ward is advocating the "*Lingua Mathematicorum*" as a model for universal character schemes (*The Works of Francis Lodwick*, 133); Ann Geneva quotes this same passage as evidence that Ward is treating the "*Lingua Mathematicorum*" as superior to ordinary language (*Astrology and the Seventeenth Century Mind*, 279). To suppose that Ward is advocating the "*Lingua Mathematicorum*" is to overlook the extent to which he is specifically concerned with differentiating algebraic symbols from numerals; it is only numerals that he calls the "*Lingua Mathemati-*

corum," and he is saying that algebraists have come to oppose their use because of this linguistic nature.

24. Ward, *Vindiciae academiarum*, 19.

25. Ward, 19.

26. Ward, 20.

27. Ward, 21. The word "Analyties" does not appear in the *Oxford English Dictionary* or in the Early English Books Online Text Creation Partnership database; it may be a misprint of *Analytics*.

28. Ward, 20.

29. Wilkins does not use quite the mechanism that Ward suggested. Ward, along with Pell, Descartes, George Dalgarno, and a number of others, thought it would be best to create symbols for the simple ideas from which more complex ones are composed and build up symbols for more complex ideas by combining them. Wilkins, on the other hand, takes a top-down approach, representing ideas by their position in a hierarchy. This difference led to a falling out between Wilkins and Dalgarno. See Maat; Blank, "Dalgarno, Wilkins, Leibniz, and the Descriptive Nature of Metaphysical Concepts."

30. Wilkins, *An Essay Towards a Real Character and a Philosophical Language*, b2r.

31. Wilkins, 13.

32. Wilkins, 19.

33. Wilkins, 20; on Wilkins and Aristotle, see Slaughter.

34. In his *Mirror of Information*, Fleming points out that, contrary to what a number of previous scholars have asserted, the language and the character are not entirely equivalent (199–208).

35. Maat, 158. See note 34.

36. Aarsleff, *From Locke to Saussure*, 262; Isermann, "Substantial vs Relational Analogy in Sixteenth and Seventeenth-Century Linguistic Thought," 108.

37. Wilkins, *Essay*, 385. Wilkins goes on to state that, on account of the difficulties of constructing signifiers that naturally correspond to things, "this Character must be by *Institution*" (386).

38. Wilkins, 20.

39. Wilkins, 235.

40. In a published letter, Wallis argues that the deaf could be taught to read without first learning to speak, offering mathematical symbols as proof that written signs can operate independently of speech ("A Letter of Dr. John Wallis to Robert Boyle Esq," 1091). Wallis was not merely speculating; he was involved in a series of efforts at deaf education, as described in Wallis, *Teaching Language to a Boy Born Deaf*.

41. Fleming, 22.

42. Wilkins, *Essay*, 20.

43. Dalgarno, *George Dalgarno on Universal Language*, 175. In the 1670 pamphlet *A Defence of the Royal Society*, John Wallis takes credit (perhaps unfairly) for turning Dalgarno against Wilkins. Wallis suggests that to write all books in Wilkins's real character "is the same thing as to Translate all Books into One Language, and to have this Language learned by All"; he takes this to undermine the whole project (16).

44. Maat.

45. Descartes, *Principles of Philosophy*, 20.

46. Descartes, "A Mersenne, Amsterdam, 20 novembre 1629," 915.

47. Leibniz, *Dissertation*.

48. Leibniz, *Dissertation*, 77.

49. On Llull, see Fidora and Sierra, *Ramon Llull*; Bonner, *The Art and Logic of Ramon Llull*.

50. Bonner, 149.

51. Leibniz, *Dissertation*, 157.

52. Leibniz, 159.

53. Leibniz, 161.

54. I am following the translation in Leibniz, 181. However, I have altered the appearance of the fractions to resemble the original Latin text more closely.

55. Leibniz, 183.

56. Leibniz, 189.

57. Leibniz, 187.

58. Leibniz, 187.

59. Leibniz, *Selections*, 20.

60. Leibniz, 22.

61. Leibniz, 23.

62. Leibniz, 23.

63. The commentary is reproduced in Leibniz, *Sämtliche Schriften und Briefe*, ser. 6, vol. 4., pt. A, pp. 27–53. On the relation between Wilkins and Leibniz, see Rutherford, "Philosophy and Language in Leibniz," 230–31; Pombo, 81; Maat; Knowlson; Blank.

64. Leibniz, *Sämtliche Schriften*, ser. 6, vol. 4, pt. A, p. 66. A similar scheme was earlier developed by Francis Lodwick; see Salmon, *Works of Francis Lodwick*, 133.

65. A few years later, Leibniz connected logical analysis to prime numbers in another way, arguing that the term *multiple of 15* is composed of the two simple terms *multiple of 3* and *multiple of 5* (*Logical Papers*, 37).

66. Leibniz, *Sämtliche Schriften*, ser. 6, vol. 4, pt. A, p. 66.

67. Leibniz, *Sämtliche Schriften*, ser. 6, vol. 4, pt. A, p. 68.

68. Couturat, *La logique de Leibniz*, 63.

69. Opitz, *Buch von der deutschen Poeterei*.

70. Leibniz, *Dissertation*, 229.

71. Leibniz, 271.

72. Leibniz, *Philosophical Papers and Letters*, 193.

73. Leibniz, 193.

74. Wilkins, *Essay*, b2r.

75. Leibniz, *Philosophical Papers*, 193; see also Leibniz, *Selections*, 16.

76. Leibniz, *Philosophical Papers*, 193.

77. Leibniz, 193.

78. The extent to which Leibniz separated the formal and informal aspects of his project is under dispute. Marcelo Dascal argues that Leibniz made room for both "hard" and "soft" forms of proof ("The Balance of Reason"). A number of scholars respond to this argument in Dascal, *Leibniz: What Kind of Rationalist?* See also Lærke, "Leibniz, the Encyclopedia, and the Natural Order of Thinking"; Heinekamp, "Ars Characteristica und natürliche Sprache bei Leibniz."

79. Leibniz, *Selections*, 50.

80. Leibniz, 51–52.

81. Leibniz, 52.

82. Leibniz, 52.

83. Leibniz, 51.

84. Leibniz, *Selections*, 15; see also Leibniz, *New Essays on Human Understanding*, 50, 85.

85. Leibniz, *Selections*, 15.

86. Leibniz, "Leibniz on His Calculating Machine," 181.

87. Leibniz, 181.

88. Leibniz, *Logical Papers*, 122–44; see Lenzen, "Arithmetical vs. 'Real' Addition."

89. Leibniz, *Logical Papers*, 143.

90. Leibniz, 124, 132, 143.

91. Leibniz, 127. This is different from Boole's system, which was designed to retain all the axioms of ordinary algebra, although some later versions of Boolean logic are closer to Leibniz's design.

92. Leibniz, 141.

93. On the (non)relation between Leibniz and Boole, see Grattan-Guinness, "Boole's Quest for the Foundations of His Logic," xliii.

94. G. Boole, *Selected Manuscripts on Logic and Its Philosophy*, 188.

95. Russell, *A Critical Exposition of the Philosophy of Leibniz*, 170; Couturat; C. I. Lewis, *Survey of Symbolic Logic*, 5–18; Cassirer, *The Philosophy of Symbolic Forms*, 1:127. Aarsleff criticizes this view in "The Eighteenth Century, Including Leibniz." On Leibniz and Russell, see Goethe, "How Did Bertrand Russell Make Leibniz Into a 'Fellow Spirit'?"; Nachtomy, "Leibniz and Russell."

96. Leibniz, *Logical Papers*, 33.

97. Leibniz, *Philosophical Texts*, 78.

98. Quoted in Capozzi and Roncaglia, "Logic and Philosophy of Logic from Humanism to Kant," 137.

99. Lambert quoted in Nöth, *Handbook of Semiotics*, 28.

100. Capozzi and Roncaglia, 139.

101. Capozzi and Roncaglia, 141.

102. Cajori, *History of Mathematical Notations*, 337.

103. Leibniz, *Dissertation*, 71.

104. To say that symbols *express* something, for Leibniz, means roughly that one can use the symbols to draw conclusions about that thing. See Mates, *The Philosophy of Leibniz*, 38; Debuiche, "La notion d'expression et ses origines mathématiques."

105. Serfati, "Mathématiques et pensée symbolique chez Leibniz," 168–73; Jones, *Good Life*, 247–48. See also Serfati, "Symbolic Inventiveness and 'Irrationalist' Practices in Mathematics."

106. See Beeley, "'Un de Mes Amis.'"

107. Leibniz, "Explication de l'arithmetique binaire," 116.

108. Leibniz, 116.

109. Leibniz, 111.

110. Van Brummelen, "Jamshīd al-Kāshī."

111. Leibniz, "Nova Methodus pro Maximis et Minimis."

112. The most obvious difference is that Leibniz uses differentials (i.e., dx) rather than derivatives (i.e., dy/dx). This aspect of calculus changed with the development of the modern notion of the function beginning in the mid-eighteenth century; see Bos, "Differentials, Higher-Order Differentials and the Derivative in the Leibnizian Calculus."

113. Leibniz, "De Geometria Recondita," 297.

114. Quoted by Cajori, *History of Mathematical Notations*, 200.

115. A number of scholars have made cases that Leibniz's calculus was rigorously founded after all. David Rabouin, for instance, argues that, around 1675–76, Leibniz had attempted to found his theory of integrals on the grounds of an Archimedean conception of mathematical rigor ("Leibniz's Rigorous Foundations of the Method of Indivisibles," 348). Hidé Ishiguro argues that Leibniz thought of infinitesimals as "a well-founded fiction" (*Leibniz's Philosophy of Logic and Language*, 92). See also Arthur, "Leibniz's Syncategorematic Infinitesimals"; Horváth, "On the Attempts Made by Leibniz to Justify His Calculus." On the basis of such a reading of Leibniz, Mikhail Katz and David Sherry argue against the received historiography in which the infinitesimal was "slain" by Berkeley (M. Katz and Sherry, "Leibniz's Infinitesimals," 593–97). This revisionist argument rests on a retrospective judgment about who was right, not on an analysis of how the ideas of Leibniz and Berkeley were received by contemporaries. The continuity that Katz and Sherry see between Leibniz and modern mathematicians such as Cantor and Weierstrass was, if it existed, disrupted by a gap of over a century in which foundational questions about the calculus were (rightly or not) considered unresolved. For more background about the early stages of the debates on calculus, see Mancosu, *Philosophy of Mathematics and Mathematical Practice in the Seventeenth Century*.

116. See Guicciardini, *Isaac Newton on Mathematical Certainty and Method*, 332.

117. See Mancosu, "The Metaphysics of the Calculus," 228–35. One of Rolle's arguments hinged, to put it simply, on treating $(x + dx, y + dy)$ as a value on the curve being differentiated; on this basis, he is able to derive the result that $dx = 0$, and thus that the infinitesimal does not exist.

118. On l'Hôpital, see Mancosu, "Metaphysics of the Calculus," 226–28; on Wolff, see Blanco, "Hermeneutics of Differential Calculus in Eighteenth-Century Northern Germany," 140.

119. Berkeley, *The Analyst*, 85. For a detailed account of Berkeley's views, see Jesseph, *Berkeley's Philosophy of Mathematics*.

120. See M. Katz and Sherry, "Leibniz's Infinitesimals," 592–93; Pycior, *Symbols*, 209–41.

121. Berkeley, *Philosophical Commentaries*, 100. On the significance of "names" and "words" in Berkeley, see Hight, "Why My Chair Is Not Merely a Congeries." Earlier editions of this text contained "never" in place of "meer."

122. Berkeley, 101.

123. Berkeley, 100. Berkeley gives a more detailed account of how signs are used in arithmetical computation in his 1710 book *A Treatise Concerning the Principles of Human Knowlege* (169–75). There, the claim that arithmetic works only with signs supports his anti-abstractionist stance by denying the need for abstract ideas of number.

124. Berkeley, 97.

125. Jesseph, *Berkeley's Philosophy of Mathematics*, 116. Elsewhere in the *Philosophical Commentaries*, Berkeley states directly that the argument that numbers are mere words does not apply to infinitesimals because the latter "are words of no use if not supposed to stand for Ideas" (42). It should be noted that Berkeley's view of geometry was far from orthodox and that he represented his doctrine as an alternative to classical geometry.

126. Newton, *Universal Arithmetick*, 227–28. Pycior has questioned whether this statement was truly representative of Newton's views (*Symbols*, 200–204). However,

this is not the only point where Newton employs this sort of standard; in the *Principia*, for instance, he claims that his solution to the Pappus problem is superior to Descartes's because it did not involve any calculations (Guicciardini, *Isaac Newton on Mathematical Certainty and Method*, 101).

127. Leibniz, "Nova Methodus pro Maximis et Minimis," 469; translation from Leibniz, "Leibniz on the Calculus," 623.

128. Leibniz, "De la chainette," 148. He glossed the term *algorithme* similarly in a 1677 manuscript: "In order to explain myself shortly and clearly, I must introduce some *fresh characters*, and give them a *new Algorithm*, that is to say, altogether special rules, for their addition, subtraction, multiplication, division, powers, roots, and also for equations" (*The Early Mathematical Manuscripts of Leibniz*, 132). On Leibniz's notion of "incomparable quantities," see Horváth, 53.

129. Elisha Coles's 1692 dictionary, for instance, states that *Algorism, Algorithme*, and *Algrim* are all "the same as *Algebra*" (*An English Dictionary*, B4r). By the eighteenth century, the idea that *algorithm* encompassed algebra as well as arithmetic was established. Ephraim Chambers's *Cyclopaedia* (1728) gives both the arithmetical and algebraic definitions of *algorithm*, although he tells us that the algebraic sense is mainly used by "the *Spaniards*" and that the meaning of the word "is properly the Art of numbering truly and readily" (1:60).

130. Harris, *A New Short Treatise of Algebra*, 118. See Guicciardini, *The Development of Newtonian Calculus in Britain, 1700–1800*, 13–14, 55–62. The usage was inconsistent, but in some instances the word seemed to suggest practical rules for problem solving as opposed to scientific theory. In a 1736 commentary on a work by Newton, for instance, its editor John Colson distinguishes the "Algorithm, or Method of Operations" from the "Principles" of arithmetic (Newton, *The Method of Fluxions and Infinite Series*, 151).

131. Bürja, "Essai d'un nouvel algorithme des logarithmes," 301; first quotation from title. Bürja introduces a different notation using a double line; I have used the modern notation here for simplicity. Expanded usages along Leibnizian lines were common in the later eighteenth century. A 1784 volume of the *Encyclopédie méthodique* (a revised version of the Diderot and d'Alembert encyclopedia) notes that the word *algorithme* has come to denote "the method & notation of all species of calculation" such as integral calculus and trigonometric functions (*Encyclopédie méthodique: Mathématiques*, 37).

132. Jolley, *Leibniz and Locke*, 162–79.

133. Jolley discusses the limits of this comparison in *Leibniz*, 211–13. On the influence of Leibniz on Kant, see Jauernig, "Kant's Critique of the Leibnizian Philosophy."

134. Quoted in Blank, 60.

135. Locke, *An Essay Concerning Human Understanding*, 489.

136. Locke, 489.

137. Locke, 364, 366.

138. Locke, 437.

139. Dawson, *Locke, Language and Early-Modern Philosophy*, 205. See Slaughter, 206.

140. Locke, *Essay*, 79–80.

141. Leibniz, *New Essays*, 49. This book is written in the form of a dialogue between Philalethes, who has just read Locke's book and who recites its claims more or less in order, and his friend Theophilus, who responds with Leibnizian rebuttals. All of my quotations are either from the preface or from Theophilus.

142. Locke, *Essay*, 69.

143. Leibniz, *New Essays*, 50.

144. Leibniz, 77. Indeed, Leibniz later argues that natural languages are not as arbitrary as Locke claims. European vernaculars, he writes, "involve a mixture of chosen features and natural and chance features of the languages upon which they are built" (278). Leibniz does, however, make an exception: "Perhaps there are some artificial languages which are wholly chosen and completely arbitrary, as that of China is believed to have been, or like those of George Dalgarno and the late Bishop Wilkins of Chester" (278).

145. Leibniz, 77.

146. Leibniz, 55.

147. On some of the religious objections to Locke's work, see Marshall, "Locke, Socinianism, 'Socinianism,' and Unitarianism"; Jolley, *Leibniz and Locke*.

148. Voltaire, *Oeuvres completes de Voltaire*, 1:848–49.

149. Locke, *Some Thoughts Concerning Education*, 39.

150. Locke, *Conduct of the Understanding*, 15. Locke clearly distinguished habit from reason; see Grant, "John Locke on Custom's Power and Reason's Authority."

151. See Cajori, 428–29.

152. Euclid, *Euclid's Elements of Geometry, from the Latin Translation of Commandine*, A4r.

153. MacLaurin, *A Treatise of Fluxions*, 2:576.

154. MacLaurin, 2:576.

155. Diderot and d'Alembert, *Encyclopédie*, 4:985.

156. Diderot and d'Alembert, 11:72–74.

157. Babbage, *Passages from the Life of a Philosopher*, 25; see Buxton, *Memoir of the Life and Labours of the Late Charles Babbage Esq. F. R. S.*, 347.

158. Histories that largely skip ahead from Leibniz to the nineteenth century include Davis, *Universal Computer*; Dasgupta, *It Began with Babbage*; Berlinski, *The Advent of the Algorithm*.

CHAPTER THREE

1. Baker, *Condorcet*, 304. A classic biographical account appears in Manuel, *The Prophets of Paris*.

2. D. Williams, *Condorcet and Modernity*, 42.

3. Condorcet, *Political Writings*, 143, 144.

4. Condorcet, 144.

5. Foucault, *The Order of Things*, 56.

6. Eco, *Search for the Perfect Language*, 283; Chartier, "Languages, Books, and Reading from the Printed Word to the Digital Text," 137.

7. Baker, 124.

8. The first published version of this text appeared in 1954 in G. G. Granger, "Langue universelle et formalisation des sciences," which omits some sections of the manuscript. A critical edition appears in the 2004 book Condorcet, *Tableau historique des progrès de l'esprit humain*. Condorcet's manuscripts contain numerous cancellations and illegible words, on account of which the two editions differ somewhat.

9. Condorcet, *Tableau*, 994–99.

10. It is here that the project falters. Whereas the objects of mathematical and

natural sciences are determinate, he writes, these four sciences are about making the objects themselves known, on account of which it is necessary to begin with some "first combinations of ideas." This passage is heavily canceled, and the manuscript ends before Condorcet even attempts this. Condorcet, *Tableau*, 1013–14.

11. Condorcet, *Tableau*, 953.

12. Condorcet, *Tableau*, 953–54. Condorcet mentions Hérigone's system in the manuscript as an example of a "hieroglyphic" notation, presumably referring to the pictographic symbols Hérigone used for some geometric ideas (995).

13. Maimieux, *Pasigraphie*.

14. Condorcet, *Tableau*, 969. The editors take this notation as equivalent to $(-1)^n a$, implying that it has a value as a whole equivalent either to a or its negation. It might, instead, be interpreted to make a statement about what the sign of a is, but this would not align with the other notation he uses elsewhere.

15. The editors of the critical edition note the inclusion of something like Boolean logic in the notation for conditionals, but they also observe that the system works differently from Boolean logic in its modern form; see Condorcet, *Tableau*, 970n.

16. Condorcet, *Tableau*, 1008.

17. Wolff, "Arbitrary, Natural, Other."

18. Condorcet, *Tableau*, 996.

19. Condorcet, *Tableau*, 982; see also 989, 993. There is an exception: Condorcet uses the first letters of Latin words for some things, such as ℝ for real numbers. He justifies this concession to linguistic convention by the fact that Latin is generally known in Europe (962–63); elsewhere he states that wholly arbitrary signs "would uselessly fatigue the memory" (994).

20. Condorcet, *Tableau*, 973–74. Granger's transcription ("Langue universelle," 213) omits the word "but" (*mais*) at the beginning of the last sentence. As the editors of the 2004 edition note, the paragraph from this point on contains a large number of canceled words and lines, which I would take to indicate an anxiety about addressing this potential flaw in the system.

21. Knowlson, 200.

22. Granger, 213.

23. Maat, 155.

24. Alexander, *Duel at Dawn*, 8. Eighteenth-century mathematicians did distinguish pure mathematics from "mixed mathematics," which involved particular applications; yet even pure mathematics was understood to be about particular aspects of physical reality, such as quantity or shape. See Daston, *Classical Probability in the Enlightenment*, 53–55.

25. Rousseau, "Discourse on the Arts and Sciences."

26. De Gérando, *Des signes et de l'art de penser considérés dans leurs rapports mutuels*, 1:xxi. De Gérando's specific target here is Condillac, *La langue des calculs*.

27. Richards, "Rigor and Clarity."

28. Heilbron, "Introductory Essay," 2.

29. See Frängsmyr, "The Mathematical Philosophy"; Manuel, 43.

30. Burke, *Reflections on the Revolution in France*, 213; see also 62, 202–3.

31. Kline, "Euler and Infinite Series."

32. Varadarajan, "Euler and His Work on Infinite Series," 526–28.

33. A translation of Euler's text appears in Barbeau and Leah, "Euler's 1760 Paper on Divergent Series"; quotation from p. 144.

34. Barbeau and Leah, 144.

35. Abel quoted in Stubhaug and Daly, *Niels Henrik Abel and His Times*, 343.

36. Barbeau and Leah, 148.

37. Condillac, *Essay on the Origin of Human Knowledge.*

38. Süßmilch, *Versuch eines Beweises.*

39. Rousseau and Herder, *On the Origin of Language.*

40. Monboddo, *Of the Origin and Progress of Language.*

41. Condorcet, *Political Writings*, 10.

42. Condorcet, 25.

43. Condorcet, 5.

44. Condorcet, 5.

45. Condorcet, 5.

46. Condorcet states this position explicitly later in the text, attributing it to Aristotle: "*even our most abstract, as it were, our most purely intellectual, ideas have their origin in our sensations*" (43).

47. Condorcet, 1.

48. Condorcet, 95–96.

49. Condorcet, 1.

50. Condillac, *Philosophical Writings of Etienne Bonnot, Abbé de Condillac*, 151.

51. Condillac, 151.

52. Condillac, 151.

53. Condillac, 410.

54. Condillac, 410. Condillac expands on this point further in the posthumously published *La langue des calculs.*

55. Baker, 114. On the relation between Condorcet and Condillac, see also Daston, *Classical Probability*, 212.

56. Diderot, *Rameau's Nephew / D'Alembert's Dream*, 221–22.

57. D'Alembert, *Oeuvres de D'Alembert*, 1:261.

58. D'Alembert, 1:261n.

59. Heeffer, "On the Nature," 13.

60. Barbeau and Leah, 147.

61. D'Alembert, 1:262.

62. D'Alembert, 1:262.

63. Condillac, *Philosophical Writings*, 410.

64. Clairaut, *Élémens d'algèbre*, 3. See Albury, introduction to *La Logique*, 23.

65. Condillac, *Philosophical Writings*, 413.

66. Condillac, 410.

67. See Hamburg, "The Theory of Equations in the 18th Century."

68. Lagrange, *Lectures on Elementary Mathematics*, 80.

69. Lagrange, 69.

70. See Ferraro and Panza, "Lagrange's Theory of Analytical Functions and His Ideal of Purity of Method."

71. Condorcet, *Essai sur l'application de l'analyse à la probabilité des décisions rendues à la pluralité des voix*, ii.

72. Condorcet, ii.

73. Baker, 117.

74. Baker, 118.

75. Quoted by Baker, 124.

76. Lifschitz, "Translation in Theory and Practice," 38. The English version was a pirated book produced without Michaelis's knowledge. It contains a number of translation errors and other anomalies, including misnumbered pages. The translations here are based on the 1769 English version, but I have corrected them with reference to the French text, which was approved by Michaelis himself. The section from which I quote is not included in the German version.

77. Michaelis, *A Dissertation on the Influence of Opinions on Language, and of Language on Opinions*, 91; Michaelis, *De l'influence des opinions sur le langage et du langage sur les opinions*, 175.

78. Michaelis, *Dissertation*, 87; Michaelis, *De l'influence*, 166.

79. Michaelis, *Dissertation*, 88; Michaelis, *De l'influence*, 168.

80. Michaelis, *Dissertation*, 88; Michaelis, *De l'influence*, 168.

81. Michaelis, *Dissertation*, 77; Michaelis, *De l'influence*, 147.

82. Herder, *Johann Gottfried Herder on World History*, 246.

83. Condorcet, *Political Writings*, 11.

84. Referring specifically to the French context, Jessica Riskin identifies these two tendencies in eighteenth-century linguistics as the "social" and the "cultural" ("Rival Idioms for a Revolutionized Science and a Republican Citizenry"). The social approach involved "deliberately orchestrated, rather than organically arising, human activity" and treated signs as "deliberately chosen" (210; 217). By contrast, in the cultural approach, "one did not deliberately invent customs, manners of thought, or sciences according to first principles, but only fostered their natural growth" (208).

85. Herder, *Philosophical Writings*, 144n.

86. Herder, 157.

87. See Berlin, *Three Critics of the Enlightenment*. Berlin is not especially sympathetic to the Counter-Enlightenment, but he does agree with this assessment of Condorcet, whose philosophy he takes to be implicitly totalitarian (*Four Essays on Liberty* 56–60, 167). Emma Rothschild has disputed this reading of Condorcet (*Economic Sentiments*). Berlin's idea of the Counter-Enlightenment has also been criticized on the grounds that the thinkers he groups together under that banner were more engaged with Enlightenment thought than he suggests; see Norton, *Herder's Aesthetics and the European Enlightenment*; Israel, *A Revolution of the Mind*.

88. See Nisbet, *Herder and Scientific Thought*, 92–93.

89. Maseres, *Occasional Essays on Various Subjects: Chiefly Political and Historical*, 168.

90. Pycior, *Symbols*, 307.

91. On Simson, see Pycior, *Symbols*, 248; I discuss Frend later in the section.

92. Maseres, *Dissertation on the Use of the Negative Sign in Algebra*, ii.

93. Maseres, ii.

94. Maseres, iii.

95. Maseres, 1.

96. Maseres, 2.

97. Richards, "Rigor and Clarity," 307.

98. Maseres, *Dissertation on the Use of the Negative Sign*, 4.

99. Maseres, 3. On the significance of subtraction for Maseres, see Lambert, "Natural History of Mathematics," 288.

100. Maseres, 20.

101. See Fisch, "'The Emergency Which Has Arrived'"; Pycior, "George Peacock and the British Origins of Symbolic Algebra."

102. Maseres, *Principles of the Doctrine of Life-Annuities*, 36; Maseres, *A Proposal for Establishing Life-annuities in Parishes for the Benefit of the Industrious Poor*. On the parliamentary effort, see D. Thomas, "Francis Maseres, Richard Price, and the Industrious Poor"; Bellhouse, *Leases for Lives*, 171–75.

103. Maseres, *Principles*, 27.

104. Maseres, 25.

105. On the history of such statistical tables, see T. Porter, *The Rise of Statistical Thinking, 1820–1900*; Hacking, *The Emergence of Probability*; Hacking, *The Taming of Chance*; Daston, *Classical Probability in the Enlightenment*; Poovey, *A History of the Modern Fact*. On annuities in particular, see Bellhouse, *Leases for Lives*.

106. Daston, *Classical Probability*, 176.

107. Daston, *Classical Probability*, xi.

108. Laplace, *Essai philosophique sur les probabilités*, 95.

109. Maseres, *Principles*, 2.

110. Maseres, *Proposal*, 10.

111. Frend, *Peace and Union Recommended to the Associated Bodies of Republicans and Anti-Republicans*, 17; Frend, *The Principles of Algebra*, xi.

112. Frend, *Principles of Algebra*, 518.

113. Frend, *Principles of Algebra*, xii.

114. Frend, *Principles of Algebra*, xii.

115. Frend, *Peace and Union*, 34.

116. See Maseres, *Occasional Essays*, 368.

117. Maseres, *A Collection of Several Commissions*, 69.

118. Maseres, *Principles*, 37.

119. Hawtrey, *The Exchequer and the Control of Expenditure*; Reitan, *Politics, Finance, and the People*.

120. Lacroix, *An Elementary Treatise on the Differential and Integral Calculus*. Richards questions this narrative by arguing that the French and English traditions remained distinct even into the nineteenth century; see Richards, "Rigor and Clarity," 299.

121. Anon., "Biographical Account of Lord Stanhope," 85. The "grave authority" is perhaps Richard Phillips, who discusses Stanhope's device in the book *Public Characters of 1800–1801* (106).

122. Gardner, *Logic Machines and Diagrams*, xiii.

123. Quoted in Wess, "The Logic Demonstrators of the 3rd Earl Stanhope (1753–1816)," 385.

124. G. Stanhope and Gooch, *The Life of Charles, Third Earl Stanhope*, 110–12; see Phillips, 90; Erdman, "Citizen Stanhope and the French Revolution."

125. G. Stanhope and Gooch, *Life*, 240–41.

126. G. Stanhope and Gooch, *Life*, 21, 113, 238.

127. The first public explanation of Stanhope's logical theory appeared in Robert Harley's 1879 article "The Stanhope Demonstrator." Another account appears in Martin Gardner's 1958 book *Logic Machines and Diagrams*, which offers an interesting analysis of Stanhope's ideas but contains little information that is not in Harley's article. Aspray (*Computing before Computers*, 106–8) and Nilsson (*The Quest for Artificial Intelligence*,

12–13) include very brief discussions of Stanhope's work in their histories of computation; both largely follow Gardner. Shilov and Silantiev have, more recently, expanded on Gardner's account, describing some early versions of the Demonstrator of which Gardner was unaware ("Logical Machines," 5). Matthew L. Jones includes an insightful study of Stanhope's work on calculating machines in his 2016 book *Reckoning with Matter*, but only briefly discusses the Demonstrator (197–99). The most detailed modern account of Stanhope's theory of logic appears in Wess, "Logic Demonstrators."

128. Howell, *Eighteenth-Century British Logic and Rhetoric*, 259. See also Franklin, "Artifice and the Natural World"; Capozzi and Roncaglia, "Logic and Philosophy of Logic."

129. Campbell, *The Philosophy of Rhetoric*, 1:164.

130. Harley, 202.

131. Anon., "Biographical Account of Lord Stanhope," 85.

132. Shilov and Silantiev, fourth unnumbered page.

133. Quoted in Wess, 381; Phillips, 106. Such thinking was common at the time; see Daston, *Classical Probability*, 198.

134. Godwin, *An Enquiry Concerning Political Justice*, 596; see also 668.

135. Godwin, 596.

136. C. Stanhope, *Observations on Mr. Pitt's Plan for the Reduction of the National Debt*, 4. On the background of Pitt's plan, see Frame, *Liberty's Apostle*, 177–86.

137. C. Stanhope, *Observations*, 5.

138. Cobbett, *Cobbett's Parliamentary History of England*, 26:31.

139. C. Stanhope, *Observations*, 14.

140. C. Stanhope, *Observations*, 14, 27; Phillips, *Public Characters*, 83.

141. Stanhope expressly defends the rationality of commoners in C. Stanhope, *A Letter from Earl Stanhope, to the Right Honourable Edmund Burke*, 10.

142. On similar efforts in the context of Revolutionary France, see Kafka, *The Demon of Writing*.

143. Condorcet, *Political Writings*, 36. Condorcet discussed the idea of "prejudice" further in a fragment entitled "Préjugés qui peuvent momentanément arrêter les progrès" (Condorcet, *Tableau*, 940–41).

144. Condorcet, *Political Writings*, 37.

145. Condorcet, *Political Writings*, 75.

146. Diderot and D'Alembert, *Encyclopédie*, 13:284.

147. Maupertuis, *Réflexions philosophiques sur l'origine des langues et la signification des mots*, 6.

148. See Condorcet, *Tableau*, 994–99.

149. Condorcet, *Political Writings*, 96–97.

150. Kant, *Foundations of the Metaphysics of Morals and What Is Enlightenment?* 91.

151. Siskin and Warner, *This Is Enlightenment*, 11. For an alternate view, see Fleischacker, *What Is Enlightenment?* Fleischacker argues that Kant employed two distinct ideas of Enlightenment: a "maximalist" one that involves replacing traditional ways of life and a "minimalist" one focused on "*how* one holds one's views, not *what* views one holds" (169). One might identify Condorcet as a maximalist to the extent that he was trying to "suppress" prejudices rather than merely open free debate.

152. Kant, 655.

153. Locke, *Conduct of the Understanding*, 93.

154. Kant, *Critique of Pure Reason*, 655.

155. *OED Online*, s.v. "*objective*, adj. 3a," s.v. "*subjective*, adj. 2," accessed February 10, 2022, https://www.oed.com/view/Entry/129634; https://www.oed.com/view/Entry/192702.

156. For an example of how these terms were defined in the nineteenth century, see Tennemann, *Manual of the History of Philosophy*, vii; see also W, Hamilton, "M. Cousin's Course of Philosophy," 196–97n. As Daston and Galison point out in *Objectivity*, it was Kant, along with, in the English-speaking world, Samuel Taylor Coleridge, who spurred this reversal in the definitions of *subjective* and *objective* (30).

157. Capozzi and Roncaglia, 147.

158. Dyck, *Novalis and Mathematics*; Schlutz, *Mind's World*, 162–213.

159. Dyck, 42.

160. Novalis, *Notes for a Romantic Encyclopaedia*, 195.

161. Novalis, *Notes*, 147–48.

162. Novalis, *Philosophical Writings*, 83.

163. Redding, "Mathematics, Computation, Language and Poetry."

164. W. Hamilton, "M. Cousin's Course of Philosophy," 197.

165. Quoted in Dyck, 81.

166. Lagrange, *Théorie des fonctions analytiques*, 80. The heavily revised 1813 version of the book attaches the heading "Fonctions dérivées; leur notation et leur algorithme" ("Derived functions; their notation and their algorithm") to the chapter explaining the procedure; Lagrange, *Théorie des fonctions analytiques*, 2nd ed., 17. See also Lagrange, *Œuvres de Lagrange*, 7:327. For a general account of Lagrange's theory, see Ferraro and Panza, "Lagrange's Theory of Analytical Functions and His Ideal of Purity of Method." Novalis cites Lagrange in *Notes*, 37.

167. See Grattan-Guinness, "Charles Babbage as an Algorithmic Thinker," 36–37. Babbage suggests that Lagrange's approach could be applied to games in *On the Influence of Signs in Mathematical Reasoning*, 19–20.

168. See Grabiner, "Who Gave You the Epsilon?"; Laugwitz, "Definite Values of Infinite Sums"; Robinson, *Non-Standard Analysis*, 267. Ferraro and Panza argue that this judgment depends on a modern conception of function alien to Lagrange's own (99); Lagrange, in their account, viewed functions in terms of algebraic formulae and would not have accepted the piecewise definitions used to construct the counterexamples (129).

169. As the philosopher Brian Rotman puts it, modern mathematics "bifurcates its discourse into a privileged formal mode and an informal one considered as supplementary and epiphenomenal" (*Ad Infinitum*, 7).

170. Bernard Bolzano, for instance, was a sharp critic of Kant; see Rusnock, "Philosophy of Mathematics."

171. Condorcet, *Political Writings*, 6.

172. For a critique of Condorcet's views on race, see Sala-Molins, *Dark Side of the Light*.

173. Condorcet, *Political Writings*, 127, 128.

CHAPTER FOUR

1. This turn affected a range of mathematical practices, not just algebra; see Daston, *Classical Probability*, 4. The symbolic turn in British algebra did not, however, parallel developments in geometry, which remained tied to classical notions until much later in the nineteenth century; see Richards, *Mathematical Visions*.

2. On Babbage and industrialism, see Babbage, *On the Economy of Machinery and Manufactures*; Schaffer, "Babbage's Intelligence"; W. Ashworth, "Memory, Efficiency, and Symbolic Analysis"; Zimmerman, "The Ideology of the Machine and the Spirit of the Factory"; Kuskey, "Math and the Mechanical Mind."

3. W. Humboldt, *On Language*, 167.

4. Coleridge wrote, in an 1800 letter to William Godwin, that he "would endeavor to destroy the old antithesis of *Words & Things*, elevating, as it were, words into Things, & living Things too" (*Letters of Samuel Taylor Coleridge*, 1:626).

5. Bopp, *Analytical Comparison of the Sanskrit, Greek, Latin and Teutonic Languages*, 14.

6. Due in large part to the influence of Raymond Williams, the nineteenth-century idea of culture is often seen as antimechanistic. More recently, a number of scholars, such as Amir Alexander and Andrea Henderson, have complicated this narrative, showing the engagement of industrialists and mathematicians with artists, poets, and fiction writers. See R. Williams, *Culture and Society, 1780–1950*; Alexander, *Duel at Dawn*; Henderson, *Algebraic Art*; Tresch, *The Romantic Machine*.

7. Boole, *Mathematical Analysis*, 5.

8. G. Boole, *Mathematical Analysis of Logic*; Boole, *An Investigation of the Laws of Thought*; Boole, "The Calculus of Logic."

9. Boole, *Mathematical Analysis*, 3.

10. As Ivor Grattan-Guinness observes, Boole never mentioned Peacock in his writings, so it is uncertain whether there was a direct influence ("Boole's Quest," xliv). However, Peacock certainly exerted an indirect influence on Boole by means of Gregory's 1840 essay "On the Real Nature of Symbolical Algebra," which is framed as an attempt to clarify and further develop Peacock's ideas. On Gregory's influence on Boole, see Laita, "The Influence of Boole's Search for a Universal Method in Analysis on the Creation of his Logic," 52; Despeaux, "'Very Full of Symbols,'" 49. Other important influences include William Rowan Hamilton and Arthur Cayley, who were developing algebraic systems that worked with arrays of numbers rather than individual numbers, and that obeyed different laws than ordinary algebra; see MacHale, *The Life and Work of George Boole*, 65–66.

11. Peacock, *Treatise of Algebra*, xiv. For an account of the context surrounding the composition of Peacock's book, see Fisch, "The Making of Peacock's Treatise." Fisch argues that, in this first edition of the book, published in 1830, Peacock held back from embracing a totally formalist position, instead trying to split the difference between the symbolic methods of Babbage and the Pauline epistemology of Francis Maseres and William Frend (168); Peacock did not, in Fisch's account, embrace fully symbolic methods until around 1840. His second edition of *Treatise of Algebra*, published in 1842, eliminates the passage quoted.

12. For an overview of the difference between Boole's system and modern Boolean logic, see Hailperin, "Boole's Algebra Isn't Boolean Algebra."

13. The requirement that categories be mutually exclusive is a notable difference between Boole's system and Leibniz's calculus of "real addition." This requirement is sometimes viewed as a mistake on Boole's part, but it is in fact necessary given his axioms. One can prove this as follows. Suppose that x and y are logical variables and $x + y$ is a logically meaningful expression. Then, by Boole's definition of *logically meaningful*, it must obey the law of duality, so that $(x + y)(x + y) = x + y$. Expanding this and

applying the law of duality again, one gets $x^2 + 2xy + y^2 = x + y \Longrightarrow x + 2xy + y = x + y \Longrightarrow 2xy = 0 \Longrightarrow xy = 0$. The latter equation means, logically interpreted, that x and y have an empty intersection, QED. Boole's exclusive interpretation of disjunction is inconvenient in practice, and most later systems of algebraic logic deviate from Boole on this point; however, in doing so, they create a greater divergence between symbolic logic and ordinary algebra than Boole intended. Boole addresses this issue directly in an 1856 manuscript titled "On the Foundations of the Mathematical Theory of Logic and on the Philosophical Interpretation of Its Methods and Processes" (in G. Boole, *Selected Manuscripts on Logic and Its Philosophy*, 91–92). See also Jevons, *Pure Logic and Other Minor Works*, 72; Hailperin, *Boole's Logic and Probability*, 87–96.

14. The interpretation of 1 and 0 is one of the few points on which Boole changed his mind between *Mathematical Analysis* and *Laws of Thought*. In the 1847 version of his theory, 1 means *true in all circumstances* and 0 means *false in all circumstances*. In the 1854 version, he interprets 1 as *true at all times* and 0 as *false at all times*. He does not explain the reason for this change very clearly, but I would venture that it is because introducing the idea of time into the interpretation enables him to connect 1 to the idea of eternity, thus making the religious implications of his theory more apparent. See G. Boole, *Mathematical Analysis*, 48–50; G. Boole, *Laws of Thought*, 162–67.

15. G. Boole, *Mathematical Analysis*, 26. Boole's inference procedure is primarily intended to determine what properties a thing must have based on a given statement; as such, it is analogous to solving an equation. Corcoran and Wood argue that this aspect of Boole's system rests on the logical fallacy of conflating roots of equations with logical consequences ("Boole's Criteria for Validity and Invalidity," 111–14); however, Frank Markham Brown has disputed this point ("George Boole's Deductive System" 307). Alonzo Church and Alan Turing would later prove, building on the work of Kurt Gödel, that a general procedure for determining the truth or falsehood of logical statements cannot exist. Boole's system did not include quantifiers such as *for all* and *there exists*, so it is not powerful enough to produce the paradoxes that lead to such problems.

16. This is how one might work out the problem in detail. We have $p(1 - g) = 0$. One can express g as a function of p in the general form $g = vp + v'(1 - p)$, where v and v' are indeterminate values. Expressing the equation in this form is possible regardless of how g and p are defined because of the law of duality, as Boole demonstrates. We substitute this into the original equation, getting $p(1 - (vp + v'(1 - p))) = 0 \Longrightarrow p - vp^2 - v'p(1 - p) = 0$. By the law of duality, the third term vanishes and the exponent disappears from the second, so we have $p - vp = 0 \Longrightarrow vp = p$. Substituting this into the general form of g gives us $g = p + v'(1 - p)$. I removed the prime mark from v' in recording the result for clarity. Boole's use of v to represent indeterminate quantities is widely regarded as problematic, since the symbol appears to be a variable even though it does not behave as one; see Hailperin, *Boole's Logic and Probability*, 97–98.

17. G. Boole, *Mathematical Analysis*, 14.

18. Grattan-Guinness, "Boole's Quest," xliii.

19. M. Boole, *Symbolical Methods of Study*, 35.

20. G. Boole, *Laws of Thought*, 417.

21. Cohen, *Equations from God*, 29–30.

22. See Greene, "A Taste for Figures," 70–71.

23. Baily, "Some Particulars Respecting the Arithmetical Powers of Zerah Colburn," 121.

24. Daston, "Enlightenment Calculations."

25. Anon., "Anecdotes of Great Memory, and Astonishing Powers of Calculation," 239.

26. See Ball, *Mathematical Recreations and Essays*, 268; Poole, *Anti-Slavery Opinions before the Year 1800*, 21.

27. Brissot de Warville, *New Travels in the United States of America*, 241.

28. Leibniz, *New Essays*, 77–78.

29. Baily, 124.

30. Colburn, *A Memoir of Zerah Colburn*, 13, 28.

31. Baily, 125. This claim is historically inaccurate.

32. Baily, 125.

33. Colburn, 164–65.

34. Colburn, 181.

35. Bishop, "The Mathematical Failure," 276.

36. Bishop, 278.

37. See Brown, "William Rowan Hamilton and William Wordsworth."

38. R. Graves, *Life of Sir William Rowan Hamilton*, 1:111.

39. Colburn, 185–89.

40. R. Graves, *Life of Hamilton*, 1:79–80.

41. Colburn, 104.

42. W. R. Hamilton, "On Quaternions," 11.

43. W. R. Hamilton, 11.

44. Ohm, *The Spirit of Mathematical Analysis, and its Relation to a Logical System*. On Ohm's relations to Kant and to Hamilton, see Martin, *Arithmetic and Combinatorics*, 39–41, 46–47.

45. Ohm, 11.

46. Ohm, 12, 13.

47. Ohm, 20.

48. Ohm, 10.

49. On the history of the Analytical Society, see Wilkes, "Herschel, Peacock, Babbage and the Development of the Cambridge Curriculum"; Becher, "Radicals, Whigs and Conservatives"; Grier, "The Inconsistent Youth of Charles Babbage." The consensus is that, in its short existence as an active organization, the society failed in its goal of reforming the Cambridge curriculum; however, some of the members later gained positions of influence at Cambridge, and symbolic methods were eventually accepted there.

50. Babbage, *Passages from the Life of a Philosopher*, 29; see Becher, 406.

51. Peacock, *Treatise on Algebra*, vii.

52. *OED Online*, s.v. "algorithm, *n*," accessed February 10, 2022, https://www.oed .com/view/Entry/4959; Hutton, *A Course of Mathematics*, 1:231.

53. Hoëné-Wroński, *Address of M. Hoene Wronski to the British Board of Longitude upon the Actual State of the Mathematics*, 9; see Hoëné-Wroński, *Introduction à la philosophie des mathématiques, et technie de l'algorithmie*. Although an advocate of infinite summation, Wroński was critical of Lagrange; see Schubring, *Conflicts between Generalization, Rigor, and Intuition*, 407.

54. Montferrier, *Encyclopédie mathématique*, 1:150.

55. Smedley, Rose, and Rose, *Encyclopædia Metropolitana*, 1:438. A similar definition appears in the French version of Lacroix's calculus text: *algorithme* sometimes

means "the system of characters that one employs to express quantities subjected to certain laws: digits are the algorithm of numeration" (Lacroix, *An Elementary Treatise on the Differential and Integral Calculus*, 3:545). This definition does not, however, refer to algebraic notation but rather to alternative systems of digits such as binary.

56. Shaw defines *algorithm* in this way: "Any system of deduction of necessary conclusions, or transmutation of laws or ideas into other forms, is an *algorithm*. Thus quaternions is an algorithm. Graphical mechanics is an algorithm." Shaw, *Mathematics*, 3.

57. Booth, "On Tangential Coordinates," 176.

58. Venn, *The Logic of Chance*, 164.

59. S. Porter, *Conversations on Arithmetic*, v, vi.

60. Lovelace, *Ada, the Enchantress of Numbers*, 68.

61. For instance, she writes in an 1850 letter to Lady Byron, "The very excitable intellect & imagination of Ralph is trained & soothed by the acquisition of true science & by the cultivation of exact reasoning power,—this last cultivation being peculiarly necessary to his sensitive & passionate nature- & best attained thro' that mathematical course of study"; Lovelace, 374.

62. See Babbage, *Babbage's Calculating Engines*, 21–50. This procedure was probably designed by Babbage.

63. Colburn, 176, 77.

64. As reported by his wife: M. Boole, "Home-Side of a Scientific Mind," 106.

65. D. Brown, *The Poetry of Victorian Scientists*; see also Forbes-MacPhail, "The Enchantress of Numbers and the Magic Noose of Poetry."

66. Lovelace, 319; Poe, *Collected Works of Edgar Allan Poe*, 3:986.

67. Weierstrass quoted in *Compte rendu du deuxième Congrès international des mathematiciens*, 149.

68. MacHale, 197–98.

69. Wordsworth and Coleridge, *Lyrical Ballads*, 105.

70. Wordsworth and Coleridge, 4.

71. Daston, *Classical Probability*, 55.

72. These lines are a part of Wordsworth's long poem *The Prelude*, which was not published in full until 1850; however, the passage from which they are extracted also appeared as a separate poem titled "French Revolution" in an 1810 edition of *The Friend* and in Wordsworth's 1815 volume of poems. See Wordsworth, *Poems by William Wordsworth*, 2:71.

73. See Sunstein, *Mary Shelley: Romance and Reality*, 39; St. Clair, *The Godwins and the Shelleys*. On James Mill's pedagogy, see J. S. Mill, *Autobiography*, 1–37. Bruce Mazlish gives a psychohistorical reading of the relationship between James and John Stuart Mill in *James and John Stuart Mill*.

74. Cleveland, *The Life and Letters of Lady Hester Stanhope*, 16. See Meryon, *Memoirs of the Lady Hester Stanhope*.

75. Wordsworth and Coleridge, 164n.

76. Wordsworth and Coleridge, 156, 157.

77. Wordsworth and Coleridge, 162n.

78. Edgeworth and Edgeworth, *Practical Education*, 1:vi.

79. Edgeworth and Edgeworth, 1:77.

80. Siskin and Warner, 170; Valenza, *Literature, Language, and the Rise of the Intellectual Disciplines in Britain*, 144.

81. Wordsworth and Coleridge, 167n.

82. Wordsworth, *The Prelude, 1799, 1805, 1850: Authoritative Texts, Context and Reception, Recent Critical Essays*, 194.

83. Wordsworth, *Prelude*, 406–8.

84. Wordsworth, *Prelude*, 476.

85. Whately, *Elements of Logic*, 10.

86. Whately, 37.

87. Mill, *A System of Logic, Ratiocinative and Inductive*, 1:439; see also 2:201–2.

88. Boole, *Laws of Thought*, 423.

89. Boole, *The Right Use of Leisure*, 14.

90. Boole, 23.

91. Boole, 23.

92. Quoted in MacHale, 111. Boole goes on to state that the mind "feels the pressures of impulses, it is conscious of the existence of powers and faculties which urge it to reduce the scattered details of its knowledge into form and order" (112); this impulse to organize knowledge must not be pursued at random but rather must serve "the true welfare of our species," which "essentially contains a moral element" (114). Whether or not there was any direct influence, the resemblance to Coleridge's "Essays on the Principles of Method" is apparent. See Coleridge, *The Collected Works of Samuel Taylor Coleridge*, vol. 4, pt. 1, pp. 448–524.

93. Mill, *Mill on Bentham and Coleridge*, 40.

94. Mill, 71, 73.

95. Mill, 140; see also 114.

96. Mill, *System of Logic*, 2:292.

97. Quoted in G. Boole, *Mathematical Analysis*, 2. Boole misquotes the passage slightly, substituting "obstacle" for "obstacles" and "mere" for "merely" and altering the punctuation in a way that does not significantly affect the meaning. The fact that Boole quotes Mill should not be taken to indicate that he endorsed Mill's views on the nature of logic. Specifically, whereas Mill made a sharp distinction between logic and psychology, Boole understood logic to be a part of psychology. Boole did not, however, construe "psychology" in the modern sense of the term, as I discuss in the final section of this chapter. See Cook, "Minds, Machines and Economic Agents"; Maas, "Mechanical Rationality."

98. G. Boole, *Mathematical Analysis*, 10.

99. G. Boole, *Mathematical Analysis*, 10.

100. M. Boole, "Home-Side," 108. Although Mary's writings should be treated with some caution in drawing conclusions about her husband, scholars have often turned to them for evidence as to his views on religion, education, and other matters that he did not address much in writing. Luis M. Laita argues, based on a comparison of her writings and his, that her representations of his views are basically accurate ("Boolean Algebra and its Extra-Logical Sources"). See also MacHale, *Life*, 25–33.

101. M. Boole, "Home-Side," 108.

102. M. Boole, "Home-Side," 108.

103. S. Porter, *Conversations on Arithmetic*, 2.

104. S. Porter, 18.

105. S. Porter, 6–7.

106. S. Porter, 6.

107. Mill, *System of Logic*, 2:261.

108. A. Humboldt, *Kosmos*, 1:42.

109. "Symbols and Notation," in *The Penny Cyclopaedia of the Society for the Diffusion of Useful Knowledge*, 23:444.

110. Whewell, *The Philosophy of the Inductive Sciences*, 1:xlix.

111. G. Boole, *Laws of Thought*, 24.

112. G. Boole, 25.

113. G. Boole, 28.

114. G. Boole, 29.

115. G. Boole, 33.

116. In "The Calculus of Logic," Boole writes, "That the forms under which propositions are actually exhibited, in accordance with the principles of this calculus, are analogous with those of a philosophical language" (184). In a footnote in *Mathematical Analysis*, he links the idea of a philosophical language to the idea that logic describes some aspects of the structure of language, including a citation to the philologist Robert Gordon Latham; see note 118.

117. G. Boole, *Laws of Thought*, 38.

118. Latham, *First Outlines of Logic, Applied to Grammar and Etymology*, 12. Boole cites this book in *Mathematical Analysis* (5n). In a note in *Laws of Thought*, Boole disagrees with Latham as to the nature of conjunctions (401n); one of Boole's innovations was to treat *and* and *or* as parts of propositions rather than (as for Latham) ways of linking propositions together.

119. G. Boole, *Laws of Thought*, 30.

120. G. Boole, *Laws of Thought*, 30–31; see also G. Boole, *Selected Manuscripts*, 70.

121. G. Boole, *Laws of Thought*, 30, 31.

122. See Kittler, *Discourse Networks 1800/1900*, 178.

123. Coleridge, *Collected Works*, vol. 4, pt. 1, pp. 512–13.

124. Coleridge, vol. 4, pt. 1, p. 513.

125. The religious aspects of Boole's thought are not incidental to his philosophy of logic; without the faith that the laws of thought are imposed by divine will, his system lacks a foundation. One is supposed to be able tell that the laws of thought are correct because they generate a correct method for reasoning. But how can one tell that the method is correct without already knowing the laws of logical validity? Since Boole has no notion of a metalogic or metamathematics in which to judge the correctness of method, he falls into circular reasoning at this point without the deus ex machina of religious faith. See Grattan-Guinness, "Boole's Quest," xli.

126. M. Boole, "Home-Side," 107; G. Boole, *Laws of Thought*, 399–424.

127. A classic study of *Paradise Lost* is S. Fish, *Surprised by Sin*; Fish argues that the text is designed to tempt readers to sympathize with Satan so as to train them to resist such temptation. For a critique of this reading, see Walker, "On Reason, Faith, and Freedom in 'Paradise Lost.'"

128. De Morgan, "A Treatise on Algebra (No. II)," 311.

129. De Morgan, "A Treatise on Algebra," 103.

130. De Morgan, 103.

131. De Morgan, 103.

132. On the relation of Boole and De Morgan to abstract algebra, see Koppelman, "The Calculus of Operations and the Rise of Abstract Algebra."

133. De Morgan, "Treatise on Algebra," 105.

134. De Morgan, 106.

135. De Morgan, 106.

136. De Morgan, "Treatise on Algebra (No. II)," 310.

137. De Morgan, 310.

138. De Morgan, 310.

139. E.g., MacHale, xvii.

140. G. Boole, *Laws of Thought*, 28.

141. G. Boole, 41.

142. G. Boole, 44.

143. G. Boole, 44.

144. G. Boole, 45.

145. G. Boole, 29.

146. G. Boole, "Calculus of Logic," 187.

147. G. Boole, *Laws of Thought*, 67.

148. G. Boole, 94–95.

149. G. Boole, 95.

150. In modern algebra, this is only true if $z + w \neq 0$. However, Boole was working in an algebraic tradition that allowed division by zero in solving equations; indeed, the values $0/0$ and $1/0$ play important roles in Boole's method. See Boole, *Laws of Thought*, 156.

151. In a manuscript written some time after *Laws of Thought*, Boole discusses an "inverse" mental operation "by which from the conception of a given class of things we ascend to the conception of some larger class from which the given class would be formed from the mental selection of those individuals which possess a given property" (*Selected Manuscripts*, 58). This "inverse operation" would seem to correspond to division in the logical calculus. Boole writes that this operation "has no verbal symbol or equivalent construction in language" and is "only conceivable by means of that operation of which it is the inverse"—that is to say, the composition of attributes (58). Thus it would seem that, at least at this later point in his career, he believed that division corresponded to an operation that can occur in the human mind but that cannot be expressed in ordinary language. This argument is problematic, however, because there is no guarantee that the "larger class" he discusses is unique; indeed, there could be infinitely many classes that satisfy this definition of p/q. Boole does not treat division as corresponding to a mental operation in *Laws of Thought*. Hailperin developed an alternative interpretation of Boolean quotients using Venn diagrams and the mathematical idea of a multiset (*Boole's Logic and Probability*, 109–12); Frank Markham Brown has criticized this interpretation for adding a layer of complexity that is not present in Boole's own work ("George Boole's Deductive System," 304).

152. Boole, *Laws of Thought*, 98.

153. Jevons, *Pure Logic*, 66.

154. Boole, *Laws of Thought*, 69

155. Boole, 67.

156. Boole, 69.

157. Boole, *Selected Manuscripts*, 53.

158. Boole, 200.

159. In his 1848 article, for instance, he states that relations such as conditional and disjunctive "are referred by Kant to distinct conditions of thought" (G. Boole, "Calculus of Logic," 197). The Kantian resonances are also apparent in his later writings on probability theory, in which he refers to a series of theorems about the properties

of statistical data as "conditions of possible experience" (G. Boole, *Studies in Logic and Probability*, 319). Boole differs from Kant in his focus on algebraic relations rather than on spatial and temporal intuitions, but he did seem to conceive of the relation between logical and conceptual truth in terms of the Kantian division between pure reason and understanding, legitimately or not.

160. Quoted in M. Boole, *Symbolical Methods of Study*, 17; see Cohen, *Equations from God*, 91.

161. Boole and Babbage did not meet in person until 1862, which was late in both of their careers (Hyman, *Charles Babbage*, 249). Although Babbage disagreed with Boole on some points (see Grattan-Guinness, "Boole's Quest," xliv), Babbage wrote on his copy of Boole's *Mathematical Analysis of Logic* "This is the work of a real *thinker*" (quoted by Hyman, 244). Boole cites Babbage in his 1860 textbook *A Treatise on the Calculus of Finite Differences*, noting that "in the state to which it has been brought, more especially by the labors of Mr Babbage," the calculus of functions "is much too extensive a branch of analysis to permit of our attempting here to give more than a general view of its objects and methods" (208). Whereas George Boole did not show much interest in Babbage's calculating machines, his wife, Mary Everest Boole, was an admirer of Babbage's computing machines and wrote enthusiastically about Babbage's theological arguments (M. Boole, *Logic Taught by Love*, 34–39).

162. M. Boole, "Home-Side," 109. Compare Novalis: "The apprentice must not yet reason. First he must become mechanically skilled, and only then may he begin to reflect and strive for insight and order concerning that which was learned" (*Notes for a Romantic Encyclopaedia*, 6).

163. M. Boole, "Home-Side," 109.

164. In the first pages of *Laws of Thought*, Boole suggests a rationale for expecting the understanding of logic to come in a sudden burst of insight. Boole's example is Aristotle's *dictum de omni et nullo*—the principle that if something is true of a category, then it is also true of any subcategory of that category. The logical truth of this principle, he argues, "is made manifest in all its generality by reflection upon a single instance of its application"; the fact that one need not offer multiple examples to convince people of its truth is evidence that it "is founded upon some general law or laws of mind" (4). Gérard Bornet argues convincingly that this passage is directed at Mill's argument in *System of Logic* that logical truths are founded on induction (Bornet, "Boole's Psychologism as a Reception Problem"). Whereas Mill's theory implies that adding more examples would strengthen the induction and thus provide further evidence of the truth of the principle, Boole holds that one can perceive the truth of the principle with certainty all at once.

165. G. Boole, *Mathematical Analysis*, 2.

166. M. Boole, *Logic Taught by Love*, 55.

167. See Barrett and Connell, "Jevons and the Logic 'Piano'"; Maas, "Mechanical Rationality."

168. See Sack, *The Software Arts*, 122. On the developments leading from Boolean logic to logic circuits, see Stanković and Astola, *From Boolean Logic to Switching Circuits and Automata*.

169. Schröder, *On the Formal Elements of the Absolute Algebra*.

170. Schröder, "Ueber Algorithmen und Calculn," 225–28. In the terminology of modern abstract algebra, these equations would make the system an Abelian group, at least if we assume the existence of an identity element and that the operation is closed

over some given set. For Schröder, an "algorithm" includes not just the stated axioms but also all other equations that can be deduced from them.

171. Schröder quoted in Peckhaus, "19th Century Logic between Philosophy and Mathematics," 442. See also Peckhaus, "Ernst Schröder on Pasigraphy."

172. Frege, "Begriffsschrift," 6. On Frege's position in the history of computation, see Von Plato, *The Great Formal Machinery Works*.

173. Frege, 7.

174. On the role of subjectivity in Romantic science, see Tresch, "Even the Tools Will Be Free."

175. On the charge of psychologism against Boole, see Bornet; Vassallo, "Psychologism in Logic"; Corcoran, "Aristotle's Prior Analytics and Boole's Laws of Thought."

176. See Henderson, *Algebraic Art*, 171.

177. G. Boole, *Laws of Thought*, 399.

178. G. Boole, 410–16.

179. W. R. Hamilton, "Theory of Conjugate Functions," 293; R. Graves, *Life of Hamilton*, 2:522.

180. "Arithmetic," in Smedley, Rose, and Rose, *Encyclopædia Metropolitana*, 1:369–523. See Lambert, "Natural History of Mathematics."

181. G. Boole, *The Social Aspect of Intellectual Culture*, 11.

182. Ball, *A History of the Study of Mathematics at Cambridge*, 123.

183. Ball, 123.

184. *Compte rendu*, 149. On Weierstrass's possible anti-Semitism, see Bair et al., "Klein vs Mehrtens."

185. A notorious example is Frege, who expressed virulently anti-Semitic views, at least later in his life. These views were not, however, shared by all other formalists. See Jacquette, *Frege: A Philosophical Biography*, 519n.

186. See Dauben, *Georg Cantor*. One of Cantor's major contributions to number theory appears in "Beiträge zur Begründung der transfiniten Mengenlehre I," 485; for an English translation, see Cantor, *Contributions to the Founding of the Theory of Transfinite Numbers*, 91.

187. Dedekind, *Was Sind und Was Sollen die Zahlen?*; Peano, *Arithmetices principia*. For English translations, see Dedekind, *Essays on the Theory of Numbers*, 31–115; Peano, *Selected Works of Giuseppe Peano*, 101–34.

188. Kronecker quoted in Gray, *Plato's Ghost*, 153.

189. Gray, 149.

190. Dedekind, *Essays*, 1.

191. Dedekind, *Essays*, 1.

192. See Janton, *Esperanto*. On the relation of mathematics to the international language movement circa 1900, see Gray, 374–88.

193. In 1903, Giuseppe Peano introduced an IAL he called Latino sine Flexione (Latin without inflections); he later began to call it Interlingua. Peano stated in his 1903 article that the construction of an artificial language was unrelated to the development of mathematical notations. See Kennedy, *Peano: Life and Works*, 107–13, 125–43. A different IAL going by the name Interlingua was developed by the International Auxiliary Language Association starting in the 1930s; see Gode and Blair, *Interlingua*.

194. See Henderson, 173–77.

195. W. Humboldt, *On Language*, 24.

196. Saussure, *A Course in General Linguistics*, 68; see also 71–74. Henderson rightly

observes that Boole anticipated Saussure by declaring that signs are arbitrary marks that should be considered primarily in terms of their formal relations (*Algebraic Art*, 7–8); however, Boole and Saussure differ on this point about individual control.

197. Welby, *What Is Meaning?*, 52. See Pietarinen, "Significs and the Origins of Analytic Philosophy."

198. Whitehead and Russell, *Principia Mathematica*, 1:364.

199. Whitehead and Russell, 1:1.

200. Cajori, *History of Mathematical Notations*, 339–40.

201. Cajori, 350.

202. Cajori, 344.

203. For instance, John Wallis was hesitant to adopt the notation $\sqrt{-a}$ until he discovered a "precedent" for it in the work of Thomas Harriot; see Pycior, *Symbols*, 107–12.

204. Cajori, 344.

CHAPTER FIVE

1. A group of American computer scientists, for instance, toured Soviet computer centers in August–September 1958; see Carr et al., "A Visit to Computation Centers in the Soviet Union." They note that a partial translation of Markov's book was available (18). A 1951 article by Markov presenting an early version of the theory was also translated into English by Edwin Hewitt in 1960; see Markov, "The Theory of Algorithms."

2. The proceedings of the conference were published in Ershov and Knuth, *Algorithms in Modern Mathematics and Computer Science*.

3. Church, "An Unsolvable Problem of Elementary Number Theory"; Post, "Finite Combinatory Processes—Formulation 1"; Turing, "On Computable Numbers, With an Application to the Entscheidungsproblem." Post does not use the exact phrase "problem solving"; he presents his model as formalizing of the activity of a "problem solver or worker" producing an answer to a problem within a given "symbol space" (103). Post's model is broadly similar to Turing's, and Post deserves credit for publishing the idea first. Turing did, however, explore the idea's implications in much greater depth than Post did. Fairly or not, Turing wound up being much more widely cited in computer science.

4. Church, 349.

5. Kleene, "Recursive Predicates and Quantifiers."

6. Kleene, 59.

7. Markov, *Theory of Algorithms*, 1; see also 58–59, 63–70.

8. Gödel, "Über formal unentscheidbare Sätze der Principia Mathematica und verwandter Systeme I." An English translation is available in Gödel, "On Formally Undecidable Propositions of Principia Mathematica and Related Systems I."

9. Markov, 1–2.

10. The phrase "Euclid's algorithm" predates "Euclidean algorithm." The earliest instance of the phrase "Euclid's algorithm" in English I have been able to find is from James Pierpont's 1898 review of Heinrich Weber's *Lehrbuch der Algebra*; see Pierpont, "Weber's Algebra," 205. Some instances in French and German appeared earlier, although the usage does not seem to have become common until the 1890s.

11. Grier, *When Computers Were Human*, 117.

12. Knuth, *The Art of Computer Programming*, 1.4–6. Another foundational work is

Niklaus Wirth's 1975 book *Algorithms + Data Structures = Programs*, whose title may be said to give an implicit definition.

13. A. Forsythe et al., *Computer Science*, 3.

14. Stone, *Introduction to Computer Organization and Data Structures*, 4.

15. Uspensky and Semenov, *Algorithms: Main Ideas and Applications*, 1. This book is a revised version of a text first presented at the symposium (see Ershov and Knuth, *Algorithms in Modern Mathematics and Computer Science*, 100–234). Uspensky and Semenov only present an informal explanation of what algorithms are; they treat *algorithm* as "a primitive notion" that is not defined within their theory (18).

16. Hilbert and Ackerman, *Principles of Mathematical Logic*.

17. On the role of intuition in computability theory, see Soare, "Formalism and Intuition in Computability."

18. Carnap, *The Logical Structure of the World*, 5.

19. Carnap, 8.

20. Carnap, 182–83.

21. See Capaldi, "The Enlightenment Project in Twentieth-Century Philosophy"; Stadler, *The Vienna Circle*.

22. Jeffrey, *Probability and the Art of Judgment*, 27–29.

23. Carnap, *Logical Structure*, 12.

24. Carnap, 276.

25. Carnap, 7.

26. Carnap, *The Logical Syntax of Language*, xv.

27. Carnap, xv.

28. Carnap, 4.

29. Tarski, *Logic, Semantics, Metamathematics*, 166. Tarski uses the term *formalized language*, which he distinguishes from "'formal' languages" that do not involve meaning.

30. On Carnap's relationship to Turing, see Floyd, "Turing on 'Common Sense,'" 111.

31. Turing, "On Computable Numbers," 250.

32. Turing, 231, 232.

33. Turing, 241.

34. Turing, 249–50.

35. Turing, "Systems of Logic Based on Ordinals," 167.

36. Wittgenstein, *Philosophical Investigations*.

37. Turing, "Systems of Logic," 172.

38. E.g., Bolter, *Turing's Man*.

39. Turing, "Systems of Logic," 214.

40. Turing, 214–15.

41. Turing, 215.

42. Turing, 215.

43. Turing, 216.

44. Turing, "On Computable Numbers," 236–65.

45. Church, "Unsolvable Problem," 351.

46. Church cites Schröder in his 1927 paper "Alternatives to Zermelo's Assumption"; see also Church, "Schröder's Anticipation of the Simple Theory of Types."

47. Priestley, *A Science of Operations*, 123.

48. Priestley, 142–45.

49. See Hoare, "The 1980 ACM Turing Award Lecture," 77.

50. Naur and Randell, *Software Engineering*; Naur and Randell, *Software Engineering Techniques*.

51. Priestley, *Science of Operations*, v.

52. Shintaro Miyazaki takes the existence of assignment operators like ALGOL's := as a point of differentiation between programming languages and algebra. While this distinction is indeed important, it should be noted that such operators are not employed in all programming languages. Purely functional languages such as Haskell forbid changing the values of variables after they are initially set; Haskell works instead through recursive functions. LISP-like programming languages, especially Scheme, include assignment features but discourage their use in favor of a recursive style. Assignment operators cannot, then, be taken as essential to algorithmic thinking. It should also be noted that some early forms of algebra, such as that of Robert of Chester, arguably do involve changing the values of variables over the course of a computation. See Miyazaki, "Algorhythmics." See also Knuth, "Algorithmic Thinking and Mathematical Thinking."

53. Sack, *The Software Arts*, 57–77.

54. See Goldstine and Von Neumann, *Planning and Coding of Problems for an Electronic Computing Instrument*.

55. Hopper, "The Education of a Computer," 272.

56. Hopper, 273.

57. Nofre, Priestley, and Alberts, "When Technology Became Language."

58. Hopper, 280.

59. Hopper, 281.

60. Sammet, "The Early History of COBOL," 210. On FLOW-MATIC, see Knuth and Pardo, *The Early Development of Programming Languages*, 89–90.

61. Quoted in Knuth and Pardo, 61.

62. Quoted in Backus, "The History of FORTRAN I, II, and III," 28.

63. Priestley, 206.

64. Bauer et al., "Proposal for a Universal Language for the Description of Computing Processes," 355.

65. Rutishauser, "Some Programming Techniques for the ERMETH," 2.

66. Zuse discusses his own contributions in *The Computer: My Life*.

67. See Miyazaki, "Algorhythmics."

68. Miyazaki points out that some early ALGOL programmers, such as John Backus, preferred to call it an *algebraic* instead of *algorithmic* language in the late 1950s; in Miyazaki's account, the fast dissemination of ALGOL 60 contributed to the transition ("Algorhythmics").

69. Hoare, "Algorithm 64."

70. "Announcement," 73.

71. Naur and Randell, *Software Engineering*, 79.

72. Ranshaw, "Certification of Algorithm 23." In 1966, the journal hosted a forum discussing whether to allow languages other than ALGOL in the "Algorithms" section; the consensus was that the publication language would have to be judged based on its clarity of expression, not its adoption for programming. See Perlis, "A New Policy for Algorithms?"; G. Forsythe, "Algorithms for Scientific Computation"; Herriot, "Algorithms Section Policy."

73. For instance, in his NATO talk, McIlroy states, "We undoubtedly produce

software by backward techniques. We undoubtedly get the short end of the stick in confrontations with hardware people because they are the industrialists and we are the crofters" (Naur and Randell, *Software Engineering*, 79).

74. Wirth, "On the Design of Programming Languages," 23.

75. Chun, "On 'Sourcery,' or Code as Fetish."

76. Wirth, "On the Design," 23.

77. Wirth, "The Essence of Programming Languages," 3. See Payette, "Hopper and Dijkstra," 67.

78. Wirth, "On the Design," 24.

79. Marcovitz and Schweppe, *An Introduction to Algorithmic Methods using the MAD Language*, 73. See Priestley, 208.

80. Wirth, "On the Design," 24. See Van Wijngaarden, "Generalized ALGOL." Babbage's credo was, as his biographer Anthony Hyman notes, "simplify and generalize" (*Charles Babbage*, 241).

81. Most important, ALGOL 60 optionally employs call by name, wherein the expressions passed in to procedures are not evaluated ahead of time but rather are literally substituted into the text where the parameter names appear in the body of the procedure.

82. Note that the ALGOL 60 language standard contains no "print" feature and, indeed, no input or output features at all. If you want to try running this program, you may have to change that line, depending on which implementation you use. For example, with the NASE A60 interpreter, you can use `outreal(1, B);`. You will also have to replace the multiplication sign with an asterisk.

83. Backus et al., "Revised Report on the Algorithmic Language ALGOL 60," 364. Since bold and italics are not generally used in modern programming languages, they deserve some explanation. ALGOL employs the practice of "stropping," which means that built-in keywords such as **if** are typographically distinguished from user-defined names; the distinction is usually made with boldface in published code and with some form of punctuation (such as the quotation marks in the example the follows later in the chapter) in code that is entered in a computer. Note that the formatting is inconsistent; in the cited version, "abs" is at one point in bold and another in italics. These details vary between reprintings of the report.

84. Backus et al., "Revised Report," 359.

85. Backus et al., "Revised Report," 351. The report quotes the original German; translation quoted from Wittgenstein, *Tractatus Logico-Philosophicus*, 55.

86. Backus, "The Syntax and Semantics of the Proposed International Algebraic Language of the Zurich ACM-GAMM Conference." See Priestley, 216.

87. Backus et al., "Revised Report," 353.

88. Backus et al., "Revised Report," 353.

89. These are not considered identifiers but rather "letter strings" that one may choose to include in the parameter list of a procedure so as to clarify the purpose of each parameter. A similar practice persists in the later programming languages Smalltalk, Objective-C, and Swift.

90. ALGOL was repeatedly compared to Esperanto in the 1960s, with various implications: to celebrate its potential to create international unity, to lament its fragmentation into multiple dialects, to mock its unrealistic ambitions. See Mounier-Kuhn, "Algol in France." For a critical perspective on the use of English in programming languages, see Golumbia, *The Cultural Logic of Computation*, 119–25.

91. Backus et al., "Revised Report," 351.

92. The international variants are discussed in Van Wijngaarden et al., "Revised Report on the Algorithmic Language Algol 68," 9. On ALGOL in Russia, see Terekhov, "ALGOL 68 and Its Impact on the USSR and Russian Programming." On the design of ALGOL 68, see Lindsey, "A History of ALGOL 68."

93. "Miscellaneous Programs," n.p.

94. Iverson, "Notation as a Tool of Thought," 445. See Iverson, *A Programming Language*.

95. Landin, "The Mechanical Evaluation of Expressions," 315. The LISP example is from the first version released in 1960; later versions eliminated the commas and sometimes used other names in place of SUM.

96. Backus et al., "Revised Report," 349.

97. Forsythe quoted in Knuth, "George Forsythe and the Development of Computer Science," 722.

98. Forsythe quoted in Knuth, 722. On the early history of academic computer science, see Misa, *Communities of Computing*.

99. Wellmon, *Organizing Enlightenment*, 249–52. Wellmon's specific example is nineteenth-century philology, as described by August Böckh.

100. Hartmanis and Stearns, "On the Computational Complexity of Algorithms."

101. Rather than giving a single statement of this thesis, Kleene gave separate definitions for "Church's thesis" and "Turing's thesis" and showed that they are equivalent (Kleene, *Introduction to Metamathematics*, 317–32, 376–81). See Soare, "Why Turing's Thesis Is Not a Thesis;" Olszewski, Wolenski, and Janusz, *Church's Thesis after 70 Years*.

102. Steele and Sussman, "Lambda: The Ultimate Imperative," 1. Steele later published a sequel to this paper titled "Lambda: The Ultimate Declarative."

103. Steele and Sussman, 1.

104. See Raymond, "Holy Wars."

105. Sammet, "The Use of English as a Programming Language," 228.

106. Sammet, 229.

107. McCarthy, "Towards a Mathematical Science of Computation," 36. Turing had proposed something similar in a 1949 lecture, although this work does not appear to have been widely known until later; see Morris and Jones, "An Early Program Proof by Alan Turing."

108. Dijkstra's preference for "computing science" over "computer science" reflected his emphasis on mathematics over machinery; see Frana and Misa, "An Interview with Edsger W. Dijkstra," 47.

109. Dijkstra, "On the Cruelty of Really Teaching Computing Science," 0. Note that the manuscript has page numbers starting with zero.

110. Dijkstra, 1.

111. Dijkstra, 1.

112. Dijkstra, 4.

113. Dijkstra, 4.

114. Dijkstra, 26.

115. Dijkstra, 16.

116. Dijkstra, 17.

117. Dijkstra, 13.

118. For a critical view of the idea of user friendliness, see Lori Emerson, *Reading Writing Interfaces*.

119. Knuth, *Literate Programming*.

120. Payette, 69.

121. Payette, 70.

122. American National Standards Institute, *Ada Programming Language*, sec. 1.3.

123. Ichbiah et al., *Rationale for the Design of the Ada Programming Language*, 5.

124. On the thinking behind CLU, see Liskov and Zilles, "Programming with Abstract Data Types."

125. Lovelace, *Ada, the Enchantress of Numbers*, 93.

126. Lovelace, 425.

127. Hoare, "1980 ACM Turing Award Lecture," 82.

128. Hoare, 82.

129. Dijkstra, "Go To Statement Considered Harmful."

130. Hoare, "1980 ACM Turing Award Lecture," 76.

131. Hudak et al., "Report on the Programming Language Haskell."

132. See Sørensen and Urzyczyn, *Lectures on the Curry-Howard Isomorphism*.

133. On the decline of the program-proving idea, see Hoare, "How Did Software Get So Reliable Without Proof?"

134. Abelson, Sussman, and Sussman, *Structure and Interpretation of Computer Programs*, xvii.

135. Abelson, Sussman, and Sussman, 1.

136. Iverson, for instance, states that "explicit procedures for calculating the exact or approximate values of various functions" are called "algorithms or *programs*" (Iverson, *A Programming Language*, vii).

137. Abelson, Sussman, and Sussman, xiii.

138. See Dourish, "Algorithms and Their Others."

139. A. Forsythe et al., *Computer Science*. The word *pseudocode* initially referred to a machine code–like notation that did not correspond directly to the operations of a real machine, but it later came to refer to notations that had more resemblance to programming languages (see Priestley, 188–93). Knuth does not use pseudocode in the latter sense but rather presents algorithms in the assembly language of a fictional computer.

140. Aho, Hopcroft, and Ullman, *Data Structures and Algorithms*, 254.

141. In Knuth's definition, *finiteness* requires that the procedure eventually come to an end, whereas *effectiveness* requires that the individual operations be performable in a finite amount of time. He discusses operations on real numbers as an example of noneffective operations; see Knuth, *Art of Computer Programming*, 1:6.

142. Aho, Hopcroft, and Ullman, 2.

143. I tried implementing this procedure several ways in Python, and it indeed concluded that the two numbers are equal. But this does not prove Stifel's equality; the program is only checking that two approximations are equal, not that the true values of the roots are equal. Things go differently if one uses the same method to test another, more elementary result: $\sqrt{2} \times \sqrt{3} = \sqrt{6}$. At least on my system, the Python program `from math import *; sqrt(2) * sqrt(3) == sqrt(6)` produces `False`, even though the equation holds in the real number field. This is because the method of approximation introduces a slight error.

144. One could also formalize algorithms in terms of Church's λ-calculus, in which it is possible to represent integers using the successor function ($S(x) = x + 1$) rather than digits; thus 2 is represented as $S(1)$, 3 as $S(S(1))$, and so forth. However, similar reasoning would apply. See Church, "Unsolvable Problem," 350. It should also be pointed

out that ALGOL had a data type called **real**, which was supposed to represent real numbers. ALGOL programs could only do so in a mathematically ideal sense, since real numbers cannot be represented exactly by computers. In later programming languages such as C, the data type was renamed *float*, indicating the floating-point representation used to approximate real numbers.

145. According to Dijkstra, this difference was a factor in programming's low status in the 1950s: "In the mathematical culture of those days you had to deal with infinity to make your topic scientifically respectable" (quoted in Frana and Misa, "Interview," 42).

146. Knuth, *Art of Computer Programming*, 1:5.

147. The idea of *online algorithm*, for instance, loosens the requirement that the input be specified at the beginning, albeit only in certain constrained ways; online algorithms receive the input piece by piece over the course of the process. For a broader alternative to the problem-solving model, see Forrest et al., "Computation in the Wild."

148. Aho, Hopcroft, and Ullman, 1.

149. Naur and Randell, *Software Engineering*, 30–31.

150. Marcovitz and Schweppe, 1.

151. Taina Bucher, for instance, argues that each sorting algorithm has its own "organizational logic": just as one can organize a list of books in a range of ways, such as "by the author's surname, by genre, or even by the color of the book jacket," algorithms such as "selection sort, merge sort, or quicksort" provide different ways of sorting a list (*If . . . Then*, 23). This comparison would seem to suggest that each algorithm orders things in its own distinctive way, which (save for the unpredictable ways sorting algorithms handle ties) is not the case. On account of the distinction between "what an algorithm does" and "how it does it," Benjamin M. Schmidt argues that humanists need not concern themselves with the technical details of algorithms but rather should focus on "understanding the transformations that algorithms attempt to bring about" ("Do Digital Humanists Need to Understand Algorithms?" 546–47). It should be noted that even if two algorithms produce the same output, they could be viewed as different in other professionally salient ways, ranging from their efficiency to their comprehensibility to other programmers; on this point, see Marino, *Critical Code Studies*, 5–8.

152. If there are ties, then the behavior of sorting algorithms does vary. So-called stable sorting algorithms will always preserve the order that the tied items had in the input. Unstable algorithms can reorder them in various ways that do not follow any straightforward pattern. Generally it is possible to avoid these issues by choosing secondary sorting criteria to fall back on in the case of ties.

153. One such caveat is that there are multiple ways of specifying the order that enable the use of different types of algorithm. The most common sorting algorithms use a pairwise function that determines where two items should be relative to each other (in effect determining whether $a < b$ or $a \geq b$, with the requirement that the transitive property holds). This is how alphabetical order is usually formalized. Other algorithms are possible if the desired position of each item is already known. This would be the case, for instance, if one is sorting a list of runners by their place in a race, with no ties. Since it is known that the runner in nth place should appear in exactly the nth position in the output list, it is possible to employ a more efficient algorithm than the standard ones. Things also become more complex if the order is intransitive, as in the circular ranking used in the rock-paper-scissors game. Such orderings would cause standard sorting algorithms to go haywire.

154. See Frana and Misa, 42.

155. Dijkstra, "A Note on Two Problems in Connexion with Graphs," 270.

156. Hart, Nilsson, and Raphael, "A Formal Basis for the Heuristic Determination of Minimum Cost Paths."

157. Johnson, "New Jersey Town to Close Streets to Non-Residents During Rush Hour."

158. My use of the word *paradigm* enmeshes me in two distinct discourses, and so it is worth some explanation. The first is Thomas Kuhn's use of the word to refer to the epistemological practices of a science in a particular context. This reference is intentional, although I do not mean to take on the full theoretical freight of Kuhn's argument; I do not, in particular, maintain that machine learning is entirely incommensurable with classical algorithms. The second is the use of the word *paradigm* within computer science, referring to styles of coding. The shift I am describing could be understood as analogous to the transition from structured to object-oriented programming, although I would argue that its implications run deeper. See Kuhn, *The Structure of Scientific Revolutions*.

159. IBM, "5 in 5 Video." It should be noted that IBM has altered its position on AI ethics since this video was published; following the protests regarding the police killing of George Floyd, it has withdrawn from the facial recognition market. See Krishna, "IBM CEO's Letter to Congress on Racial Justice Reform."

160. On this point, see Ed Finn, *What Algorithms Want*.

161. Sack, 145.

162. PageRank is sometimes described as an unsupervised machine learning algorithm, but I would consider it a classical algorithm because (unlike in, say, unsupervised clustering tasks) the desired output is fully specified by an equation.

163. Page et al., "The PageRank Citation Ranking"; Brin and Page, "The Anatomy of a Large-Scale Hypertextual Web Search Engine." For a contextual account of PageRank's development, see Rieder, *Engines of Order*, 265–304.

164. See M. Jones, "Querying the Archive."

165. Bavelas, "Communication Patterns in Task-Oriented Groups"; Freeman, "A Set of Measures of Centrality Based on Betweenness."

166. L. Katz, "A New Status Index Derived from Sociometric Analysis"; Bonacich, "Factoring and Weighting Approaches to Status Scores and Clique Identification"; Kleinberg, "Authoritative Sources in a Hyperlinked Environment."

167. Page et al., 1.

168. Page et al., 1.

169. Daston and Galison, *Objectivity*, 115–90.

170. Page et al., 3.

171. Page et al., 3.

172. Page et al., 4.

173. Page et al., 4.

174. Page et al., 5.

175. Altman and Tennenholtz, "Ranking Systems."

176. Bánky, Iván, and Grolmusz, "Equal Opportunity for Low-Degree Network Nodes."

177. A classic, albeit now dated, article on the idea of *relevance* is Van Couvering, "Is Relevance Relevant?"

178. One of the algorithms Knuth designed is the LR(k) parser generator, which he described in a 1965 article; this method can mechanically translate a grammar in

Backus–Naur form into an algorithm for parsing the language, albeit subject to some constraints. See Knuth, "On the Translation of Languages from Left to Right." A classic textbook on this topic is Aho, Sethi, and Ullman, *Compilers*, also known as the "dragon book."

CODA

1. Priestley argues that the practice of data abstraction led to "The Unification of Data and Algorithms" during the 1970s (*A Science of Operations*, 277–96). However, he is referring more to data structures than to the data themselves, so this argument does not conflict with mine.

2. The account given here primarily focuses on the approach known as *deep learning*. For overviews, see Goodfellow, Bengio, and Courville, *Deep Learning*; Skansi, *Guide to Deep Learning Basics*.

3. The idea of a neural network is often traced to a 1943 article by Warren S. McCulloch and Walter Pitts, titled "A Logical Calculus of the Ideas Immanent in Neural Activity." Their theory, however, is based on propositional logic, which gives the model an "all-or-none" character that conflicts with how modern ML works (118). The ANN as we know it depends on the backpropagation algorithm, which provides a means of adjusting network parameters to improve predictions; see Rumelhart, Hinton, and Williams, "Learning Representations by Back-Propagating Errors."

4. See Minsky and Papert, *Perceptrons*.

5. One exception is the Rectified Linear Unit (ReLU) function, commonly used in neural networks; it contains one undifferentiable point, whose derivative is simply chosen arbitrarily. While differentiability does not fit with Alan Turing's model of computation, Alex Graves, Greg Wayne, and Ivo Danihelka have proposed a "Neural Turing Machine," which aims to reconcile the two by making Turing Machines differentiable ("Neural Turing Machines"; see also Kaiser and Sutskever, "Neural GPUs Learn Algorithms"). It should also be noted that, while modern Boolean algebra is not differentiable, George Boole's own algebra system arguably was; he did not treat logical statements as resolvable to one or zero but rather manipulated them as symbolic equations that behaved analogously to linear differential equations. This connection to calculus is totally lost in the version of Boolean logic used in programming languages.

6. The disconnect between real numbers and their binary representations leads to well-known practical difficulties that are dealt with in the subfield of numerical computing. On such issues in particular relation to deep learning, see Goodfellow Bengio, and Courville, 78–95.

7. Generalization is usually defined in relation to a test data set that is specifically prepared for the purpose of evaluation. This test data set serves as an imperfect stand-in for the range of data that may be encountered in production, about which it is usually hard to reason with certainty.

8. Goodfellow, Bengio, and Courville, 109.

9. Breiman, "Statistical Modeling," 199. For a contextual account of Breiman's work, see M. Jones, "How We Became Instrumentalists (Again)."

10. The idea of a "black box" has become a figure for a broader form of social organization; see Pasquale, *The Black Box Society*.

11. Breiman, "Statistical Modeling," 199.

12. Rieder, *Engines of Order*, 13.

13. Breiman, "Statistical Modeling," 205.

14. Breiman, "Random Forests."

15. Breiman, "Statistical Modeling," 205.

16. Breiman, 218, 219.

17. Breiman, 230.

18. For a critique of this approach, see Raji et al., "AI and the Everything in the Whole Wide World Benchmark."

19. Goodfellow, Bengio, and Courville, 97.

20. Goodfellow, Bengio, and Courville, 151.

21. Goodfellow, Bengio, and Courville, 105.

22. For one attempt at an alternative definition, see Amoore, *Cloud Ethics*.

23. Goodfellow, Bengio, and Courville, 208.

24. In a 2015 piece in the *Atlantic*, Ian Bogost argues that the cultural fixation on algorithms has taken on a quasi-religious character as people have allowed computers "to replace gods in their minds"; the repeated mantra that we live in an age of algorithms, he argues, contributes to an irrational belief in their power ("The Cathedral of Computation," n.p.). What it means to study algorithms from a social perspective has become a matter of controversy. Paul Dourish has warned of the importance of being clear about what algorithms are and are not ("Algorithms and Their Others"). Nick Seaver, on the other hand, has argued that social scientists should avoid defining the term *algorithm* and instead study how "algorithmic systems" function in practical contexts ("Algorithms as Culture"). Seaver is right to note that an exclusive focus on technicalities can distract from social issues, but his approach bears the danger of reinforcing the tendency Bogost describes by turning the word *algorithm* into a floating signifier. I would argue, in particular, that an uncritical acceptance of the claim that algorithms are "able to learn" can serve the interests of those who profit from ML technology by assigning agency to an ambiguously defined entity.

25. Vaswani et al., "Attention Is All You Need."

26. Technically, it is token by token, not word by word; the GPT models break some words down into parts. On BERT, see Devlin et al., "BERT."

27. Raffel et al., "Exploring the Limits of Transfer Learning with a Unified Text-to-Text Transformer."

28. Shannon, "A Mathematical Theory of Communication," 387.

29. Shannon, 388.

30. Shannon, 379.

31. Shannon and Weaver, *The Mathematical Theory of Communication*, 24.

32. Shannon and Weaver, 27.

33. Weaver, *Translation*, 11.

34. Weaver, 10.

35. Radford et al., "Better Language Models and Their Implications."

36. Radford et al., "Language Models Are Unsupervised Multitask Learners," sixth unnumbered page.

37. Rajpurkar et al., "SQuAD: 100,000+ Questions for Machine Comprehension of Text"; Rajpurkar, Jia, and Liang, "Know What You Don't Know: Unanswerable Questions for SQuAD."

38. Stanford NLP Group, "Economic_inequality."

39. There are, to be sure, language models trained on historical English. The model

MacBERTh, released in December 2021, was trained on English texts from the years 1450–1950. The training data for this model are approximately a hundredth of the size of GPT-3's training data. See Manjavacas.

40. Bender et al., "On the Dangers of Stochastic Parrots," 614.

41. On the politics of labeling in image recognition training data, see Crawford and Paglen, "Excavating AI."

42. I am not the first to draw this comparison; Ted Underwood made a similar point while responding to the question of how one could explain machine learning to an eighteenth century person: Ted Underwood (@Ted_Underwood), "I'd say, turns out Locke (1689) was right that we form general Ideas by abstracting detail from particular sensations, and Hartley (1749) was right that this happens through the mathematical concord of electrical vibrations in nerves. We can do the same thing in tiny wires," Twitter, May 15, 2021, 6:11 a.m., https://twitter.com/Ted_Underwood/status/1393524414646521861.

43. Mentor, "The Conscience of a Hacker," n.p. For a critique of this form of utopianism, see Chun, Discriminating Data.

44. Mentor, n.p.

45. Mentor, n.p.

46. O'Neil, Weapons of Math Destruction; Noble, Algorithms of Oppression.

47. T. Brown et al., "Language Models Are Few-Shot Learners," 35–38.

48. See Barratt, "InterpNET"; Došilović, Brčić, and Hlupić, "Explainable Artificial Intelligence: A Survey."

49. Bender et al., "On the Dangers of Stochastic Parrots."

50. For an examples in the popular press, see Vincent, "OpenAI's Latest Breakthrough Is Astonishingly Powerful, but Still Fighting Its Flaws."

51. I generated these predictions myself using the HuggingFace transformer library.

52. In the mid-twentieth century, a common position among literary critics was that meaning is created by the reader; this was taken furthest in the school called reader-response criticism. A famous critique of this position came from Steven Knapp and Walter Benn Michaels, who ask what one would think if waves spontaneously carved some lines from Wordsworth into the sand: one would either suppose some supernatural power was at work or else view the shapes as mere meaningless accidents ("Against Theory" 727–28).

53. Collins and Skover, Robotica, 42. For a critique of this argument, see Siraganian, "Against Theory, Now with Bots!"

54. Bender et al., 611.

55. Bender et al., 616.

56. Metzler et al., "Rethinking Search."

57. Radford et al., "Language Models Are Unsupervised Multitask Learners," seventh unnumbered page.

58. I did this by entering the text "Is liberty, equality, or fraternity the most important? The most important one is" and generating a completion. To be sure, I was not using the system in quite the right way—it should be fine-tuned or in some other way primed for this particular type of task. But the problem goes beyond the issues of where the information is coming from and how the model is set up.

59. Andersen, A Theory of Computer Semiotics, 10.

60. N. Katherine Hayles adopted the term skeuomorph, which originated in archae-

ology, to refer to an element of an older technology that persists in a newer one, such as the use of an image of a floppy disk to indicate *save*; see Hayles, *How We Became Posthuman*, 17.

61. Branwen, "GPT-3 Creative Fiction."

62. Brown et al., "Language Models Are Few-Shot Learners," 56.

63. Esser, Rombach, and Ommer, "Taming Transformers for High-Resolution Image Synthesis." This program can be run online at https://colab.research.google.com/github/eps696/aphantasia/blob/master/CLIP_VQGAN.ipynb

64. On the potential spuriousness of some of these differences, see Veitch et al., "Counterfactual Invariance to Spurious Correlations."

65. Schick and Schütze, "It's Not Just Size That Matters," fourth unnumbered page.

66. See Cappelen, *Fixing Language*; Burgess, Cappelen, and Plunkett, *Conceptual Engineering and Conceptual Ethics*. Carnap's work, to be clear, is only one of a number of reference points for this intellectual program, and not all advocates are strongly influenced by his approach.

Bibliography

Aarsleff, Hans. "The Eighteenth Century, Including Leibniz." *Current Trends in Linguistics* 13 (1975): 383–479.

———. *From Locke to Saussure: Essays on the Study of Language and Intellectual History.* Minneapolis: University of Minnesota Press, 1982.

Abelson, Harold, Gerald Jay Sussman, and Julie Sussman. *Structure and Interpretation of Computer Programs.* 2nd ed. Cambridge, MA: MIT Press, 1996.

Aho, Alfred V., John E. Hopcroft, and Jeffrey D. Ullman. *Data Structures and Algorithms.* 1983. Reprinted with corrections, Reading, MA: Addison-Wesley, 1987.

Aho, Alfred V., Ravi Sethi, and Jeffrey D. Ullman. *Compilers: Principles, Techniques, and Tools.* Reading, MA: Addison-Wesley, 1987.

Aksoy, Asuman Güven. "Al-Khwārizmī and the Hermeneutic Circle: Reflections on a Trip to Samarkand." *Journal of Humanistic Mathematics* 6, no. 2 (July 2016): 114–27. https://doi.org/10.5642/jhummath.201602.09.

Albury, W. R. Introduction to *La Logique*, by Étienne Bonnot de Condillac, 7–34. New York: Abaris Books, 1979.

Alexander, Amir. *Duel at Dawn: Heroes, Martyrs, and the Rise of Modern Mathematics.* Cambridge, MA: Harvard University Press, 2010.

———. *Infinitesimal: How a Dangerous Mathematical Theory Shaped the Modern World.* New York: Scientific American / Farrar, Straus, and Giroux, 2014.

al-Khwārizmī, Muḥammad Ibn Mūsā. *The Algebra of Mohammed Ben Musa.* Translated by Friedrich Rosen. London: Oriental Translation Fund, 1831.

———. *Algoritmi de numero indorum.* Edited by Baldassarre Boncompagni. Trattati d'arithmetica 1. Rome, 1857.

———. *Mohammed Ibn Musa Alchwarizmi's Algorismus: Das Früheste Lehrbuch Zum Rechnen Mit Indischen Ziffern.* Edited by Kurt Vogel. Milliaria 3. Aalen, Ger.: O. Zeller, 1963.

———. *Robert of Chester's Latin Translation of Al-Khwārizmī's Al-Jabr: A New Critical Edition.* Edited by Barnabas Bernard Hughes. Translated by Robert of Chester. Stuttgart: F. Steiner Verlag Wiesbaden, 1989.

Alsted, Johann Heinrich. *Encyclopaedia septem tomis distincta.* Herborn, Ger., 1630.

Altman, Alon, and Moshe Tennenholtz. "Ranking Systems: The PageRank Axioms." In *EC '05: Proceedings of the 6th ACM Conference on Electronic Commerce*, 1–8. Vancouver, BC: Association for Computing Machinery, 2005. https://doi.org/10.1145/1064009.1064010.

American National Standards Institute. *Ada Programming Language*. ANSI/MIL STD 1815A-1983. Approved February 17, 1983; canceled June 7, 1997. Quick Search Assist. Data updated November 30, 2021. https://quicksearch.dla.mil/qsDocDetails.aspx?ident_number=37152.

Amoore, Louise. *Cloud Ethics: Algorithms and the Attributes of Ourselves and Others*. Durham, NC: Duke University Press, 2020.

Andersen, P. B. *A Theory of Computer Semiotics: Semiotic Approaches to Construction and Assessment of Computer Systems*. Cambridge: Cambridge University Press, 1990.

"Anecdotes of Great Memory, and Astonishing Powers of Calculation." *Universal Magazine* 98 (April 1796): 239–40.

An Introduction for to Lerne to Recken with the Pen or with the Counters Accordynge to the Trewe Cast of Algorysme [. . .]. London, 1539. Early English Books, 1475–1640, microfilm reel position 68:07.

"Announcement." *Communications of the ACM* 3, no. 2 (February 1960): 73.

Antognazza, Maria Rosa. *Leibniz: An Intellectual Biography*. Cambridge: Cambridge University Press, 2009.

Aristotle and St. Thomas Aquinas. *On Interpretation / Commentary by St. Thomas and Cajetan (Peri Hermeneias)*. Translated by Jean T. Oesterle. Milwaukee: Marquette University Press, 1962.

Arthur, Richard T. W. "Leibniz's Syncategorematic Infinitesimals." *Archive for History of Exact Sciences* 67, no. 5 (2013): 553–93.

Ashworth, E. J. "'Do Words Signify Ideas or Things?' The Scholastic Sources of Locke's Theory of Language." *Journal of the History of Philosophy* 19, no. 3 (July 1981): 299–326. https://doi.org/10.1353/hph.2008.0250.

Ashworth, William J. "Memory, Efficiency, and Symbolic Analysis: Charles Babbage, John Herschel, and the Industrial Mind." *Isis* 87, no. 4 (1996): 629–53.

Aspray, William. *Computing before Computers*. Ames: Iowa State University Press, 1990.

Augustine. *Confessions*. Translated by R. S. Pine-Coffin. London: Penguin, 1961.

Aydin, Nuh, and Lakhdar Hammoudi. "Root Extraction by Al-Kashi and Stevin." *Archive for History of Exact Sciences* 69, no. 3 (May 2015): 291–310. https://doi.org/10.1007/s00407-015-0150-3.

Babbage, Charles. *Babbage's Calculating Engines: Being a Collection of Papers Relating to Them; Their History, and Construction*. Edited by Henry P. Babbage. London: E. and F. N. Spon, 1889.

———. *On the Economy of Machinery and Manufactures*. London, 1832.

———. *On the Influence of Signs in Mathematical Reasoning*. Cambridge: Printed by J. Smith, 1826.

———. *Passages from the Life of a Philosopher*. London: Longman, Green, Longman, Roberts, and Green, 1864.

Backus, John W., F. L. Bauer, J. Green, C. Katz, J. McCarthy, P. Naur, A. J. Perlis, et al. "Revised Report on the Algorithmic Language ALGOL 60." *Computer Journal* 5, no. 4 (January 1, 1963): 349–67. https://doi.org/10.1093/comjnl/5.4.349.

———. "The History of FORTRAN I, II, and III." In Wexelblat, *History of Programming Languages*, 25–45.

———. "The Syntax and Semantics of the Proposed International Algebraic Language of the Zurich ACM-GAMM Conference." In *Proceedings of the International Conference on Information Processing, UNESCO, Paris, 1959,* 125–32, 1960.

Bacon, Francis. *The Advancement of Learning.* Edited by Joseph Devey. New York: P. F. Collier, 1901.

———. *The New Organon: And Related Writings.* Translated by F. H. Anderson. Indianapolis: Bobbs-Merrill, 1960.

Baily, Francis. "Some Particulars Respecting the Arithmetical Powers of Zerah Colburn, a Child under Eight Years of Age." *Philosophical Magazine and Journal* 40 (July–December 1812): 119–25.

Bair, Jacques, Piotr Błaszczyk, Peter Heinig, Mikhail G. Katz, Jan Peter Schäfermeyer, and David Sherry. "Klein vs Mehrtens: Restoring the Reputation of a Great Modern." *Matematychni Studii* 48, no. 2 (December 19, 2016): 189–219. https://doi.org/10.15330/ms.48.2.189-219.

Baker, Keith Michael. *Condorcet: From Natural Philosophy to Social Mathematics.* Chicago: University of Chicago Press, 1975.

Ball, W. W. Rouse. *A History of the Study of Mathematics at Cambridge.* Cambridge: Cambridge University Press, 1889.

———. *Mathematical Recreations and Essays.* 10th ed. London: Macmillan, 1937.

Bánky, Dániel, Gábor Iván, and Vince Grolmusz. "Equal Opportunity for Low-Degree Network Nodes: A PageRank-Based Method for Protein Target Identification in Metabolic Graphs." *PLOS One* 8, no. 1 (2013): e54204. https://doi.org/10.1371/journal.pone.0054204.

Barbeau, E. J., and P. J. Leah. "Euler's 1760 Paper on Divergent Series." *Historia Mathematica* 3, no. 2 (May 1, 1976): 141–60. https://doi.org/10.1016/0315-0860(76)90030-6.

Barbin, E., J. Borowczyk, J.-L. Chabert, M. Guillemot, A. Michel-Pajus, A. Djebbar, and J.-C. Martzloff. *A History of Algorithms: From the Pebble to the Microchip.* Edited by Jean-Luc Chabert. Translated by C. Weeks. Berlin: Springer-Verlag, 1999.

Barratt, Shane. "InterpNET: Neural Introspection for Interpretable Deep Learning." Revised November 16, 2017. Cornell University. http://arxiv.org/abs/1710.09511.

Barrett, Lindsay, and Matthew Connell. "Jevons and the Logic 'Piano.'" *Rutherford Journal* 1 (2005–2006). http://www.rutherfordjournal.org/article010103.html.

Bashmakova, I. G., G. S. Smirnova, and Abe Shenitzer. "The Birth of Literal Algebra." *American Mathematical Monthly* 106, no. 1 (1999): 57–66. https://doi.org/10.2307/2589589.

Bauer, F. L., H. Bottenbruch, H. Rutishauser, and K. Samelson. "Proposal for a Universal Language for the Description of Computing Processes." In *Computer Programming and Artificial Intelligence,* edited by John Weber Carr, 355–73. [Ann Arbor, MI]: University of Michigan College of Engineering, 1958. http://www.softwarepreservation.org/projects/ALGOL/report/BauerBRS-Proposal_for_a_Universal_Language-1958.pdf.

Bavelas, Alex. "Communication Patterns in Task-Oriented Groups." *Journal of the Acoustical Society of America* 22, no. 6 (November 1, 1950): 725–30. https://doi.org/10.1121/1.1906679.

Becher, Harvey W. "Radicals, Whigs and Conservatives: The Middle and Lower Classes in the Analytical Revolution at Cambridge in the Age of Aristocracy." *British Journal for the History of Science* 28, no. 4 (1995): 405–26.

Beeley, Philip. "'Un de Mes Amis': On Leibniz's Relation to the English Mathematician and Theologian John Wallis." In Phemister and Brown, *Leibniz and the English-Speaking World*, 63–81.

Bellhouse, D. R. *Leases for Lives: Life Contingent Contracts and the Emergence of Actuarial Science in Eighteenth-Century England.* Cambridge: Cambridge University Press, 2017.

Bender, Emily M., Timnit Gebru, Angelina McMillan-Major, and Shmargaret Shmitchell. "On the Dangers of Stochastic Parrots: Can Language Models Be Too Big? 🦜 " In *Proceedings of the 2021 ACM Conference on Fairness, Accountability, and Transparency*, 610–23. New York: Association for Computing Machinery, 2021. https://doi.org/10.1145/3442188.3445922.

Berkeley, George. *The Analyst: Or, a Discourse Addressed to an Infidel Mathematician.* London, 1734.

———. *Philosophical Commentaries.* Edited by George H. Thomas. Routledge Library Editions: 18th Century Philosophy 12. London: Routledge, 1989.

———. *A Treatise Concerning the Principles of Human Knowlege.* Dublin, 1710. Eighteenth Century Collections Online. https://link.gale.com/apps/doc/CW0118263402/ECCO?u=psucic&sid=bookmark-ECCO&xid=90512785&pg=1.

Berlin, Isaiah. *Four Essays on Liberty.* Oxford: Oxford University Press, 1979.

———. *Three Critics of the Enlightenment: Vico, Hamann, Herder.* 2nd ed. Princeton, NJ: Princeton University Press, 2013.

Berlinski, David. *The Advent of the Algorithm: The 300-Year Journey from an Idea to the Computer.* San Diego, New York, and London: Harcourt, 2001.

Besteman, Catherine, and Hugh Gusterson. *Life by Algorithms: How Roboprocesses Are Remaking Our World.* Chicago: University of Chicago Press, 2019.

Binder, Jeffrey M. "Romantic Disciplinarity and the Rise of the Algorithm," *Critical Inquiry* 46, no. 4 (Summer 2020): 813–34.

"Biographical Account of Lord Stanhope." *Annals of Philosophy, Or, Magazine of Chemistry, Mineralogy, Mechanics, Natural History, Agriculture, and the Arts* 11, no. 62 (February 1818): 81–86.

Bishop, Coleman E. "The Mathematical Failure." *Chautauquan* 4, no. 5 (February 1884): 275–78.

Bisk, Yonatan, Ari Holtzman, Jesse Thomason, Jacob Andreas, Yoshua Bengio, Joyce Chai, Mirella Lapata, et al. "Experience Grounds Language." Cornell University. Revised November 2, 2020. http://arxiv.org/abs/2004.10151.

Blair, Ann M. "Revisiting Renaissance Encyclopaedism." In *Encyclopaedism from Antiquity to the Renaissance*, edited by Jason König and Greg Woolf, 377–97. Cambridge: Cambridge University Press, 2013. https://doi.org/10.1017/CBO9781139814683.

Blanco, Mónica. "Hermeneutics of Differential Calculus in Eighteenth-Century Northern Germany." *Sudhoffs Archiv* 92, no. 2 (2008): 133–64.

Blank, Andreas. "Dalgarno, Wilkins, Leibniz, and the Descriptive Nature of Metaphysical Concepts." In Phemister and Brown, *Leibniz and the English-Speaking World*, 51–61.

Bliss, Charles Kasiel. *Semantography (Blissymbolics): A Simple System of 100 Logical Pictorial Symbols, Which Can Be Operated and Read like 1+2=3 in All Languages.* 2nd ed. Sydney: Semantography (Blissymbolics) Publications, 1965.

Boethius. *Boethian Number Theory: A Translation of the De Institutione Arithmetica (With Introduction and Notes).* Translated by Michael Masi. Amsterdam: Rodopi, 1983.

Bogost, Ian. "The Cathedral of Computation." *Atlantic*, January 15, 2015. https://www
.theatlantic.com/technology/archive/2015/01/the-cathedral-of-computation/
384300/.

Bolter, J. David. *Turing's Man: Western Culture in the Computer Age*. Chapel Hill: University of North Carolina Press, 1984.

Bonacich, Phillip. "Factoring and Weighting Approaches to Status Scores and Clique Identification." *Journal of Mathematical Sociology* 2, no. 1 (January 1, 1972): 113–20. https://doi.org/10.1080/0022250X.1972.9989806.

Bonner, Anthony. *The Art and Logic of Ramon Llull: A User's Guide*. Leiden, Neth.: Brill, 2007.

Boole, George. "The Calculus of Logic." *Cambridge and Dublin Mathematical Journal* 3 (1848): 183–98.

———. *An Investigation of the Laws of Thought: On Which Are Founded the Mathematical Theories of Logic and Probabilities*. London: Macmillan, 1854.

———. *The Mathematical Analysis of Logic, Being an Essay Towards a Calculus of Deductive Reasoning*. Cambridge: Macmillan, Barclay, & Macmillan, 1847.

———. *The Right Use of Leisure: An Address, Delivered Before the Members of the Lincoln Early Closing Association, February 9, 1847*. London: J. Nisbet, 1847.

———. *Selected Manuscripts on Logic and Its Philosophy*. Edited by Ivor Grattan-Guinness and Gérard Bornet. Boston: Birkhäuser, 1997.

———. *The Social Aspect of Intellectual Culture: An Address Delivered in the Cork Athenæum, May 29th, 1855 : At the Soirée of the Cuvierian Society*. Cork, Ire.: George Purcell, 1855.

———. *Studies in Logic and Probability*. Mineola, NY: Dover, 2012.

———. *A Treatise on the Calculus of Finite Differences*. Cambridge: Macmillan, 1860.

Boole, Mary Everest. "Home-Side of a Scientific Mind." *Dublin University Magazine: A Literary and Political Journal* 91, no. 1 (January 1878): 105–14.

———. *Logic Taught by Love*. Boston: Alfred Mudge & Son, 1890.

———. *Symbolical Methods of Study*. London: Kegan, Paul, Trench, 1884.

Booth, James. "On Tangential Coordinates." *Proceedings of the Royal Society of London* 9 (January 1, 1859): 175–88. https://doi.org/10.1098/rspl.1857.0043.

Bopp, Franz. *Analytical Comparison of the Sanskrit, Greek, Latin and Teutonic Languages, Shewing the Original Identity of Their Grammatical Structure*. Edited by Joseph-Daniel Guigniaut. Amsterdam: Benjamins, 1989.

Bornet, Gérard. "Boole's Psychologism as a Reception Problem." In G. Boole, *Selected Manuscripts on Logic and Its Philosophy*, xlvii–lviii.

Bos, Henk J. M. "Differentials, Higher-Order Differentials and the Derivative in the Leibnizian Calculus." *Archive for History of Exact Sciences* 14, no. 1 (1974): 1–90.

———. *Redefining Geometrical Exactness: Descartes' Transformation of the Early Modern Concept of Construction*. New York: Springer-Verlag, 2001.

Branwen, Gwern. "GPT-3 Creative Fiction." Modified July 1, 2021. https://www.gwern
.net/GPT-3.

Breiman, Leo. "Random Forests." *Machine Learning* 45, no. 1 (October 2001): 5–32.

———. "Statistical Modeling: The Two Cultures (With Comments and a Rejoinder by the Author)." *Statistical Science* 16, no. 3 (August 2001): 199–231. https://doi.org/
10.1214/ss/1009213726.

Bright, Timothie. *Charac[terie.] An Ar[te] of Shorte, Swift[e], and Secrete Writing by*

Character [. . .]. London, 1588. Early English Books, 1475–1640, microfilm reel position 198:06.

Brin, Sergey, and Lawrence Page. "The Anatomy of a Large-Scale Hypertextual Web Search Engine." *Computer Networks and ISDN Systems* 30, no. 1–7 (April 1998): 107–17. https://doi.org/10.1016/S0169-7552(98)00110-X.

Brissot de Warville, Jacques-Pierre. *New Travels in the United States of America [. . .]*. Translated by Joel Barlow. Dublin, 1792.

Brown, Daniel. *The Poetry of Victorian Scientists*. Cambridge: Cambridge University Press, 2013.

———. "William Rowan Hamilton and William Wordsworth: The Poetry of Science." *Studies in Romanticism* 51, no. 4 (2012): 475–501.

Brown, Frank Markham. "George Boole's Deductive System." *Notre Dame Journal of Formal Logic* 50, no. 3 (2009): 303–30. https://doi.org/10.1215/00294527-2009-013.

Brown, Tom B., Benjamin Mann, Nick Ryder, Melanie Subbiah, Jared Kaplan, Prafulla Dhariwal, Arvind Neelakantan, et al. "Language Models Are Few-Shot Learners." Cornell University. Last revised July 22, 2020. http://arxiv.org/abs/2005.14165.

Bucher, Taina. *If . . . Then: Algorithmic Power and Politics*. Oxford Studies in Digital Politics. Oxford: Oxford University Press, 2018.

Burgess, Alexis, Herman Cappelen, and David Plunkett, eds. *Conceptual Engineering and Conceptual Ethics*. Oxford: Oxford University Press, 2020.

Bürja, Abel. "Essai d'un nouvel algorithme des logarithmes." *Mémoires de l'Académie royale des sciences et belles-lettres, 1788–89*, 300–325.

Burke, Edmund. *Reflections on the Revolution in France*. Library of Liberal Arts 46. New York: Liberal Arts Press, Bobbs-Merrill, 1955.

Buxton, Harry W. *Memoir of the Life and Labours of the Late Charles Babbage Esq. F. R. S.* Cambridge, MA: MIT Press, 1988.

Cajori, Florian. *A History of Mathematical Notations*. Mineola, NY: Dover, 1993.

Campbell, George. *The Philosophy of Rhetoric*. 2 vols. London, 1776.

Cantor, Georg. "Beiträge Zur Begründung Der Transfiniten Mengenlehre" I. *Mathematische Annalen* 46, no. 4 (November 1895): 481–512.

———. "Beiträge Zur Begründung Der Transfiniten Mengenlehre" II. *Mathematische Annalen* 49 (1897): 207–246.

———. *Contributions to the Founding of the Theory of Transfinite Numbers*. Translated by Philip E. B. Jourdain. Chicago: Open Court, 1915.

Capaldi, Nicholas. "The Enlightenment Project in Twentieth-Century Philosophy." In *Modern Enlightenment and the Rule of Reason*, edited by John C. McCarthy, 257–82. Washington, DC: Catholic University of America Press, 1998. https://doi.org/10.2307/j.ctt22h6r4z.15.

Capozzi, Mirella, and Gino Roncaglia. "Logic and Philosophy of Logic from Humanism to Kant." In *The Development of Modern Logic*, edited by Leila Haaparanta, 78–147. Oxford: Oxford University Press, 2009.

Cappelen, Herman. *Fixing Language: An Essay on Conceptual Engineering*. Oxford: Oxford University Press, 2018.

Cardano, Girolamo. *The Rules of Algebra: Ars Magna*. Translated by T. Richard Witmer. Mineola, NY: Dover, 1968.

Carnap, Rudolf. *The Logical Structure of the World: Pseudoproblems in Philosophy*. Translated by Rolf A. George. Berkeley: University of California Press, 1967.

————. *The Logical Syntax of Language.* Translated by Amethe Smeaton, Countess von Zeppelin. London: Routledge & Kegan Paul, 1937.

Carr, John W., Alan J. Perlis, James E. Robertson, and Norman R. Scott. "A Visit to Computation Centers in the Soviet Union." *Communications of the ACM* 2, no. 6 (June 1959): 8–20.

Cassirer, Ernst. *The Philosophy of Symbolic Forms.* Edited by John Michael Krois. Translated by Ralph Manheim. 4 vols. New Haven, CT: Yale University Press, 1953.

Chaitin, Gregory J. Information, *Randomness and Incompleteness: Papers on Algorithmic Information Theory.* Hackensack, NJ: World Scientific, 1990.

Chambers, Ephraim. *Cyclopaedia: Or, An Universal Dictionary of Arts and Sciences.* 2 vols. London, 1728.

Chartier, Roger. "Languages, Books, and Reading from the Printed Word to the Digital Text." Translated by Teresa Lavender Fagen. *Critical Inquiry* 31, no. 1 (Autumn 2004): 133–52. https://doi.org/10.1086/427305.

Chaucer, Geoffrey. *The Riverside Chaucer.* Edited by Larry Dean Benson. Boston: Houghton Mifflin, 1987.

Chrisomalis, Stephen. *Reckonings: Numerals, Cognition, and History.* Cambridge, MA: MIT Press, 2020.

Christianidis, Jean, and Jeffrey Oaks. "Practicing Algebra in Late Antiquity: The Problem-Solving of Diophantus of Alexandria." *Historia Mathematica* 40, no. 2 (May 1, 2013): 127–63. https://doi.org/10.1016/j.hm.2012.09.001.

Chun, Wendy Hui Kyong. *Discriminating Data: Correlation, Neighborhoods, and the New Politics of Recognition.* Cambridge, MA: MIT Press, 2021.

————. "On 'Sourcery,' or Code as Fetish." *Configurations* 16, no. 3 (2008): 299–324.

Church, Alonzo. "Alternatives to Zermelo's Assumption." *Transactions of the American Mathematical Society* 29, no. 1 (January 1927): 178–208. https://doi.org/10.2307/1989285.

————. "Schröder's Anticipation of the Simple Theory of Types." *Erkenntnis* 10, no. 3 (1976): 407–11.

————. "An Unsolvable Problem of Elementary Number Theory." *American Journal of Mathematics* 58, no. 2 (1936): 345–63.

Cifoletti, Giovanna. "From Valla to Viète: The Rhetorical Reform of Logic and Its Use in Early Modern Algebra." *Early Science and Medicine* 11, no. 4 (2006): 390–423.

————. "La question de l'algèbre: Mathématiques et rhétorique des hommes de droit dans la France du XVIe siècle." *Annales: Histoire, Sciences Sociales* 50, no. 6 (1995): 1385–1416. https://doi.org/10.3406/ahess.1995.279438.

Clairaut, Alexis-Claude. *Élémens d'algèbre.* Paris, 1746.

Cleveland, Catherine Lucy Wilhelmina Powlett. *The Life and Letters of Lady Hester Stanhope.* London: John Murray, 1914.

Cobbett, William, ed. *Cobbett's Parliamentary History of England: From the Norman Conquest, in 1066, to the Year, 1803.* 36 vols. London: Printed by T. Curson Hansard, Published by R. Bagshaw, 1806.

Cohen, Daniel J. *Equations from God: Pure Mathematics and Victorian Faith.* Baltimore: JHU Press, 2008.

Colburn, Zerah. *A Memoir of Zerah Colburn: Written by Himself [. . .].* Springfield, MA: G. and C. Merriam, 1833.

Coleridge, Samuel Taylor. *The Collected Works of Samuel Taylor Coleridge.* 16 vols.

London: Routledge and Kegan Paul; Princeton, NJ: Princeton University Press, 1969.

————. *Letters of Samuel Taylor Coleridge*. Edited by Earl Leslie Griggs. 6 vols. Oxford: Clarendon, 1966.

Coles, Elisha. *An English Dictionary [. . .]*. London: Printed for Peter Parker, 1692. Early English Books, 1641–1700, microfilm reel position 450:28.

Collins, Ronald K. L., and David M. Skover. *Robotica: Speech Rights and Artificial Intelligence*. Cambridge: Cambridge University Press, 2018.

Compte rendu du deuxième Congrès international des mathematiciens tenu à Paris du 6 au 12 août 1900. Paris: Gauthier-Villars, 1902. http://gallica.bnf.fr/ark:/12148/bpt6k994421.

Condillac, Étienne Bonnot de. *Essay on the Origin of Human Knowledge*. Translated by Hans Aarsleff. Cambridge: Cambridge University Press, 2001.

————. *La langue des calculs*. Paris: Librarie Sandoz et Fischbacher, 1877. http://gallica.bnf.fr/ark:/12148/bpt6k6532532j.

————. *Philosophical Writings of Etienne Bonnot, Abbé de Condillac*. Translated by Lawrence Erlbaum. Hillsdale, NJ: L. Erlbaum Associates, 1982.

Condorcet, Marie-Jean-Antoine-Nicolas Caritat, Marquis de. *Essai sur l'application de l'analyse à la probabilité des décisions rendues à la pluralité des voix*. Paris: De l'Imprimerie royale, 1785.

————. *Political Writings*. Edited by Steven Lukes and Nadia Urbinati. Cambridge: Cambridge University Press, 2012.

————. *Tableau historique des progrès de l'esprit humain: Projets, esquisse, fragments et notes (1772–1794)*. Edited by Groupe Condorcet. Paris: INED, 2004.

Cook, Simon. "Minds, Machines and Economic Agents: Cambridge Receptions of Boole and Babbage." *Studies in History and Philosophy of Science* 36A, no. 2 (June 2005): 331–50. https://doi.org/10.1016/j.shpsa.2005.04.001.

Corcoran, John. "Aristotle's Prior Analytics and Boole's Laws of Thought." *History and Philosophy of Logic* 24 (2003): 261–88.

Corcoran, John, and Susan Wood. "Boole's Criteria for Validity and Invalidity." In Gasser, *Boole Anthology*, 101–28.

Cormack, Lesley B. "The Role of Mathematical Practitioners and Mathematical Practice in Developing Mathematics as the Language of Nature" in Gorham et al., *Language of Nature*, 205–28.

Cormack, Lesley B., Steven A. Walton, and John A. Schuster, eds. *Mathematical Practitioners and the Transformation of Natural Knowledge in Early Modern Europe*. Cham, Switz.: Springer International, 2017.

Couturat, Louis. *La logique de Leibniz*. Hildesheim, Ger.: Georg Olms, 1985.

Cram, David, Andrew R. Linn, and Elke Nowak, eds. *From Classical to Contemporary Linguistics*. Vol. 2 of *History of Linguistics 1996*. Amsterdam: John Benjamins, 1999.

Cram, David, and Jaap Maat. "Universal Language Schemes in the 17th Century." In *History of the Language Sciences*, edited by Sylvian Auroux, E. F. K. Koerner, Hans-Josef Niederehe, and Kees Versteegh, 1030–43. Berlin: Walter de Gruyter, 2000.

Crawford, Kate, and Trevor Paglen. "Excavating AI: The Politics of Images in Machine Learning Training Sets." AI Now Institute, New York University. Published September 19, 2019. https://excavating.ai.

Creel, Herrlee Glessner. "On the Nature of Chinese Ideography." *T'oung Pao* 32, no. 2/3 (1936): 85–161.

Crossley, John N., and Alan S. Henry. "Thus Spake Al-Khwārizmī: A Translation of the Text of Cambridge University Library Ms. Ii.vi.5." *Historia Mathematica* 17, no. 2 (May 1, 1990): 103–31. https://doi.org/10.1016/0315-0860(90)90048-I.

d'Alembert, Jean Le Rond. *Oeuvres de D'Alembert*. 5 vols. Paris, 1821.

Dalgarno, George. *George Dalgarno on Universal Language: "The Art of Signs" (1661), "The Deaf and Dumb Man's Tutor" (1680), and the Unpublished Papers*. Translated by David Cram and Jaap Maat. Oxford: Oxford University Press, 2001.

Dary, Michael. *Interest Epitomized Both Compound and Simple [. . .]*. London, 1677. Early English Books, 1641–1700, microfilm reel position 1307:05.

Dascal, Marcelo. "The Balance of Reason." In *Logic, Thought and Action*, edited by Daniel Vanderveken, 27–47. Dordrecht, Neth.: Springer, 2005.

———. *Leibniz, Language, Signs, and Thought: A Collection of Essays*. Amsterdam: John Benjamins, 1987.

———, ed. *Leibniz: What Kind of Rationalist?* Logic, Epistemology, and the Unity of Science 13. Dordrecht, Neth.: Springer Science+Business Media, 2009.

———, ed. *The Practice of Reason: Leibniz and His Controversies*. Amsterdam: John Benjamins, 2010.

Dasgupta, Subrata. *It Began with Babbage: The Genesis of Computer Science*. New York: Oxford University Press, 2014.

Daston, Lorraine. *Classical Probability in the Enlightenment*. Princeton, NJ: Princeton University Press, 1988.

———. "Enlightenment Calculations." *Critical Inquiry* 21, no. 1 (1994): 182–202.

Daston, Lorraine, and Peter Galison. *Objectivity*. New York: Zone Books, 2010.

Dauben, Joseph Warren. *Georg Cantor: His Mathematics and Philosophy of the Infinite*. Princeton, NJ: Princeton University Press, 1979.

Davis, Martin. *The Universal Computer: The Road from Leibniz to Turing*. 3rd ed. Boca Raton, FL: CRC Press, 2018.

Dawson, Hannah. *Locke, Language and Early-Modern Philosophy*. Ideas in Context. Cambridge: Cambridge University Press, 2007.

Dear, Peter. *Discipline and Experience: The Mathematical Way in the Scientific Revolution*. Chicago: University of Chicago Press, 1995.

———. *The Intelligibility of Nature: How Science Makes Sense of the World*. Chicago: University of Chicago Press, 2006.

———, ed. *The Literary Structure of Scientific Argument: Historical Studies*. Philadelphia: University of Pennsylvania Press, 1991.

Debuiche, Valerie. "La notion d'expression et ses origines mathématiques." *Studia Leibnitiana* 41, no. 1 (2009): 88–117.

Dedekind, Richard. *Essays on the Theory of Numbers*. Translated by Wooster Woodruff Beman. Chicago: Open Court, 1901.

———. *Was Sind und Was Sollen die Zahlen?* Braunschweig, Ger.: Vieweg, 1888.

Dee, John. Preface to Euclid, *Elements of Geometrie*.

de Gérando, Joseph Marie. *Des signes et de l'art de penser considérés dans leurs rapports mutuels*. 4 vols. Paris, 1799. http://gallica.bnf.fr/ark:/12148/bpt6k840388.

De Morgan, Augustus. "A Treatise on Algebra." *Quarterly Journal of Education* 9, no. 17 (January 1835): 91–110.

———. "A Treatise on Algebra (No. II)." *Quarterly Journal of Education* 9, no. 18 (April 1835): 293–311.

Descartes, René. "A Mersenne, Amsterdam, 20 novembre 1629." In *Œuvres et lettres*, edited by André Bridoux, 911–15. Bibliothèque de la Pléiade 40. Paris: Gallimard, 2004.

———. *A Discourse on Method*. Translated by John Veitch. London: Everyman's Library, 1992.

———. *The Geometry of René Descartes*. Translated by David Eugene Smith and Marcia L. Latham. Chicago: Open Court, 1925.

———. *The Philosophical Works of Descartes*. Translated by Elizabeth S. Haldane and G. R. T. Ross. London: Cambridge University Press, 1967.

———. *Principles of Philosophy*. Translated by Valentine Rodger Miller and Reese P. Miller. Dordrecht, Neth.: Kluwer, 1982.

Despeaux, Sloan Evans. "'Very Full of Symbols': Duncan F. Gregory, the Calculus of Operations, and the Cambridge Mathematical Journal." In *Episodes in the History of Modern Algebra (1800–1950)*, edited by Jeremy J. Gray and Karen Hunger Parshall, 49–72. Providence, RI: American Mathematical Society, 2011.

Devlin, Jacob, Ming-Wei Chang, Kenton Lee, and Kristina Toutanova. "BERT: Pre-Training of Deep Bidirectional Transformers for Language Understanding." Cornell University. Last revised May 24, 2019. http://arxiv.org/abs/1810.04805.

Diderot, Denis. *Rameau's Nephew / D'Alembert's Dream*. Translated by Leonard Tancock. London: Penguin UK, 1976.

Diderot, Denis, and Jean Le Rond d'Alembert, eds. *Encyclopédie, ou Dictionnaire raisonné des sciences, des arts et des métiers*. 17 vols. Paris: Briasson, 1751–72.

Dijkstra, Edsger W. "Go To Statement Considered Harmful." Letter to the editor. *Communications of the ACM* 11, no. 3 (March 1968): 147–48. https://doi.org/10.1145/362929.362947.

———. "A Note on Two Problems in Connexion with Graphs." *Numerische Mathematik* 1, no. 1 (December 1959): 269–71. https://doi.org/10.1007/BF01386390.

———. "On the Cruelty of Really Teaching Computing Science." E. W. Dijkstra Archive: The manuscripts of Edsger W. Dijkstra, 1930–2002. https://www.cs.utexas.edu/~EWD/transcriptions/EWD10xx/EWD1036.html.

Diophantus. *Books IV to VII of Diophantus' Arithmetica: In the Arabic Translation Attributed to Qustā Ibn Lūqā*. Translated by Jacques Sesiano. New York: Springer-Verlag, 1982.

Došilović, Filip Karlo, Mario Brčić, and Nikica Hlupić. "Explainable Artificial Intelligence: A Survey." In *2018 41st International Convention on Information and Communication Technology, Electronics and Microelectronics (MIPRO)*, 0210–15. IEEE Xplore, July 2, 2018. https://doi.org/10.23919/MIPRO.2018.8400040.

Dourish, Paul. "Algorithms and Their Others: Algorithmic Culture in Context." *Big Data & Society* 3, no. 2 (July–December 2016): 1–11. https://doi.org/10.1177/2053951716665128.

Dreyfus, Hubert L. *What Computers Still Can't Do: A Critique of Artificial Reason*. Cambridge, MA: MIT Press, 1992.

Dyck, Martin. *Novalis and Mathematics: A Study of Friedrich von Hardenberg's "Fragments on Mathematics" and Its Relation to Magic, Music, Religion, Philosophy, Language, and Literature*. New York: AMS Press, 1969.

Eco, Umberto. "The Language of the Austral Land." In *From the Tree to the Labyrinth: Historical Studies on the Sign and Interpretation*, translated by Anthony Oldcorn, 424–39. Cambridge, MA: Harvard University Press, 2014.

————. *The Search for the Perfect Language.* Translated by James Fentress. Cambridge, MA: Blackwell, 1995.

Edgeworth, Maria, and Richard Lovell Edgeworth. *Practical Education.* 2 vols. London, 1798.

Emerson, Lori. *Reading Writing Interfaces: From the Digital to the Bookbound.* Minneapolis: University of Minnesota Press, 2014.

Encyclopédie méthodique: Mathématiques. Vol. 1. Paris, 1784.

Erdman, David V. "Citizen Stanhope and the French Revolution." *Wordsworth Circle* 15, no. 1 (1984): 8–17.

Erickson, Paul, Judy L. Klein, Lorraine Daston, Rebecca Lemov, Thomas Sturm, and Michael D. Gordin. *How Reason Almost Lost Its Mind: The Strange Career of Cold War Rationality.* Chicago: University of Chicago Press, 2013.

Ershov, Andrei P., and Donald E. Knuth. *Algorithms in Modern Mathematics and Computer Science: Proceedings, Urgench, Uzbek SSR September 16–22, 1979.* Berlin: Springer-Verlag, 1981.

Esser, Patrick, Robin Rombach, and Björn Ommer. "Taming Transformers for High-Resolution Image Synthesis." Last revised June 23, 2021. https://arxiv.org/abs/2012.09841v3.

Euclid. *Elements, Books I–XIII.* Translated by Thomas L. Heath. New York: Barnes & Noble, 2006.

————. *The Elements of Geometrie of the Most Auncient Philosopher Euclide of Megara.* Translated by Henry Billingsley. With preface by John Dee. London: John Daye, 1570. Early English Books, 1475–1640, microfilm reel 342:05.

————. *Euclid's Elements of Geometry, from the Latin Translation of Commandine [. . .].* London, 1754.

Evans, Gillian R. "From Abacus to Algorism: Theory and Practice in Medieval Arithmetic." *British Journal for the History of Science* 10, no. 2 (1977): 114–31.

Feingold, Mordechai. *The Mathematicians' Apprenticeship: Science, Universities and Society in England, 1560–1640.* Cambridge: Cambridge University Press, 1984.

Ferraro, Giovanni, and Marco Panza. "Lagrange's Theory of Analytical Functions and His Ideal of Purity of Method." *Archive for History of Exact Sciences* 66, no. 2 (2012): 95–197.

Fibonacci, Leonardo. *Fibonacci's Liber Abaci: A Translation into Modern English of Leonardo Pisano's Book of Calculation.* Translated by Laurence E. Sigler. New York: Springer-Verlag, 2002.

Fidora, Alexander, and Carles Sierra, eds. *Ramon Llull: From the Ars Magna to Artificial Intelligence.* Barcelona: Artificial Intelligence Research Institute, 2011.

Finn, Ed. *What Algorithms Want: Imagination in the Age of Computing.* Cambridge, MA: MIT Press, 2017.

Fisch, Menachem. "'The Emergency Which Has Arrived': The Problematic History of Nineteenth-Century British Algebra; A Programmatic Outline." *British Journal for the History of Science* 27, no. 3 (September 1994): 247–76.

————. "The Making of Peacock's Treatise on Algebra: A Case of Creative Indecision." *Archive for History of Exact Sciences* 54, no. 2 (1999): 137–79. https://doi.org/10.1007/s004070050037.

Fish, John Charles Lounsbury. *Mathematics of the Paper Location of a Railroad.* New York: M. C. Clark, 1905.

Fish, Stanley Eugene. *Surprised by Sin: The Reader in "Paradise Lost."* 2nd ed. Cambridge, MA: Harvard University Press, 1997.

Fleischacker, Samuel. *What Is Enlightenment?* Abingdon, UK: Routledge, 2013.

Fleming, James Dougal. *The Mirror of Information in Early Modern England: John Wilkins and the Universal Character.* London: Palgrave Macmillan, 2017.

Floyd, Juliet. "Turing on 'Common Sense': Cambridge Resonances." In *Philosophical Explorations of the Legacy of Alan Turing: Turing 100*, edited by Juliet Floyd and Alisa Bokulich, 103–49. Boston Studies in the Philosophy and History of Science. Cham, Switz.: Springer International, 2017. https://doi.org/10.1007/978-3-319-53280-6_5.

Forbes-Macphail, Imogen. "The Enchantress of Numbers and the Magic Noose of Poetry: Literature, Mathematics, and Mysticism in the Nineteenth Century." *Journal of Language, Literature and Culture* 60, no. 3 (December 2013): 138–56. https://doi.org/10.1179/2051285613Z.00000000016.

Forrest, Stephanie, Justin Balthrop, Matthew Glickman, and David Ackley. "Computation in the Wild." In *Robust Design: A Repertoire of Biological, Ecological, and Engineering Case Studies*, edited by Erica Jen, 207–30. New York: Oxford University Press, 2005.

Forster, Peter G. *The Esperanto Movement.* The Hague, Neth.: Mouton, 1982.

Forsythe, Alexandra I., Thomas A. Keenan, Elliot I. Organick, and Warren Stenberg. *Computer Science: A First Course.* New York: Wiley, 1969.

Forsythe, George E. "Algorithms for Scientific Computation." *Communications of the ACM* 9, no. 4 (April 1966): 255–56. https://doi.org/10.1145/365278.365294.

Foucault, Michel. *The Order of Things: An Archaeology of the Human Sciences.* Reissue ed. New York: Vintage, 1994.

Fowler, David. "Dedekind's Theorem: $\sqrt{2} \times \sqrt{3} = \sqrt{6}$." *American Mathematical Monthly* 99, no. 8 (October 1992): 725–33. https://doi.org/10.2307/2324238.

Frame, Paul. *Liberty's Apostle: Richard Price, His Life and Times.* Wales and the French Revolution. Cardiff: University of Wales Press, 2015.

Frana, Philip L., and Thomas J. Misa. "An Interview with Edsger W. Dijkstra." *Communications of the ACM* 53, no. 8 (August 2010): 41–47. https://doi.org/10.1145/1787234.1787249.

Frängsmyr, Tore. "The Mathematical Philosophy." In Frängsmyr, Heilbron, and Rider, *Quantifying Spirit in the 18th Century*, 27–44.

Frängsmyr, Tore, J. L. Heilbron, and Robin E. Rider, eds. *The Quantifying Spirit in the 18th Century.* Berkeley: University of California Press, 1990.

Franklin, James. "Artifice and the Natural World: Mathematics, Logic, Technology." In *The Cambridge History of Eighteenth-Century Philosophy*, edited by Knut Haakonssen, 2.817-53. Cambridge: Cambridge University Press, 2006.

Freeman, Linton C. "A Set of Measures of Centrality Based on Betweenness." *Sociometry* 40, no. 1 (March 1977): 35–41. https://doi.org/10.2307/3033543.

Frege, Gottlob. "Begriffsschrift." In *From Frege to Gödel: A Source Book in Mathematical Logic, 1879–1931*, 1–82. Cambridge, MA: Harvard University Press, 1967.

———. "Sense and Reference." Translated by Max Black. *Philosophical Review* 57, no. 3 (May 1948): 209–30. https://doi.org/10.2307/2181485.

Frend, William. *Peace and Union Recommended to the Associated Bodies of Republicans and Anti-Republicans.* St. Ives, UK, 1792.

———. *The Principles of Algebra.* London, 1796.

Galilei, Galileo. *Discoveries and Opinions of Galileo: Including The Starry Messenger*

(1610), Letter to the Grand Duchess Christina (1615), and Excerpts from Letters on Sunspots (1613), The Assayer (1623). Translated by Stillman Drake. New York: Anchor, 1957.

Gardner, Martin. *Logic Machines and Diagrams*. 2nd ed. Chicago: University of Chicago Press, 1982.

Gasser, James, ed. *A Boole Anthology: Recent and Classical Studies in the Logic of George Boole*. Synthese Library. Dordrecht, Neth.: Kluwer Academic Publishers, 2000. https://doi.org/10.1007/978-94-015-9385-4.

Gelb, Ignace J. *A Study of Writing: The Foundations of Grammatology*. London: Routledge and Kegan Paul, 1952.

Geneva, Ann. *Astrology and the Seventeenth Century Mind: William Lilly and the Language of the Stars*. Manchester, UK: Manchester University Press, 1995.

Gleick, James. *The Information: A History, a Theory, a Flood*. New York: Vintage Books, 2012.

Gnanadesikan, Amalia E. *The Writing Revolution: Cuneiform to the Internet*. Chichester, UK: Wiley-Blackwell, 2009.

Gode, Alexander, and Hugh E. Blair. *Interlingua: A Grammar of the International Language*. New York: Frederick Ungar, 1971.

Gödel, Kurt. "On Formally Undecidable Propositions of Principia Mathematica and Related Systems I." In *From Frege to Gödel: A Source Book in Mathematical Logic, 1879–1931*, edited by Jean van Heijenoort, 596–616. Cambridge, MA: Harvard University Press, 2002.

———. "Über formal unentscheidbare Sätze der Principia Mathematica und verwandter Systeme I." *Monatshefte für Mathematik und Physik* 38, no. 1 (December 1, 1931): 173–98. https://doi.org/10.1007/BF01700692.

Godwin, William. *An Enquiry Concerning Political Justice, and Its Influence on General Virtue and Happiness*. 2 vols. London, 1793.

Goethe, Norma B. "How Did Bertrand Russell Make Leibniz Into a 'Fellow Spirit'?" In Phemister and Brown, *Leibniz and the English Speaking World*, 195–205.

Goldstine, Herman H., and John Von Neumann. *Planning and Coding of Problems for an Electronic Computing Instrument*. 3 vols. Report on the Mathematical and Logical Aspects of an Electronic Computing Instrument 2. Princeton, NJ: Institute for Advanced Study, 1947.

Golumbia, David. *The Cultural Logic of Computation*. Cambridge, MA: Harvard University Press, 2009.

Goodfellow, Ian, Yoshua Bengio, and Aaron Courville. *Deep Learning*. Cambridge, MA: MIT Press, 2016.

Gorham, Geoffrey, Benjamin Hill, Edward Slowik, and C. Kenneth Waters, eds. *The Language of Nature: Reassessing the Mathematization of Natural Philosophy in the Seventeenth Century*. Minnesota Studies in the Philosophy of Science. Minneapolis: University of Minnesota Press, 2016.

Grabiner, Judith V. "Who Gave You the Epsilon? Cauchy and the Origins of Rigorous Calculus." *American Mathematical Monthly* 90, no. 3 (1983): 185–94. https://doi.org/10.2307/2975545.

Granger, G. G. "Langue universelle et formalisation des sciences: Un fragment inédit de Condorcet." *Revue d'histoire des sciences et leurs applications* 7 (1954): 197–219.

Grant, Ruth W. "John Locke on Custom's Power and Reason's Authority." *Review of Politics* 74, no. 4 (2012): 607–29.

Grattan-Guinness, Ivor. "Boole's Quest for the Foundations of His Logic." In G. Boole, *Selected Manuscripts on Logic and Its Philosophy*, xiii–xlvii.

———. "Charles Babbage as an Algorithmic Thinker." *IEEE Annals of the History of Computing* 14, no. 3 (1992): 34–48. https://doi.org/10.1109/85.150067.

Graves, Alex, Greg Wayne, and Ivo Danihelka. "Neural Turing Machines." Cornell University. Last revised December 10, 2014. http://arxiv.org/abs/1410.5401.

Graves, Robert Perceval. *Life of Sir William Rowan Hamilton [. . .]*. 3 vols. Dublin: Hodges, Figgis; London: Longmans, Green, 1882.

Gray, Jeremy. *Plato's Ghost: The Modernist Transformation of Mathematics*. Princeton, NJ: Princeton University Press, 2008.

Greene, Stephen. "A Taste for Figures." *New England Quarterly* 26, no. 1 (March 1953): 65–77. https://doi.org/10.2307/362336.

Gregory, Duncan F. "On the Real Nature of Symbolical Algebra." *Transactions of the Royal Society of Edinburgh* 14 (1840): 208–16.

Grier, David Alan. "The Inconsistent Youth of Charles Babbage." *IEEE Annals of the History of Computing* 32, no. 4 (December 2010): 18–31. https://doi.org/10.1109/MAHC.2010.67.

———. *When Computers Were Human*. Princeton, NJ: Princeton University Press, 2005.

Guicciardini, Niccolò. *The Development of Newtonian Calculus in Britain, 1700–1800*. Cambridge: Cambridge University Press, 1989.

———. *Isaac Newton on Mathematical Certainty and Method*. Cambridge, MA: MIT Press, 2009.

Guillaumin, Jean-Yves. "Boethius's De Institutione Arithmetica and Its Influence on Posterity." In *A Companion to Boethius in the Middle Ages*, edited by Noel Howard Kaylor and Philip Edward Phillips, 135–61. Brill's Companions to the Christian Tradition 30. Leiden, Ger.: Brill, 2012.

Guillory, John. "The Memo and Modernity." *Critical Inquiry* 31, no. 1 (2004): 108–32.

Hacking, Ian. *The Emergence of Probability: A Philosophical Study of Early Ideas about Probability, Induction and Statistical Inference*. 2nd ed. Cambridge: Cambridge University Press, 2006.

———. *The Taming of Chance*. Cambridge: Cambridge University Press, 1990.

Hailperin, Theodore. "Boole's Algebra Isn't Boolean Algebra." *Mathematics Magazine* 54, no. 4 (1981): 173–84. https://doi.org/10.2307/2689628.

———. *Boole's Logic and Probability: A Critical Exposition from the Standpoint of Contemporary Algebra, Logic, and Probability Theory*. Amsterdam: North-Holland, 1986.

Hamburg, Robin Rider. "The Theory of Equations in the 18th Century: The Work of Joseph Lagrange." *Archive for History of Exact Sciences* 16, no. 1 (1976): 17–36.

Hamilton, William Rowan. "On Quaternions; or On a New System of Imaginaries in Algebra (Part 1)." *London, Edinburgh and Dublin Philosophical Magazine and Journal of Science* 25 (1844): 10–13.

———. "Theory of Conjugate Functions, or Algebraic Couples; with a Preliminary and Elementary Essay on Algebra as the Science of Pure Time." *Transactions of the Royal Irish Academy* 17 (1831): 293–423.

Hamilton, William Stirling. "M. Cousin's Course of Philosophy." *Edinburgh Review, or Critical Journal* 50 (October 1829): 194–221.

Harkness, Deborah E. *The Jewel House: Elizabethan London and the Scientific Revolution*. New Haven, CT: Yale University Press, 2007.

Harley, Robert. "The Stanhope Demonstrator." *Mind* 4, no. 14 (1879): 192–210.

Harriot, Thomas. *Artis Analyticae Praxis: An English Translation with Commentary*. Translated by Muriel Seltman and Robert Goulding. New York: Springer Science+Business Media, 2007.

Harris, John. *A New Short Treatise of Algebra: With the Geometrical Construction of Equations, as Far as the Fourth Power Or Dimension. Together with a Specimen of the Nature and Algorithm of Fluxions*. London, 1702.

Hart, Peter E., Nils J. Nilsson, and Bertram Raphael. "A Formal Basis for the Heuristic Determination of Minimum Cost Paths." *IEEE Transactions on Systems Science and Cybernetics* 4, no. 2 (July 1968): 100–107. https://doi.org/10.1109/TSSC.1968 .300136.

Hartmanis, J., and R. E. Stearns. "On the Computational Complexity of Algorithms." *Transactions of the American Mathematical Society* 117 (1965): 285–306. https://doi .org/10.2307/1994208.

Haugeland, John. *Artificial Intelligence: The Very Idea*. Cambridge, MA: MIT Press, 1985.

Hawtrey, R. G. *The Exchequer and the Control of Expenditure*. London: Humphrey Milford; Oxford University Press, 1921.

Hayles, N. Katherine. *How We Became Posthuman: Virtual Bodies in Cybernetics, Literature, and Informatics*. University of Chicago Press, 2008.

Heeffer, Albrecht. "On the Nature and Origin of Algebraic Symbolism." In *New Perspectives on Mathematical Practices: Essays in Philosophy and History of Mathematics*, edited by Bart Van Kerkhove, 1 27. Singapore: World Scientific, 2008.

Heeffer, Albrecht, and Maarten Van Dyck, eds. *Philosophical Aspects of Symbolic Reasoning in Early Modern Mathematics*. London: College Publications, 2010.

Heilbron, J. L. "Introductory Essay." In Frängsmyr, Heilbron, and Rider, *Quantifying Spirit in the 18th Century*, 1–23.

Heinekamp, Albert. "Ars characteristica und natürliche sprache bei Leibniz." *Tijdschrift Voor Filosofie* 34, no. 3 (1972): 446–88.

Henderson, Andrea K. *Algebraic Art: Mathematical Formalism and Victorian Culture*. Oxford: Oxford University Press, 2018.

Herder, Johann Gottfried. *Johann Gottfried Herder on World History: An Anthology*. Edited by Hans Adler and Ernest A. Menze. Translated by Michael Palma and Ernest A. Menze. London: Routledge, 2015.

———. *Philosophical Writings*. Translated by Michael N. Forster. Cambridge: Cambridge University Press, 2002.

Hérigone, Pierre. *Cvrsvs mathematicvs, nova, brevi et clara methodo demonstratvs, per notas reales & vniuersales, citra vsum, cuiuscunque idiomatis intellectu faciles [. . .]*. 5 vols. Paris, 1634. https://gallica.bnf.fr/ark:/12148/bpt6k618506.

Herriot, J. G. "Algorithms Section Policy." *Communications of the ACM* 9, no. 4 (April 1966): 256. https://doi.org/10.1145/365278.365295.

Hesse, Mary B. "Hooke's Philosophical Algebra." *Isis* 57, no. 1 (1966): 67–83.

Hicks, Mar. *Programmed Inequality: How Britain Discarded Women Technologists and Lost Its Edge in Computing*. Cambridge, MA: MIT Press, 2017.

Hight, Marc A. "Why My Chair Is Not Merely a Congeries: Berkeley and the Single-Idea Thesis." In *Reexamining Berkeley's Philosophy*, edited by Stephen H. Daniel, 82–107. Toronto: University of Toronto Press, 2007.

Hilbert, David, and Wilhelm Ackermann. *Principles of Mathematical Logic*. Edited by

Robert E. Luce. Translated by Lewis M. Hammond, George G. Leckie, and F. Stein-
    hardt. Providence, RI: American Mathematical Society, 1999.
Hill, Katherine. "'Juglers or Schollers?': Negotiating the Role of a Mathematical Practi-
    tioner." *British Journal for the History of Science* 31, no. 3 (1998): 253–74.
Hoare, C. A. R. "Algorithm 64: Quicksort." *Communications of the ACM* 4, no. 7 (July
    1961): 321.
———. "How Did Software Get So Reliable without Proof?" In *Proceedings of the
    Third International Symposium of Formal Methods Europe on Industrial Benefit and
    Advances in Formal Methods*, 1–17. Berlin: Springer-Verlag, 1996.
———. "The 1980 ACM Turing Award Lecture." *Communications of the ACM* 24, no. 2
    (February 1981): 75–83.
Hobbes, Thomas. *Elements of Philosophy the First Section, Concerning Body*. London,
    1656. Early English Books, 1641–1700, microfilm reel position 634:08.
———. *Leviathan, Or, The Matter, Form, and Power of a Common-Wealth Ecclesiasti-
    cal and Civil [. . .]*. London: 1651. Early English Books, 1641–1700, microfilm reel
    position 670:04.
———. *Six Lessons to the Professors of the Mathematiques One of Geometry the Other
    of Astronomy, in the Chaires Set Up by the Noble and Learned Sir Henry Savile in the
    University of Oxford*. London, 1656. Early English Books, 1641–1700, microfilm reel
    position 1360:01.
Hoëné-Wroński, Józef Maria. *Address of M. Hoene Wronski to the British Board of Lon-
    gitude upon the Actual State of the Mathematics [. . .]*. Translated by W. Gardiner.
    London: T. Egerton, 1820.
———. *Introduction à la philosophie des mathématiques, et technie de l'algorithmie*.
    Paris: Courcier, 1811.
Hooke, Robert. *Lectures de Potentia Restitutiva, or, Of Spring [. . .]*. London: Printed
    for John Martyn, 1678. Early English Books, 1641–1700, microfilm reel position
    1125:02.
———. *The Posthumous Works of Robert Hooke [. . .]*. London: Richard Waller, 1705.
———. "Some Observations, and Conjectures Concerning the Chinese Characters."
    Philosophical Transactions of the Royal Society of London 16, no. 180 (January 1,
    1687): 63–78. https://doi.org/10.1098/rstl.1686.0011.
Hope, Jonathan. *Shakespeare and Language: Reason, Eloquence and Artifice in the
    Renaissance*. London: Arden Shakespeare, 2010.
Hopper, Grace Murray. "The Education of a Computer." *Annals of the History of Com-
    puting* 9, no. 3/4 (July–September/October–November 1987): 271–81. https://doi
    .org/10.1109/MAHC.1987.10032.
Horváth, Miklós. "On the Attempts Made by Leibniz to Justify His Calculus." *Studia
    Leibnitiana* 18, no. 1 (1986): 60–71.
Hotson, Howard. *Commonplace Learning: Ramism and Its German Ramifications,
    1543–1630*. Oxford: Oxford University Press, 2007.
Howell, Wilbur Samuel. *Eighteenth-Century British Logic and Rhetoric*. Princeton, NJ:
    Princeton University Press, 1971.
———. *Logic and Rhetoric in England, 1500–1700*. Princeton, NJ: Princeton University
    Press, 1956.
Hudak, Paul, Simon Peyton Jones, Philip Wadler, Brian Boutel, Jon Fairbairn, Joseph
    Fasel, María M. Guzmán, et al. "Report on the Programming Language Haskell: A

Non-Strict, Purely Functional Language Version 1.2." *ACM SIGPLAN Notices* 27, no. 5 (May 1992): 1–164. https://doi.org/10.1145/130697.130699.

Humboldt, Alexander von. *Kosmos: A General Survey of Physical Phenomena of the Universe*. 2 vols. London: Hippolyte Baillière, 1845.

Humboldt, Wilhelm von. *On Language: On the Diversity of Human Language Construction and Its Influence on the Mental Development of the Human Species*. Translated by Michael Losonsky. Cambridge Texts in the History of Philosophy. Cambridge: Cambridge University Press, 1999.

Hutton, Charles. *A Course of Mathematics*. 11th ed. 2 vols. London, 1836.

Hyman, Anthony. *Charles Babbage: Pioneer of the Computer*. Princeton, NJ: Princeton University Press, 1982.

IBM. "5 in 5 Video: AI Bias Will Explode; But Only the Unbiased AI Will Survive." IBM Newsroom, August 9, 2021. Video, 1:40. https://newsroom.ibm.com/IBM -research?item=30310.

Ichbiah, J., J. Barnes, R. Firth, and M. Woodger. *Rationale for the Design of the Ada Programming Language*. Cambridge: Cambridge University Press, 1991.

Ifrah, Georges. *The Universal History of Computing: From the Abacus to the Quantum Computer*. Translated by E. F. Harding. New York: Wiley, 2001.

———. *The Universal History of Numbers: From Prehistory to the Invention of the Computer*. Translated by David Bellos. New York: Wiley, 2000.

Isermann, Michael M. "Letters, Sounds and Things: Orthography, Phonetics and Metaphysics in Wilkins's 'Essay' (1668)." *Historiographia Linguistica* 34, no. 2–3 (January 2007): 213–56. https://doi.org/10.1075/hl.34.2.03ise.

———. "Substantial vs Relational Analogy in Sixteenth and Seventeenth-Century Linguistic Thought." In Cram, Linn, and Nowak, *From Classical to Contemporary Linguistics*, 105–12.

Ishiguro, Hidé. *Leibniz's Philosophy of Logic and Language*. 2nd ed. Cambridge: Cambridge University Press, 1990.

Israel, Jonathan. *A Revolution of the Mind: Radical Enlightenment and the Intellectual Origins of Modern Democracy*. Princeton, NJ: Princeton University Press, 2009.

Iverson, Kenneth E. "Notation as a Tool of Thought." *Communications of the ACM* 23, no. 8 (August 1980): 444–65. https://doi.org/10.1145/358896.358899.

———. *A Programming Language*. New York: Wiley, 1962.

Jacquette, Dale. *Frege: A Philosophical Biography*. Cambridge: Cambridge University Press, 2019.

Jajdelska, Elspeth. *Silent Reading and the Birth of the Narrator*. Toronto: University of Toronto Press, 2007.

Janton, Pierre. *Esperanto: Language, Literature, and Community*. Edited by Humphrey Tonkin. Translated by Humphrey Tonkin, Jane Edwards, and Karen Johnson-Weiner. Albany, NY: SUNY Press, 1993.

Jauernig, Anja. "Kant's Critique of the Leibnizian Philosophy: Contra the Leibnizians, but Pro Leibniz." In *Kant and the Early Moderns*, edited by Daniel Garber and Béatrice Longuenesse, 41–63. Princeton, NJ: Princeton University Press, 2008.

Jeffrey, Richard. *Probability and the Art of Judgment*. Cambridge: Cambridge University Press, 1992.

Jesseph, Douglas M. *Berkeley's Philosophy of Mathematics*. Chicago: University of Chicago Press, 1993.

———. "Geometry, Religion and Politics: Context and Consequences of the Hobbes–

Wallis Dispute." *Notes and Records: The Royal Society Journal of the History of Science* 72, no. 4 (December 20, 2018): 469–86. https://doi.org/10.1098/rsnr.2018.0026.

———. "The 'Merely Mechanical' vs. the 'Scab of Symbols': Seventeenth Century Disputes over the Criteria of Mathematical Rigor," in Heeffer and Van Dyck, *Philosophical Aspects of Symbolic Reasoning in Early Modern Mathematics*, 273–88.

———. "Ratios, Quotients, and the Language of Nature." In Gorham et al., *Language of Nature*, 160–77.

———. *Squaring the Circle: The War Between Hobbes and Wallis*. Chicago: University of Chicago Press, 1999.

Jevons, William Stanley. *Pure Logic and Other Minor Works*. Edited by Robert Adamson and Harriet A. Jevons. London: Macmillan, 1890.

Johnson, Anthony. "New Jersey Town to Close Streets to Non-Residents During Rush Hour." ABC7 New York, December 6, 2017. https://abc7ny.com/2747744/.

Jolley, Nicholas. *Leibniz*. London: Routledge, 2005.

———. *Leibniz and Locke: A Study of the New Essays on Human Understanding*. Oxford: Clarendon Press, 1984.

Jones, Matthew L. *The Good Life in the Scientific Revolution: Descartes, Pascal, Leibniz, and the Cultivation of Virtue*. Chicago: University of Chicago Press, 2006.

———. "How We Became Instrumentalists (Again): Data Positivism since World War II." *Historical Studies in the Natural Sciences* 48, no. 5 (November 2018): 673–84. https://doi.org/10.1525/hsns.2018.48.5.673.

———. "Querying the Archive: Data Mining from Apriori to PageRank." In *Science in the Archives: Pasts, Presents, Futures*, edited by Lorraine Daston, 311–28. Chicago: University of Chicago Press, 2017.

———. *Reckoning with Matter: Calculating Machines, Innovation, and Thinking about Thinking from Pascal to Babbage*. Chicago: University of Chicago Press, 2016.

Jones, Richard Foster. *The Triumph of the English Language: A Survey of Opinions Concerning the Vernacular from the Introduction of Printing to the Restoration*. Stanford, CA: Stanford University Press, 1953.

Kafka, Ben. *The Demon of Writing: Powers and Failures of Paperwork*. New York: Zone Books, 2012.

Kaiser, Łukasz, and Ilya Sutskever. "Neural GPUs Learn Algorithms." Cornell University. Last revised March 15, 2016. http://arxiv.org/abs/1511.08228.

Kant, Immanuel. *Critique of Pure Reason*. Translated by Marcus Weigelt. New York: Penguin Books, 2007.

———. *Foundations of the Metaphysics of Morals and What Is Enlightenment?* Translated by Lewis White Beck. New York: Liberal Arts Press, 1959.

Katz, Karin Usadi, and Mikhail G. Katz. "Stevin Numbers and Reality." *Foundations of Science* 17, no. 2 (June 2012): 109–23. https://doi.org/10.1007/s10699-011-9228-9.

Katz, Leo. "A New Status Index Derived from Sociometric Analysis." *Psychometrika* 18, no. 1 (March 1953): 39–43. https://doi.org/10.1007/BF02289026.

Katz, Mikhail G., and David Sherry. "Leibniz's Infinitesimals: Their Fictionality, Their Modern Implementations, and Their Foes from Berkeley to Russell and Beyond." *Erkenntnis* 78, no. 3 (2013): 571–625.

Katz, Victor J., and Karen Hunger Parshall. *Taming the Unknown: A History of Algebra from Antiquity to the Early Twentieth Century*. Princeton, NJ: Princeton University Press, 2014.

Kauffunger, Nicolaus. *Plenaria Arithmetica; Oder, Rechen Buch Auff Linien Vnd Ziffern [. . .].* Cassel, 1647.

Kennedy, Hubert C. *Peano: Life and Works of Giuseppe Peano.* Dordrecht, Neth.: D. Reidel, 1980.

Kittler, Friedrich A. *Discourse Networks 1800/1900.* Translated by Michael Metteer. Stanford, CA: Stanford University Press, 1990.

Kleene, Stephen Cole. *Introduction to Metamathematics.* New York: ISHI, 2009.

———. "Recursive Predicates and Quantifiers." *Transactions of the American Mathematical Society* 53, no. 1 (1943): 41–73. https://doi.org/10.2307/1990131.

Klein, Jacob. *Greek Mathematical Thought and the Origin of Algebra.* Translated by Eva Brann. New York: Dover, 1992.

Kleinberg, Jon M. "Authoritative Sources in a Hyperlinked Environment." *Journal of the ACM* 46, no. 5 (September 1999): 604–32. https://doi.org/10.1145/324133.324140.

Kline, Morris. "Euler and Infinite Series." *Mathematics Magazine* 56, no. 5 (1983): 307–14. https://doi.org/10.2307/2690371.

Knapp, Steven, and Walter Benn Michaels. "Against Theory." *Critical Inquiry* 8, no. 4 (1982): 723–42.

Knowlson, James. *Universal Language Schemes in England and France, 1600–1800.* Toronto: University of Toronto Press, 1975.

Knuth, Donald E. "Algorithmic Thinking and Mathematical Thinking." *American Mathematical Monthly* 92, no. 3 (1985): 170–81.

———. "Ancient Babylonian Algorithms." *Communications of the ACM* 15, no. 7 (1972): 671–77.

———. *The Art of Computer Programming.* 3rd ed. 4 vols. Boston: Addison-Wesley, 2011.

———. "George Forsythe and the Development of Computer Science." *Communications of the ACM* 15, no. 8 (1972): 721–26.

———. *Literate Programming.* Chicago: University of Chicago Press, 1992.

———. "On the Translation of Languages from Left to Right." *Information and Control* 8, no. 6 (December 1965): 607–39.

Knuth, Donald E., and Luis Trabb Pardo. *The Early Development of Programming Languages.* [Stanford, CA]: Computer Science Department, Stanford University, 1976. https://apps.dtic.mil/sti/citations/ADA032123.

Koetsier, Teun, and Karin Reich. "Michael Stifel and His Numerology." In *Mathematics and the Divine*, edited by T. Koetsier and L. Bergmans, 291–310. Amsterdam: Elsevier Science, 2005. https://doi.org/10.1016/B978-044450328-2/50017-5.

Koppelman, Elaine. "The Calculus of Operations and the Rise of Abstract Algebra." *Archive for History of Exact Sciences* 8, no. 3 (1971): 155–242.

Krishna, Arvind. "IBM CEO's Letter to Congress on Racial Justice Reform." *THINK-Policy Blog*, IBM, June 8, 2020. https://www.ibm.com/blogs/policy/facial -recognition-sunset-racial-justice-reforms/.

Kuhn, Thomas S. *The Structure of Scientific Revolutions: 50th Anniversary Edition.* Chicago: University of Chicago Press, 2012.

Kuskey, Jessica. "Math and the Mechanical Mind: Charles Babbage, Charles Dickens, and Mental Labor in *Little Dorrit*." *Dickens Studies Annual: Essays on Victorian Fiction* 45, no. 1 (July 14, 2014): 247–74. https://doi.org/10.7756/dsa.045.012/247 -274.

Lacroix, Silvestre-François. *An Elementary Treatise on the Differential and Integral Calculus*. Translated by Charles Babbage, George Peacock, and John Frederick William Herschel. Cambridge, 1816.

Lærke, Mogens. "Leibniz, the Encyclopedia, and the Natural Order of Thinking." *Journal of the History of Ideas* 75, no. 2 (April 2014): 237–59. https://doi.org/10.1353/jhi.2014.0009.

Lagrange, Joseph-Louis. *Lectures on Elementary Mathematics*. Translated by Thomas J. McCormack. Chicago: Open Court, 1898.

———. *Œuvres de Lagrange*. 14 vols. Paris: Gauthier-Villars, 1867.

———. *Théorie des fonctions analytiques*. Paris: De l'imprimerie de la république, 1797.

———. *Théorie des fonctions analytiques*. 2nd ed. Paris: Courcier, 1813.

Laita, Luis M. "Boolean Algebra and Its Extra-Logical Sources: The Testimony of Mary Everest Boole." *History and Philosophy of Logic* 1, no. 1–2 (1980): 37–60. https://doi.org/10.1080/01445348008837004.

———. "The Influence of Boole's Search for a Universal Method in Analysis on the Creation of His Logic." In Gasser, *Boole Anthology*, 45–57.

Lambert, Kevin. "A Natural History of Mathematics: George Peacock and the Making of English Algebra." *Isis* 104, no. 2 (June 2013): 278–302. https://doi.org/10.1086/670948.

Lamy, Bernard. *The Art of Speaking*. In *The Rhetorics of Thomas Hobbes and Bernard Lamy*. Edited by John T. Harwood. Carbondale, IL: SIU Press, 1986.

Landin, P. J. "The Mechanical Evaluation of Expressions." *Computer Journal* 6, no. 4 (January1964): 308–20. https://doi.org/10.1093/comjnl/6.4.308.

Lantz, Johann. *Institutionum arithmeticarum libri quatuor [. . .]*. Munich, 1616.

Laplace, Pierre-Simon marquis de. *Essai philosophique sur les probabilités*. Paris: Courcier, 1814.

Larson, Max. "Optimizing Chess: Philology and Algorithmic Culture." *Diacritics* 46, no. 1 (2018): 30–53. https://doi.org/10.1353/dia.2018.0001.

Latham, Robert Gordon. *First Outlines of Logic, Applied to Grammar and Etymology*. London: Taylor and Walton, 1847.

Laugwitz, Detlef. "Definite Values of Infinite Sums: Aspects of the Foundations of Infinitesimal Analysis around 1820." *Archive for History of Exact Sciences* 39, no. 3 (September 1989): 195–245. https://doi.org/10.1007/BF00329867.

Leibniz, Gottfried Wilhelm. "De Geometria Recondita [. . .]." *Acta Eruditorum*, June 1686, 292–300.

———. "De la chainette [. . .]." *Journal des sçavans*, March 31, 1692, 147–53.

———. *Dissertation on Combinatorial Art*. Translated by Massimo Mugnai, Han van Ruler, and Martin Wilson. Oxford: Oxford University Press, 2020.

———. *The Early Mathematical Manuscripts of Leibniz*. Translated by J. M. Child. Mineola, NY: Dover, 2005.

———. "Explication de l'arithmetique binaire [. . .]." *Memoires de l'Académie Royale Des Sciences* (1703): 110–16.

———. "Leibniz on His Calculating Machine." Translated by Mark Kormes. In *A Source Book in Mathematics*, edited by David Eugene Smith, 173–81. New York: McGraw-Hill, 1929.

———. "Leibniz on the Calculus." Translated by Evelyn Walker. In *A Source Book in Mathematics*, edited by David Eugene Smith, 619–26. New York: McGraw-Hill, 1929.

———. *Logical Papers: A Selection*. Edited by George Henry Radcliffe Parkinson. Oxford: Clarendon, 1966.

———. *New Essays on Human Understanding*. Translated by Peter Remnant and Jonathan Bennett. Cambridge: Cambridge University Press, 1981.

———. "Nova Methodus pro Maximis et Minimis." *Acta Eruditorum*, October 1684, 467–73.

———. *Philosophical Papers and Letters*. Edited by Leroy E. Loemker. Dordrecht, Neth.: D. Reidel, 1970.

———. *Philosophical Texts*. Translated by R. S. Woolhouse and Richard Francks. Oxford Philosophical Texts. Oxford: Oxford University Press, 1998.

———. *Sämtliche Schriften und Briefe*. 56 vols. Berlin: Akademie Verlag, 1998.

———. *Selections*. Edited by Philip P. Wiener. New York: Charles Scribner's Sons, 1979.

Lennon, Brian. "Machine Translation: A Tale of Two Cultures." In *A Companion to Translation Studies*, edited by Sandra Bermann and Catherine Porter, 133–46. Chichester, UK: John Wiley & Sons, 2014.

———. *Passwords: Philology, Security, Authentication*. Cambridge, MA: Harvard University Press, 2018.

Lenoir, Timothy, ed. *Inscribing Science: Scientific Texts and the Materiality of Communication*. Stanford, CA: Stanford University Press, 1998.

Lenzen, Wolfgang. "Arithmetical vs. 'Real' Addition: A Case Study of the Relation between Logic, Mathematics, and Metaphysics in Leibniz." In *Leibnizian Inquiries: A Group of Essays*, edited by Nicholas Rescher, 149–57. CPS Publications in Philosophy of Science. Lanham, MD: University Press of America, 1989.

Lewis, C. I. *A Survey of Symbolic Logic*. Berkeley: University of California Press, 1918.

Lewis, Rhodri. *Language, Mind and Nature: Artificial Languages in England from Bacon to Locke*. Cambridge: Cambridge University Press, 2007.

Lifschitz, Avi S. "Translation in Theory and Practice: The Case of Johann David Michaelis's Prize Essay on Language and Opinions." In *Cultural Transfer through Translation: The Circulation of Enlightened Thought in Europe by Means of Translation*, edited by Stefanie Stockhorst, 29–44. Amsterdam: Rodopi, 2010.

Lindsey, C. H. "A History of ALGOL 68." In *History of Programming Languages—II*, edited by Thomas J. Bergin Jr. and Richard G. Gibson Jr., 27–96. New York: ACM Press, 1996. https://doi.org/10.1145/234286.1057810.

Liskov, Barbara, and Stephen Zilles. "Programming with Abstract Data Types." In *Proceedings of the ACM SIGPLAN Symposium on Very High Level Languages*, 50–59. New York: Association for Computing Machinery, 1974. https://doi.org/10.1145/800233.807045.

Locke, John. *Conduct of the Understanding*. Edited by Thomas Fowler. Oxford: Clarendon Press, 1901.

———. *An Essay Concerning Human Understanding*. Edited by Roger Woolhouse. Penguin Classics. London: Penguin Books, 2004.

———. *Some Thoughts Concerning Education*. Edited by Robert Herbert Quick. Cambridge: Cambridge University Press, 1889.

Loemker, Leroy E. "Leibniz and the Herborn Encyclopedists." *Journal of the History of Ideas* 22, no. 3 (1961): 323–38. https://doi.org/10.2307/2708128.

Long, Pamela O. *Artisan/Practitioners and the Rise of the New Sciences, 1400–1600*. Corvallis: Oregon State University Press, 2011.

Lovelace, Ada King, Countess of. *Ada, the Enchantress of Numbers: A Selection from the Letters of Lord Byron's Daughter and Her Description of the First Computer.* Edited by Betty Alexandra Toole. Mill Valley, CA: Strawberry Press, 1992.

Maas, Harro. "Mechanical Rationality: Jevons and the Making of Economic Man." *Studies in History and Philosophy of Science Part A* 30, no. 4 (December 1999): 587–619. https://doi.org/10.1016/S0039-3681(99)00030-8.

Maat, Jaap. *Philosophical Languages in the Seventeenth Century: Dalgarno, Wilkins, Leibniz.* Dordrecht, Neth.: Kluwer Academic Publishers, 2004.

MacHale, Desmond. *The Life and Work of George Boole: A Prelude to the Digital Age.* Cork, Ire.: Cork University Press, 2014.

MacLaurin, Colin. *A Treatise of Fluxions: In Two Books.* 2 vols. Edinburgh, 1742.

Mahoney, Michael Sean. "The History of Computing in the History of Technology." *Annals of the History of Computing* 10, no. 2 (1988): 113–23.

———. *The Mathematical Career of Pierre de Fermat, 1601–1665.* 2nd ed. Princeton, NJ: Princeton University Press, 1994.

Maimieux, Joseph de. *Pasigraphie, ou Premiers élémens du nouvel art-science d'écrire ou imprimer en une langue de manière à être lu et entendu dans toute autre langue sans traduction [. . .].* Paris: Bureau de la Pasigraphie, 1797. https://gallica.bnf.fr/ark:/12148/bpt6k5401094b.

Malcolm, Noel, and Jacqueline Stedall. *John Pell (1611–1685) and His Correspondence with Sir Charles Cavendish: The Mental World of an Early Modern Mathematician.* Oxford: Oxford University Press, 2005.

Mancosu, Paolo. "The Metaphysics of the Calculus: A Foundational Debate in the Paris Academy of Sciences, 1700–1706." *Historia Mathematica* 16, no. 3 (August 1989): 224–48. https://doi.org/10.1016/0315-0860(89)90019-0.

———. *Philosophy of Mathematics and Mathematical Practice in the Seventeenth Century.* Oxford: Oxford University Press, 1996.

Manjavacas, Enrique. "MacBERTh." Accessed February 21, 2022. https://github.com/emanjavacas/macberth-eval.

Manuel, Frank Edward. *The Prophets of Paris: Turgot, Condorcet, Saint-Simon, Fourier, and Comte.* Harper Torchbook. New York: Harper & Row, 1965.

Marcovitz, Alan B., and Earl Justin Schweppe. *An Introduction to Algorithmic Methods Using the MAD Language.* New York: Macmillan, 1966.

Marino, Mark C. *Critical Code Studies.* Cambridge, MA: MIT Press, 2020.

Markley, Robert. *Fallen Languages: Crises of Representation in Newtonian England, 1660–1740.* Ithaca, NY: Cornell University Press, 1993.

Markov, A. A. Jr. *Theory of Algorithms.* Translated by Jacques J. Schorr-Kon and PST Staff. Washington, DC: Israel Program for Scientific Translations, 1961.

———. "The Theory of Algorithms." Translated by Edwin Hewitt. *American Mathematical Society Translations,* 2nd ser., 15 (1960): 1–14.

Marshall, John. "Locke, Socinianism, 'Socinianism,' and Unitarianism." In *English Philosophy in the Age of Locke,* edited by M. A. Stewart, 111–82. Oxford Studies in the History of Philosophy 3. Oxford: Clarendon Press, 2000.

Martin, Gottfried. *Arithmetic and Combinatorics: Kant and His Contemporaries.* Translated by Judy Wubnig. Carbondale, IL: SIU Press, 1985.

Maseres, Francis. *A Collection of Several Commissions, and Other Public Instruments [. . .].* London, 1772.

———. *Dissertation on the Use of the Negative Sign in Algebra [. . .].* London, 1758.

https://catalog.lindahall.org/permalink/01LINDAHALL_INST/1nrd3ls/
alma991430113405961.

———. *Occasional Essays on Various Subjects: Chiefly Political and Historical [. . .].*
London, 1809.

———. *The Principles of the Doctrine of Life-Annuities [. . .].* London, 1783.

———. *A Proposal for Establishing Life-Annuities in Parishes for the Benefit of the Industrious Poor.* London, 1772.

Mason, William Albert. *A History of the Art of Writing.* New York: Macmillan, 1920.

Massa Esteve, Mª Rosa. "Symbolic Language in Early Modern Mathematics: The Algebra of Pierre Hérigone (1580–1643)." *Historia Mathematica* 35, no. 4 (November 2008): 285–301. https://doi.org/10.1016/j.hm.2008.05.003.

Mates, Benson. *The Philosophy of Leibniz: Metaphysics and Language.* New York: Oxford University Press, 1986.

Mattheier, Klaus J. "German." In *Germanic Standardizations: Past to Present,* edited by Ana Deumert and Wim Vandenbussche, 211–44. Amsterdam: John Benjamins, 2003.

Maupertuis, Pierre Louis Moreau de. *Réflexions philosophiques sur l'origine des langues et la signification des mots.* [Paris, 1748]. http://gallica.bnf.fr/ark:/12148/btv1b8626902w.

Mazlish, Bruce. *James and John Stuart Mill.* Abingdon, UK: Routledge, 2017.

McCarthy, J. "Towards a Mathematical Science of Computation." In *Program Verification: Fundamental Issues in Computer Science,* edited by Timothy R. Colburn, James H. Fetzer, and Terry L. Rankin, 35–56. Studies in Cognitive Systems. Dordrecht, Neth.: Springer Science+Business Media, 1993. https://doi.org/10.1007/978-94-011-1793-7_2.

McCulloch, Warren S., and Walter Pitts. "A Logical Calculus of the Ideas Immanent in Nervous Activity." *Bulletin of Mathematical Biophysics* 5, no. 4 (December 1943): 115–33.

McMahon, William E. "The Semantics of Johann Alsted." In Cram, Linn, and Nowak, *From Classical to Contemporary Linguistics,* 123–29.

———. "The Semantics of Post-Medieval Lullism." *Henry Sweet Society for the History of Linguistic Ideas Bulletin* 38, no. 1 (May 2002): 43–53. https://doi.org/10.1080/02674971.2002.11745552.

Mentor, The [Lloyd Blankenship]. "The Conscience of a Hacker." Internet Archive. Published April 24, 2005. http://archive.org/details/The_Conscience_of_a_Hacker.

Meryon, Charles Lewis. *Memoirs of the Lady Hester Stanhope.* 3 vols. London: Henry Colburn, 1845.

Metzler, Donald, Yi Tay, Dara Bahri, and Marc Najork. "Rethinking Search: Making Domain Experts Out of Dilettantes." *ACM SIGIR Forum* 55, no. 1 (June 2021): 1–27. https://doi.org/10.1145/3476415.3476428.

Michaelis, Johann David. *De l'influence des opinions sur le langage et du langage sur les opinions [. . .].* Brême: G. L. Förster, 1762. http://gallica.bnf.fr/ark:/12148/bpt6k75491v.

———. *A Dissertation on the Influence of Opinions on Language, and of Language on Opinions [. . .].* 2nd ed. London, 1771.

Mill, John Stuart. *Autobiography.* London: Longmans, Green, Reader, and Dyer, 1873.

———. *Mill on Bentham and Coleridge.* Edited by F. R. Leavis. Westport, CT: Greenwood, 1983.

———. *A System of Logic, Ratiocinative and Inductive.* 2 vols. London: John W. Parker, 1843.

Minsky, Marvin, and Seymour A. Papert. *Perceptrons: An Introduction to Computational Geometry.* Cambridge, MA: MIT Press, 2017.

Misa, Thomas J., ed. *Communities of Computing: Computer Science and Society in the ACM.* New York: Morgan & Claypool, 2016.

"Miscellaneous Programs." Dick Grune, n.d. https://dickgrune.com/CS/Algol68/MiscellaneousPrograms.

Miyazaki, Shintaro. "Algorhythmics: Understanding Micro-Temporality in Computational Cultures." *Computational Culture,* no. 2 (September 28, 2012). http://computationalculture.net/algorhythmics-understanding-micro-temporality-in-computational-cultures/.

Monboddo, Lord James Burnett. *Of the Origin and Progress of Language.* 2nd ed. 6 vols. Edinburgh, 1774.

Montferrier, A. S. de. *Encyclopédie mathématique ou exposition complète de toutes les branches des mathématiques d'après les principes de la philosophie des mathématiques de Hoëné Wronski.* 4 vols. Paris: Amyot, 1856.

Moore, Jonas. *Moores Arithmetick Discovering the Secrets of That Art in Numbers and Species [. . .].* London, 1650. Early English Books, 1641–1700, microfilm reel position 112:04.

Morris, F. K., and C. B. Jones. "An Early Program Proof by Alan Turing." *Annals of the History of Computing* 6, no. 2 (April 1984): 139–43.

Mounier-Kuhn, Pierre. "Algol in France: From Universal Project to Embedded Culture." *IEEE Annals of the History of Computing* 36, no. 4 (December 2014): 6–25. https://doi.org/10.1109/MAHC.2014.50.

Nachtomy, Ohad. "Leibniz and Russell." In Phemister and Brown, *Leibniz and the English-Speaking World,* 207–18.

Napier, John. *Mirifici logarithmorum canonis descriptio ejusque usus, in utraque trigonometria [. . .].* Edinburgh, 1614. Early English Books, 1475–1640, microfilm reel position 1147:03.

Naur, Peter, and Brian Randell, eds. *Software Engineering: Report on a Conference Sponsored by the NATO Science Committee, Garmisch, Germany, 7th to 11th October 1968.* Published January 1969. http://homepages.cs.ncl.ac.uk/brian.randell/NATO/nato1968.PDF.

———. *Software Engineering Techniques: Report on a Conference Sponsored by the NATO Science Committee, Rome, Italy, 27th to 31st October 1969.* Published April 1970. http://homepages.cs.ncl.ac.uk/brian.randell/NATO/nato1969.PDF.

Neal, Katherine. *From Discrete to Continuous: The Broadening of Number Concepts in Early Modern England.* Dordrecht, Neth.: Springer Science+Business Media, 2002.

Nesselmann, Georg Heinrich Ferdinand. *Versuch einer kritischen Geschichte der Algebra.* Berlin: G. Reimer, 1842.

Newton, Isaac. *The Method of Fluxions and Infinite Series: With Its Application to the Geometry of Curve-Lines.* Edited by John Colson. London, 1736.

———. *Universal Arithmetick: Or, A Treatise of Arithmetical Composition and Resolution [. . .].* Translated by Joseph Ralphson. 2nd ed. London, 1728.

Nilsson, Nils J. *The Quest for Artificial Intelligence.* Cambridge: Cambridge University Press, 2010.

Nisbet, H. B. *Herder and Scientific Thought*. [Cambridge]: Cambridge Modern Humanities Research Association, 1970.

Noble, Safiya Umoja. *Algorithms of Oppression: How Search Engines Reinforce Racism*. New York: NYU Press, 2018.

Nofre, David, Mark Priestley, and Gerard Alberts. "When Technology Became Language: The Origins of the Linguistic Conception of Computer Programming, 1950–1960." *Technology and Culture* 55, no. 1 (January 2014): 40–75. https://doi.org/10.1353/tech.2014.0031.

Norton, Robert Edward. *Herder's Aesthetics and the European Enlightenment*. Ithaca, NY: Cornell University Press, 1991.

Nöth, Winfried. *Handbook of Semiotics*. Advances in Semiotics. Bloomington: Indiana University Press, 1990.

Novalis. *Notes for a Romantic Encyclopaedia: Das Allgemeine Brouillon*. Translated by David W. Wood. Albany, NY: SUNY Press, 2012.

———. *Philosophical Writings*. Translated by Margaret Mahoney Stoljar. Albany, NY: SUNY Press, 1997.

Oaks, Jeffrey A. "François Viète's Revolution in Algebra." *Archive for History of Exact Sciences* 72, no. 3 (May 2018): 245–302. https://doi.org/10.1007/s00407-018-0208-0.

Ohm, Martin. *The Spirit of Mathematical Analysis, and Its Relation to a Logical System*. Translated by Alexander John Ellis. London: John W. Parker, 1843.

Olszewski, Adam, Jan Wolenski, and Robert Janusz, eds. *Church's Thesis after 70 Years*. Berlin: Walter de Gruyter, 2006.

O'Neil, Cathy. *Weapons of Math Destruction: How Big Data Increases Inequality and Threatens Democracy*. New York: Crown, 2016.

O'Neill, Timothy Michael. *Ideography and Chinese Language Theory: A History*. Berlin: Walter de Gruyter GmbH, 2016.

Ong, Walter J. *Ramus, Method, and the Decay of Dialogue: From the Art of Discourse to the Art of Reason*. Cambridge, MA: Harvard University Press, 1958.

Opitz, Martin. *Buch von der Deutschen Poeterei*. Neudrucke Deutscher Litteraturwerke des XVI. und XVII. Jahrhunderts 1. Halle: M. Niemeyer, 1913.

Otis, Jessica. "'Set Them to the Cyphering Schoole': Reading, Writing, and Arithmetical Education, circa 1540–1700." *Journal of British Studies* 56, no. 3 (July 2017): 453–82. https://doi.org/10.1017/jbr.2017.59.

Ottenheimer, Harriet Joseph, and Judith M. S. Pine. *The Anthropology of Language: An Introduction to Linguistic Anthropology*. Boston: Cengage Learning, 2019.

Oughtred, William. *The Circles of Proportion and the Horizontall Instrument [. . .]*. Translated by William Forster. London, 1632. Early English Books, 1475–1640, microfilm reel position 1322:17, 1323:01.

———. *The Key of the Mathematicks New Forged and Filed [. . .]*. London, 1647. Early English Books, 1641–1700, microfilm reel position 1023:12.

Padley, G. A. *Grammatical Theory in Western Europe, 1500–1700: The Latin Tradition*. Cambridge: Cambridge University Press, 1976.

———. *Grammatical Theory in Western Europe, 1500–1700: Trends in Vernacular Grammar*. 2 vols. Cambridge: Cambridge University Press, 1985.

Page, Lawrence, Sergei Brin, Rajeev Motwani, and Terry Winograd. "The Page-Rank Citation Ranking: Bringing Order to the Web." Stanford University InfoLab.

Published 1999. Last modified December 28, 2008. http://ilpubs.stanford.edu: 8090/422/.

Panaccio, Claude. *Mental Language: From Plato to William of Ockham*. Translated by Joshua P. Hochschild and Meredith K. Ziebart. New York: Fordham University Press, 2017.

Pantin, Isabelle. "La représentation des mathématiques chez Jacques Peletier Du Mans: Cosmos hiéroglyphique ou ordre rhétorique?" *Rhetorica: A Journal of the History of Rhetoric* 20, no. 4 (2002): 375–89.

Parshall, Karen Hunger. "A Plurality of Algebras, 1200–1600: Algebraic Europe from Fibonacci to Clavius." *BSHM Bulletin: Journal of the British Society for the History of Mathematics* 32, no. 1 (2017): 2–16. https://doi.org/10.1080/17498430.2016.1225340.

Pasquale, Frank. *The Black Box Society: The Secret Algorithms That Control Money and Information*. Cambridge, MA: Harvard University Press, 2015.

Payette, Sandy. "Hopper and Dijkstra: Crisis, Revolution, and the Future of Programming." *IEEE Annals of the History of Computing* 36, no. 4 (October–December 2014): 64–73. https://doi.org/10.1109/MAHC.2014.54.

Peacock, George. *A Treatise on Algebra*. Cambridge: J. & J. J. Deighton, 1830.

———. *A Treatise on Algebra*. 2nd ed. Cambridge: J. & J. J. Deighton, 1842.

Peano, Giuseppe. *Arithmetices principia: Nova methodo*. Rome: Fratres Bocca, 1889.

———. *Selected Works of Giuseppe Peano*. Edited by Hubert C. Kennedy. Toronto: University of Toronto Press, 1973.

Peckhaus, Volker. "Ernst Schröder on Pasigraphy." *Revue d'histoire des sciences* 67, no. 2 (2014): 207–30.

———. "19th Century Logic Between Philosophy and Mathematics." *Bulletin of Symbolic Logic* 5, no. 4 (1999): 433–50. https://doi.org/10.2307/421117.

The Penny Cyclopaedia of the Society for the Diffusion of Useful Knowledge. 27 vols. London: Charles Knight, 1833–44.

Perlis, Alan J. "A New Policy for Algorithms?" *Communications of the ACM* 9, no. 4 (April 1966): 255. https://doi.org/10.1145/365278.365293.

Pesic, Peter. "Hearing the Irrational: Music and the Development of the Modern Concept of Number." *Isis* 101, no. 3 (September 2010): 501–30. https://doi.org/10.1086/655790.

———. *Music and the Making of Modern Science*. Cambridge, MA: MIT Press, 2014.

———. "Secrets, Symbols, and Systems: Parallels between Cryptanalysis and Algebra, 1580–1700." *Isis* 88, no. 4 (1997): 674–92.

Phemister, Pauline, and Stuart Brown. *Leibniz and the English-Speaking World*. Dordrecht, Neth.: Springer, 2007.

Phillips, Richard. *Public Characters of 1800–1801*. Dublin, 1801.

Pierpont, James. "Weber's Algebra." *Bulletin of the American Mathematical Society* 4 (February 1898): 200–234.

Pietarinen, Ahti-Veikko. "Significs and the Origins of Analytic Philosophy." *Journal of the History of Ideas* 70, no. 3 (2009): 467–90.

Plato. *Plato's Phaedrus*. Translated and with introduction and commentary by R. Hackforth. Cambridge: Cambridge University Press, 1952.

Plofker, Kim. *Mathematics in India*. Princeton, NJ: Princeton University Press, 2009.

Poe, Edgar Allan. *Collected Works of Edgar Allan Poe*. Edited by Thomas Ollive Mabbott. 3 vols. Cambridge, MA: Belknap Press of Harvard University Press, 1969.

Pombo, Olga. *Leibniz and the Problem of a Universal Language*. Münster, Ger.: Nodus, 1987.

Poole, William Frederick. *Anti-Slavery Opinions before the Year 1800*. Cincinnati: Robert Clarke, 1873.

Poovey, Mary. *A History of the Modern Fact: Problems of Knowledge in the Sciences of Wealth and Society*. Chicago: University of Chicago Press, 1998.

Porter, Sarah. *Conversations on Arithmetic*. London: Charles Knight, 1835.

Porter, Theodore M. *The Rise of Statistical Thinking, 1820–1900*. Princeton, NJ: Princeton University Press, 1986.

Post, Emil L. "Finite Combinatory Processes—Formulation 1." *Journal of Symbolic Logic* 1, no. 3 (September 1936): 103–5.

Powell, Barry B. *Writing: Theory and History of the Technology of Civilization*. Chichester, UK: Wiley-Blackwell, 2009.

Priestley, Mark. *A Science of Operations: Machines, Logic and the Invention of Programming*. History of Computing. London: Springer-Verlag, 2011.

Pycior, Helena M. "George Peacock and the British Origins of Symbolical Algebra." *Historia Mathematica* 8, no. 1 (February 1981): 23–45. https://doi.org/10.1016/0315-0860(81)90003-3.

———. *Symbols, Impossible Numbers, and Geometric Entanglements: British Algebra through the Commentaries on Newton's Universal Arithmetick*. Cambridge: Cambridge University Press, 1997.

Rabouin, David. "'Analytica Generalissima Humanorum Cognitionum': Some Reflections on the Relationship between Logical and Mathematical Analysis in Leibniz." *Studia Leibnitiana* 45, no. 1 (2013): 109–30.

———. "Leibniz's Rigorous Foundations of the Method of Indivisibles." In *Seventeenth-Century Indivisibles Revisited*, edited by Vincent Jullien, 347–64. Cham, Switz.: Springer International, 2015.

Radford, Alec, Karthik Narasimhan, Tim Salimans, and Ilya Sutskever. *Improving Language Understanding by Generative Pre-Training*. OpenAI, 2018. https://cdn.openai.com/research-covers/language-unsupervised/language_understanding_paper.pdf.

Radford, Alec, Jeffrey Wu, Dario Amodei, Daniela Amodei, Jack Clark, Miles Brundage, Ilya Sutskever, Amanda Askell, David Lansky, and Danny Hernandez. "Better Language Models and Their Implications." *OpenAI* (blog), February 14, 2019. https://openai.com/blog/better-language-models/.

Radford, Alec, Jeffrey Wu, Rewon Child, David Luan, Dario Amodei, and Ilya Sutskever. *Language Models Are Unsupervised Multitask Learners*. OpenAI, 2019. https://cdn.openai.com/better-language-models/language_models_are_unsupervised_multitask_learners.pdf.

Raffel, Colin, Noam Shazeer, Adam Roberts, Katherine Lee, Sharan Narang, Michael Matena, Yanqi Zhou, Wei Li, and Peter J. Liu. "Exploring the Limits of Transfer Learning with a Unified Text-to-Text Transformer." Cornell University. Last revised July 28, 2020. http://arxiv.org/abs/1910.10683.

Raji, Inioluwa Deborah, Emily M. Bender, Amandalynne Paullada, Emily Denton, and Alex Hanna. "AI and the Everything in the Whole Wide World Benchmark." Cornell University. Published November 26, 2021. https://arxiv.org/abs/2111.15366.

Rajpurkar, Pranav, Robin Jia, and Percy Liang. "Know What You Don't Know: Un-

answerable Questions for SQuAD." Cornell University. Published June 11, 2018.
http://arxiv.org/abs/1806.03822.

Rajpurkar, Pranav, Jian Zhang, Konstantin Lopyrev, and Percy Liang. "SQuAD:
100,000+ Questions for Machine Comprehension of Text." Cornell University. Last
revised October 10, 2016. http://arxiv.org/abs/1606.05250.

Ranshaw, Russell W. "Certification of Algorithm 23: MATHSORT." *Communications of
the ACM* 4, no. 5 (May 1961): 238. https://doi.org/10.1145/366532.366568.

Raymond, Eric S. "Holy Wars." In *The Jargon File (version 4.4.7)*, December 29, 2003.
http://www.catb.org/jargon/html/H/holy-wars.html.

Recorde, Robert. *The Whetstone of Witte Whiche Is the Seconde Parte of Arithmetike
[. . .]*. London, 1557. Early English Books, 1475–1640, microfilm reel position
550:13.

Redding, Paul. "Mathematics, Computation, Language, and Poetry: The Novalis
Paradox." In *The Relevance of Romanticism: Essays on German Romantic Philosophy*,
edited by Dalia Nassar, 221–38. New York: Oxford University Press, 2014.

Reitan, Earl. *Politics, Finance, and the People: Economic Reform in England in the Age of
the American Revolution, 1770–92*. Basingstoke, UK: Palgrave Macmillan, 2007.

Richards, Joan L. *Mathematical Visions: The Pursuit of Geometry in Victorian England*.
Waltham, MA: Academic Press, 1988.

———. "Rigor and Clarity: Foundations of Mathematics in France and England, 1800–
1840." *Science in Context* 4, no. 2 (October 1991): 297–319. https://doi.org/10.1017/
S0269889700000983.

Rieder, Bernhard. *Engines of Order: A Mechanology of Algorithmic Techniques*. Recur-
sions: Theories of Media, Materiality, and Cultural Techniques. Amsterdam: Amsterdam University Press, 2020.

Riskin, Jessica. "Rival Idioms for a Revolutionized Science and a Republican Citizenry."
Isis 89, no. 2 (1998): 203–32.

Robinson, Abraham. *Non-Standard Analysis*. Rev. ed. Princeton, NJ: Princeton University Press, 2016.

Rodal, Jocelyn. "Patterned Ambiguities: Virginia Woolf, Mathematical Variables, and
Form." *Configurations* 26, no. 1 (2018): 73–101.

Rodet, Léon. *Sur les notations numériques et algébriques antérieurement au XVIe siècle*.
Paris: Ernest Leroux, Éditeur, 1881.

Rosenfeld, Sophia A. *A Revolution in Language: The Problem of Signs in Late Eighteenth-
Century France*. Stanford, CA: Stanford University Press, 2001.

Rothschild, Emma. *Economic Sentiments: Adam Smith, Condorcet, and the Enlightenment*. 2nd ed. Cambridge, MA: Harvard University Press, 2002.

Rotman, Brian. *Ad Infinitum: The Ghost in Turing's Machine; Taking God Out of Mathematics and Putting the Body Back In; An Essay in Corporeal Semiotics*. Stanford, CA:
Stanford University Press, 1993.

Rousseau, Jean-Jacques. "Discourse on the Arts and Sciences." In *The Discourses and
Other Early Political Writings*, translated by Victor Gourevitch, 1–28. Cambridge:
Cambridge University Press, 1997.

Rousseau, Jean-Jacques, and Johann Gottfried Herder. *On the Origin of Language*.
Translated by John H. Moran and Alexander Gode. Chicago: University of Chicago
Press, 1966.

Rudolff, Christoff. *Behend unnd hubsch Rechnung durch die kunstreichen regeln Algebre,
so gemeincklich die Coss genennt werden [. . .]*. Strasbourg, 1525.

Rumelhart, David E., Geoffrey E. Hinton, and Ronald J. Williams. "Learning Representations by Back-Propagating Errors." *Nature* 323 (October 1986): 533–36.

Rusnock, Paul. "Philosophy of Mathematics: Bolzano's Responses to Kant and Lagrange." *Revue d'histoire des sciences* 52, no. 3/4 (1999): 399–427.

Russell, Bertrand. *A Critical Exposition of the Philosophy of Leibniz, with an Appendix of Leading Passages.* Cambridge: Cambridge University Press, 1900.

Rutherford, Donald. "Philosophy and Language in Leibniz." In *The Cambridge Companion to Leibniz*, edited by Nicholas Jolley, 224–69. Cambridge: Cambridge University Press, 1994.

Rutishauser, Heinz. "Some Programming Techniques for the ERMETH." *Journal of the ACM* 2, no. 1 (January 1955): 1–4.

Sack, Warren. *The Software Arts.* Cambridge, MA: MIT Press, 2019.

Sadler, John Edward. *J. A. Comenius and the Concept of Universal Education.* Routledge Library Editions: History of Education 32. London: Routledge, 2013.

Saenger, Paul. *Space between Words: The Origins of Silent Reading.* Stanford, CA: Stanford University Press, 1997.

Sala-Molins, Louis. *Dark Side of the Light: Slavery and the French Enlightenment.* Translated by John Conteh-Morgan. Minneapolis: University of Minnesota Press, 2006.

Salmon, Vivian. *The Study of Language in 17th-Century England.* 2nd ed. Amsterdam: John Benjamins, 1988.

———. *The Works of Francis Lodwick: A Study of His Writings in the Intellectual Context of the Seventeenth Century.* London: Longman, 1972.

Sammet, Jean E. "The Early History of COBOL." In Wexelblat, *History of Programming Languages*, 199–243.

———. "The Use of English as a Programming Language." *Communications of the ACM* 9, no. 3 (March 1966): 228–30.

Sarton, George. "Simon Stevin of Bruges (1548–1620)." *Isis* 21, no. 2 (1934): 241–303.

Sasaki, Chikara. *Descartes's Mathematical Thought.* Dordrecht, Neth.: Springer Science+Business Media, 2003.

Saussure, Ferdinand de. *Course in General Linguistics.* Edited by Charles Bally, Albert Sechehaye, and Albert Riedlinger. Translated by Roy Harris. Chicago: Open Court, 1986.

Schaffer, Simon. "Babbage's Intelligence: Calculating Engines and the Factory System." *Critical Inquiry* 21, no. 1 (1994): 203–27.

Schick, Timo, and Hinrich Schütze. "It's Not Just Size That Matters: Small Language Models Are Also Few-Shot Learners." Cornell University. Last revised April 12, 2021. http://arxiv.org/abs/2009.07118.

Schlutz, Alexander M. *Mind's World: Imagination and Subjectivity from Descartes to Romanticism.* Seattle: University of Washington Press, 2010.

Schmidt, Benjamin M. "Do Digital Humanists Need to Understand Algorithms?" In *Debates in the Digital Humanities 2016*, edited by Matthew K. Gold and Lauren F. Klein, 546–55. Minneapolis: University of Minnesota Press, 2016. https://doi.org/10.5749/j.cttlcn6thb.51.

Schmitter, Amy M. "Mind and Sign: Method and the Interpretation of Mathematics in Descartes's Early Work." *Canadian Journal of Philosophy* 30, no. 3 (2000): 371–411.

Schröder, Ernst. *On the Formal Elements of the Absolute Algebra.* Edited by Davide Bondoni. Milan: LED Edizioni Universitarie, 2012.

————. "Ueber Algorithmen und Calculn." *Archiv der Mathematik und Physik* 2, no. 5 (1887): 225–78.

Schubring, Gert. *Conflicts between Generalization, Rigor, and Intuition: Number Concepts Underlying the Development of Analysis in 17th–19th Century France and Germany*. New York: Springer Science+Business Media, 2005.

Searle, John R. "Minds, Brains, and Programs." *Behavioral and Brain Sciences* 3 (1980): 417–57.

Seaver, Nick. "Algorithms as Culture: Some Tactics for the Ethnography of Algorithmic Systems." *Big Data & Society* 4, no. 2 (July–December 2017): 1-12.

Serfati, Michel. *La révolution symbolique: La constitution de l'écriture symbolique mathématique*. Paris: Pétra, 2005.

————. "Mathématiques et pensée symbolique chez Leibniz." *Revue d'histoire des sciences* 54, no. 2 (2001): 165–221.

————. "Symbolic Inventiveness and 'Irrationalist' Practices in Leibniz's Mathematics." In Dascal, *Leibniz: What Kind of Rationalist?* 125–39.

Shannon, Claude E. "A Mathematical Theory of Communication." *Bell System Technical Journal* 27 (July–October 1948): 379–423.

Shannon, Claude E., and Warren Weaver. *The Mathematical Theory of Communication*. Urbana: University of Illinois Press, 1998.

Shaw, James Byrnie. *Mathematics: The Science of Algorithms*. Jacksonville, IL: Henderson & Depew, 1895.

Shilov, V. V., and S. A. Silantiev. "Logical Machines: Predecessors of Modern Intellectual Technologies." In *History of High-Technologies and Their Socio-Cultural Contexts Conference (HIS^{TEL}CON), 2015 ICOHTEC/IEEE International*, 1–20. IEEE *Xplore*, October 29, 2015. https://doi.org/10.1109/HISTELCON.2015.7307312.

Siraganian, Lisa. "Against Theory, Now with Bots! On the Persistent Fallacy of Intentionless Speech." *Nonsite.Org*, no. 36 (August 2, 2021). https://nonsite.org/against-theory-now-with-bots-on-the-persistent-fallacy-of-intentionless-speech/.

Siskin, Clifford, and William Warner, eds. *This Is Enlightenment*. Chicago: University of Chicago Press, 2010.

Skansi, Sandro, ed. *Guide to Deep Learning Basics: Logical, Historical and Philosophical Perspectives*. Cham, Switz.: Springer Nature, 2020.

Slaughter, M. M. *Universal Languages and Scientific Taxonomy in the Seventeenth Century*. Cambridge: Cambridge University Press, 1982.

Smedley, Rev. Edward, Rev. Hugh James Rose, and Rev. Henry John Rose, eds. *Encyclopædia Metropolitana*. 26 vols. London, 1817–45.

Smith, David Eugene. *Rara Arithmetica: A Catalogve of the Arithmetics Written Before the Year MDCI [. . .]*. Boston: Ginn and Company, 1908.

Smith, Fenny, and Gareth Ffowc Roberts, eds. *Robert Recorde: The Life and Times of a Tudor Mathematician*. Cardiff: University of Wales Press, 2012.

Soare, Robert Irving. "Formalism and Intuition in Computability." *Philosophical Transactions: Mathematical, Physical and Engineering Sciences* 370, no. 1971 (2012): 3277–3304.

————. "Why Turing's Thesis Is Not a Thesis." In Sommaruga and Strahm, *Turing's Revolution*, 297–310.

Sommaruga, Giovanni, and Thomas Strahm. *Turing's Revolution: The Impact of His Ideas about Computability*. Cham, Switz.: Springer International, 2015.

Sørensen, Morten Heine, and Pawel Urzyczyn. *Lectures on the Curry-Howard Isomorphism*. Amsterdam: Elsevier, 2006.

Stadler, Friedrich. *The Vienna Circle: Studies in the Origins, Development, and Influence of Logical Empiricism*. Vienna Circle Institute Library 4. Cham, Switz.: Springer International, 2015.

Stanford Natural Language Processing (NLP) Group. "Economic_inequality." SQuAD 2.0: The Stanford Question Answering Dataset. https://rajpurkar.github.io/SQuAD -explorer/explore/v2.0/dev/Economic_inequality.html.

Stanhope, Charles. *A Letter from Earl Stanhope, to the Right Honourable Edmund Burke: Containing a Short Answer to His Late Speech on the French Revolution*. 2nd ed. London, 1790.

———. *Observations on Mr. Pitt's Plan for the Reduction of the National Debt*. London, 1786.

Stanhope, Ghita, and G. P. Gooch. *The Life of Charles, Third Earl Stanhope*. London: Longmans, Green, 1914.

Stanković, Radomir S., and Jaakko Astola. *From Boolean Logic to Switching Circuits and Automata: Towards Modern Information Technology*. Berlin: Springer-Verlag, 2011.

St. Clair, William. *The Godwins and the Shelleys: A Biography of a Family*. Baltimore: JHU Press, 1989.

Stedall, Jacqueline A. *From Cardano's Great Art to Lagrange's Reflections: Filling a Gap in the History of Algebra*. Zürich: European Mathematical Society, 2011.

———. "Notes Made by Thomas Harriot on the Treatises of François Viète." *Archive for History of Exact Sciences* 62, no. 2 (2008): 179–200.

———. *The Greate Invention of Algebra: Thomas Harriot's Treatise on Equations*. Oxford: Oxford University Press, 2003.

Steele, Guy Lewis Jr. "Lambda: The Ultimate Declarative." MIT AI Memo No. 379. November 1, 1976. https://dspace.mit.edu/handle/1721.1/6091.

Steele, Guy Lewis Jr., and Gerald Jay Sussman. "Lambda: The Ultimate Imperative." MIT AI Memo No. 353. March 1, 1976. https://dspace.mit.edu/handle/1721.1/5790.

Stevin, Simon. *L'arithmetique*. Leiden, 1585.

———. *Les oeuvres mathematiques de Simon Stevin de Bruges [. . .]*. Leiden, Ger.: Bonaventure & Abraham Elsevier, 1634.

———. "On Decimal Fractions." In *Source Book in Mathematics*, edited by David Eugene Smith, translated by Vera Sanford, 20–34. New York: McGraw-Hill, 1929.

Stifel, Michael. *Arithmetica integra*. Nuremburg, 1544.

Stone, Harold S. *Introduction to Computer Organization and Data Structures*. New York: McGraw-Hill, 1972.

Struik, D. J. "Simon Stevin and the Decimal Fractions." *Mathematics Teacher* 52, no. 6 (1959): 474–78.

Stubhaug, Arild, and Richard H. Daly. *Niels Henrik Abel and His Times: Called Too Soon by Flames Afar*. Berlin: Springer-Verlag, 2000.

Sunstein, Emily W. *Mary Shelley: Romance and Reality*. Baltimore: JHU Press, 1989.

Süßmilch, Johann Peter. *Versuch eines Beweises, daß die erste Sprache ihren Ursprung nicht vom Menschen, sondern allein vom Schöpfer erhalten habe*. Berlin, 1766.

Tarski, Alfred. *Logic, Semantics, Metamathematics: Papers from 1923 to 1938*. Translated by J. H. Woodger. 2nd ed. Indianapolis: Hackett, 1983.

Taylor, Eva Germaine Rimington. *The Mathematical Practitioners of Tudor and Stuart England*. Cambridge: Cambridge University Press, 1954.

Tennemann, Wilhelm Gottlieb. *A Manual of the History of Philosophy*. Translated by Arthur Johnson. London: Henry G. Bohn, 1852.

Terekhov, Andrey. "ALGOL 68 and Its Impact on the USSR and Russian Programming." In *2014 Third International Conference on Computer Technology in Russia and in the Former Soviet Union*, 97–106. IEEE *Xplore*, February 9, 2015. https://doi.org/10.1109/SoRuCom.2014.29.

Thomas, D. O. "Francis Maseres, Richard Price, and the Industrious Poor." *Enlightenment and Dissent* 4 (1985): 65–82.

Thomas, Wolfgang. "Algorithms: From Al-Khwarizmi to Turing and Beyond." In Sommaruga and Strahm, *Turing's Revolution: The Impact of His Ideas about Computability*, 29–42.

Tillery, Denise. "Engendering the Language of the New Science: The Subject of John Wilkins's Language Project." *Eighteenth Century* 46, no. 1 (2005): 59–79.

Tresch, John. "Even the Tools Will Be Free: Humboldt's Romantic Technologies." In *The Heavens on Earth: Observatories and Astronomy in Nineteenth Century Science and Culture*, edited by John Aubin, Charlotte Bigg, and Otto Sibum, 253–85. Durham, NC: Duke University Press, 2010.

———. *The Romantic Machine: Utopian Science and Technology after Napoleon*. Chicago: University of Chicago Press, 2012.

Turing, Alan M. "Computing Machinery and Intelligence." *Mind* 59, no. 236 (1950): 433–60. https://doi.org/10.1093/mind/LIX.236.433.

———. "On Computable Numbers, With an Application to the Entscheidungsproblem." *Proceedings of the London Mathematical Society* 2, no. 42 (1936): 230–65.

———. "Systems of Logic Based on Ordinals." *Proceedings of the London Mathematical Society*, 2, 45, no. 1 (1939): 161–228. https://doi.org/10.1112/plms/s2-45.1.161.

Ulivi, Elisabetta. "Masters, Questions and Challenges in the Abacus Schools." *Archive for History of Exact Sciences* 69, no. 6 (2015): 651–70.

Unger, J. Marshall. *Ideogram: Chinese Characters and the Myth of Disembodied Meaning*. Honolulu: University of Hawai'i Press, 2004.

Uspensky, Vladimir, and Alexei Semenov. *Algorithms: Main Ideas and Applications*. Translated by A. Shen. Mathematics and Its Applications. Dordrecht, Neth.: Springer Science+Business Media, 1993.

Valenza, Robin. *Literature, Language, and the Rise of the Intellectual Disciplines in Britain, 1680–1820*. Cambridge: Cambridge University Press, 2009.

Van Brummelen, Glen. "Jamshīd Al-Kāshī: Calculating Genius." *Mathematics in School* 27, no. 4 (1998): 40–44.

Van Couvering, Elizabeth. "Is Relevance Relevant? Market, Science, and War: Discourses of Search Engine Quality." *Journal of Computer-Mediated Communication* 12, no. 3 (April 1, 2007): 866–87. https://doi.org/10.1111/j.1083-6101.2007.00354.x.

Van Wijngaarden, A. "Generalized ALGOL." *Annual Review in Automatic Programming* 3 (1963): 17–26. https://doi.org/10.1016/S0066-4138(63)80002-6.

Van Wijngaarden, A., B. J. Mailloux, J. E. L. Peck, C. H. A. Koster, M. Sintzoff, C. H. Lindsey, L. G. L. T. Meertens, and R. G. Fisker, eds. "Revised Report on the Algorithmic Language Algol 68." *ALGOL Bulletin*, no. 47 (1981): supplement.

Varadarajan, V. S. "Euler and His Work on Infinite Series." *Bulletin of the American Mathematical Society* 44, no. 4 (2007): 515–39.

Vassallo, Nicla. "Psychologism in Logic: Some Similarities between Boole and Frege." In Gasser, *Boole Anthology*, 311–25.

Vaswani, Ashish, Noam Shazeer, Niki Parmar, Jakob Uszkoreit, Llion Jones, Aidan N. Gomez, Lukasz Kaiser, and Illia Polosukhin. "Attention Is All You Need." Cornell University. Last revised December 6, 2017. http://arxiv.org/abs/1706.03762.

Veitch, Victor, Alexander D'Amour, Steve Yadlowsky, and Jacob Eisenstein. "Counterfactual Invariance to Spurious Correlations: Why and How to Pass Stress Tests." Cornell University. November 2, 2021. http://arxiv.org/abs/2106.00545.

Venn, John. *The Logic of Chance*. London: Macmillan, 1866.

Viète, François. *The Analytic Art*. Translated by T. Richard Witmer. Mineola, NY: Dover, 2006.

———. *In artem analyticem isagoge*. Tours, Fr., 1591.

———. *Opera mathematica, In unum Volumen congesta, ac recognita [. . .]*. Leiden, Ger., 1646.

Vincent, James. "OpenAI's Latest Breakthrough Is Astonishingly Powerful, but Still Fighting Its Flaws." The Verge, July 30, 2020. https://www.theverge.com/21346343/gpt-3-explainer-openai-examples-errors-agi-potential.

Voltaire. *Oeuvres completes de Voltaire*. 3 vols. Paris: Sautelet, 1827.

Von Plato, Jan. *The Great Formal Machinery Works: Theories of Deduction and Computation at the Origins of the Digital Age*. Princeton, NJ: Princeton University Press, 2017.

Waerden, Bartel L. van der. *A History of Algebra: From al-Khwārizmī to Emmy Noether*. Berlin: Springer-Verlag, 1985.

Walker, William. "On Reason, Faith, and Freedom in 'Paradise Lost.'" *Studies in English Literature, 1500–1900* 47, no. 1 (2007): 143–59.

Wallis, John. *A Defence of the Royal Society [. . .]*. London: Printed by T. S. for Thomas Moore, 1670.

———. "A Letter of Dr. John Wallis to Robert Boyle Esq [. . .]." *Philosophical Transactions of the Royal Society of London* 5, no. 61 (July 18, 1670): 1087–99. https://doi.org/10.1098/rstl.1670.0011.

———. *Teaching Language to a Boy Born Deaf: The Popham Notebook and Associated Texts*. Edited by David Cram and Jaap Maat. Oxford: Clarendon Press, 2017.

———. *A Treatise of Algebra, Both Historical and Practical [. . .]*. London, 1685. Early English Books, 1641–1700, microfilm reel position 402:29, 854:14.

Ward, Seth. *Idea trigonometriae demonstratae (in usum juventutus Oxoniensis) [. . .]*. Oxford: Leonard Lichfield, 1654. Early English Books, 1641–1700, microfilm reel position 1135:06.

———. *Vindiciae academiarum: Containing, Some Briefe Animadversions upon Mr Websters Book, Stiled, The Examination of Academies [. . .]*. Oxford, 1654. Thompson Tracts, microfilm reel position 113:E.738[5].

Weaver, Warren. *Translation*. Memorandum, July 15, 1949. Internet Archive Machine Translation Archive. https://web.archive.org/web/20210126031922/http://www.mt-archive.info/Weaver-1949.pdf.

Webster, John. *Academiarum Examen, Or The Examination of Academies [. . .]*. London, 1653. Thompson Tracts, microfilm reel position 111:E.724[14].

Weinberg, Justin. "Philosophers on GPT-3 (Updated with Replies by GPT-3)." Daily Nous, July 30, 2020. http://dailynous.com/2020/07/30/philosophers-gpt-3/.

Welby, Victoria. *What Is Meaning? Studies in the Development of Significance*. London: Macmillan, 1903.

Wellmon, Chad. *Organizing Enlightenment: Information Overload and the Invention of the Modern Research University*. Baltimore: JHU Press, 2015.

Wess, Jane. "The Logic Demonstrators of the 3rd Earl Stanhope (1753–1816)." *Annals of Science* 54, no. 4 (July 1997): 375–95.

Wexelblat, Richard L., ed. *History of Programming Languages*, ACM Monograph Series. New York: Academic Press, 1981.

Whately, Richard. *Elements of Logic [. . .]*. London: Mawman, 1826.

Whewell, William. *The Philosophy of the Inductive Sciences: Founded upon Their History*. 2 vols. London: John W. Parker, 1840.

Whitehead, Alfred North, and Bertrand Russell. *Principia Mathematica*. 3 vols. Cambridge: Cambridge University Press, 1910.

Wiener, Norbert. *Cybernetics, Or Control and Communication in the Animal and the Machine*. 2nd ed. Cambridge: MIT Press, 1961.

Wilde, Lisa Jennifer. "English Numeracy and the Writing of New Worlds, 1543–1622." PhD diss., Princeton University, 2014. https://dataspace.princeton.edu/handle/88435/dsp01g732d9137.

Wilkes, M. V. "Herschel, Peacock, Babbage and the Development of the Cambridge Curriculum." *Notes and Records of the Royal Society of London* 44, no. 2 (1990): 205–19.

Wilkins, John. *An Essay Towards a Real Character and a Philosophical Language*. London: S. Gellibrand, 1668. Early English Books, 1641–1700, microfilm reel position 196:12.

———. *Mathematicall Magick: Or, the VVonders That May Be Performed by Mechanicall Geometry [. . .]*. 2 vols. London, 1648. Early English Books, 1641–1700, microfilm reel position 443:09.

———. *Mercury, or the Secret and Svvift Messenger [. . .]*. London, 1641. Thompson Tracts, microfilm reel position 163:E.1100[4].

Williams, David. *Condorcet and Modernity*. Cambridge: Cambridge University Press, 2004.

Williams, Raymond. *Culture and Society, 1780–1950*. New York: Columbia University Press, 1983.

Williams, Travis D. "Mathematical Enargeia: The Rhetoric of Early Modern Mathematical Notation." *Rhetorica* 34, no. 2 (Spring 2016): 163–211. https://doi.org/10.1525/rh.2016.34.2.163.

———. "Procrustean Marxism and Subjective Rigor: Early Modern Arithmetic and Its Readers." In *"Raw Data" Is an Oxymoron*, edited by Lisa Gitelman, 41–59. Cambridge, MA: MIT Press, 2013.

Wilson, Peter H. *Heart of Europe: A History of the Holy Roman Empire*. Cambridge, MA: Belknap Press of Harvard University Press, 2016.

Wirth, Niklaus. *Algorithms + Data Structures = Programs*. Englewood Cliffs, NJ: Prentice-Hall, 1976.

———. "The Essence of Programming Languages." In *Modular Programming Languages*, edited by László Böszörményi and Peter Schojer, 1–11. Lecture Notes in Computer Science. Berlin: Springer-Verlag, 2003. https://doi.org/10.1007/978-3-540-45213-3_1.

———. "On the Design of Programming Languages." In *Programming Languages, a Grand Tour*, edited by Ellis Horowitz, 23–30. Rockville, MD: Computer Sciences Press, 1987.

Wittgenstein, Ludwig. *Philosophical Investigations*. Translated by G. E. M. Anscombe, P. M. S. Hacker, and Joachim Schulte. 4th ed. Chichester, UK: John Wiley & Sons, 2009.

———. *Tractatus Logico-Philosophicus*. Edited by Marc A. Joseph. Peterborough, ON: Broadview Press, 2014.

Wolff, Tristram. "Arbitrary, Natural, Other: J. G. Herder and Ideologies of Linguistic Will." *European Romantic Review* 27, no. 2 (March 3, 2016): 259–80. https://doi.org/10.1080/10509585.2016.1140044.

Wootton, David. *Galileo: Watcher of the Skies*. New Haven, CT: Yale University Press, 2010.

Wordsworth, William. *The Prelude, 1799, 1805, 1850: Authoritative Texts, Context and Reception, Recent Critical Essays*. Edited by Jonathan Wordsworth, Meyer Howard Abrams, and Stephen Gill. New York: Norton, 1979.

———. *Poems by William Wordsworth*. 2 vols. London, 1815.

Wordsworth, William, and Samuel Taylor Coleridge. *Lyrical Ballads*. Edited by W. J. B. Owen. 2nd ed. London: Oxford University Press, 1969.

Zetterberg, J. Peter. "The Mistaking of 'the Mathematicks' for Magic in Tudor and Stuart England." *Sixteenth Century Journal* 11, no. 1 (Spring 1980): 83–97. https://doi.org/10.2307/2539477.

Zimmerman, Andrew. "The Ideology of the Machine and the Spirit of the Factory: Remarx on Babbage and Ure." *Cultural Critique*, no. 37 (Autumn 1997): 5–29. https://doi.org/10.2307/1354539.

Zuse, Konrad. *The Computer: My Life*. Translated by Patricia McKenna and J. Andrew Ross. Berlin: Springer Science+Business Media, 1993.

Index

abaci, 22

abacus schools, 28

Abbasid Caliphate, 28

Abel, Niels Henrik, 95, 121

Abelson, Harold, 192–93

abstraction: algorithmic thinking and, 3, 7, 14, 33, 164, 197–99, 206; Diderot on, 98; Leibniz on, 82; in linguistic thought, 61; in modern mathematics, 13; of quantity, 29, 69; symbolic algebra and, 35–42; Wirth on, 180–81

academic disciplines. *See* disciplinarity

Ackerman, Wilhelm, 169

ACM. *See* Association for Computing Machinery

aconventionality, 173, 183–84, 188, 223–25

Ada (programming language), 190–91

Alexandria, 27, 37–38, 42

algebra: abstract, 146; algorithmic thinking, significance for, 5, 34, 39, 78, 175–76; al-Khwārizmī on, 27–28; Berkeley on, 76–77; Boole's logic system and, 125–26; conceptual problems in, 29–33, 98–109, 131–32; Condillac on, 97; Condorcet's universal language and, 87–89, 96; early history of, 27; historiography of, 28; Lagrange's contributions, 99–100, 120; Leibnizian calculus and, 77–78; low prestige in early modern period, 16, 28–29, 77, 84; nineteenth-century symbolic turn, 123, 125, 130–32; notation (*see*

mathematical notation); Novalis on, 119–20; word, origin of, 21

ALGOL (programming language), 6, 11, 168–69, 187; history of, 178–80; naming of, 179; technical design, 181–84

algorism. *See* calculation

algorithm (word), 3–5; algebraic systems, usage to refer to, 9, 78, 133, 246n129; Church's usage, 175–76; computer science, adoption in, 11, 168, 179–80, 192–93; definition, controversy about in social sciences, 272n24; definitions in computer science textbooks, 168, 193–97; Knuth's definition, 195–96; Lagrange's usage, 120–21; Leibniz's usage, 53, 77–78; machine learning, usage in, 207–10; modern sense, emergence of, 132–33, 163, 166–69; origins and early usage, 16, 22; as a primitive notion, 264n15; *program*, usage as a synonym of, 193, 268n136; Schröder's usage, 154; Stifel's usage, 30

algorithms: A*, 197; abstraction and, 181; algebra and, 5, 77–78, 133; applications in the early twentieth century, 167–68; bubble sort, 194, 197; classical paradigm, 194–203, 206; *Communications of the ACM*, section in, 179–80, 216; for compound interest, 41–42; for differentiation in calculus, 53; Dijkstra's algorithm, 197–99; for

311